Lecture Notes in Networks and Systems 774

The series "Lecture Notes in Networks and Systems" publishes the latest developments in Networks and Systems—quickly, informally and with high quality. Original research reported in proceedings and post-proceedings represents the core of LNNS.

Volumes published in LNNS embrace all aspects and subfields of, as well as new challenges in, Networks and Systems.

The series contains proceedings and edited volumes in systems and networks, spanning the areas of Cyber-Physical Systems, Autonomous Systems, Sensor Networks, Control Systems, Energy Systems, Automotive Systems, Biological Systems, Vehicular Networking and Connected Vehicles, Aerospace Systems, Automation, Manufacturing, Smart Grids, Nonlinear Systems, Power Systems, Robotics, Social Systems, Economic Systems and other. Of particular value to both the contributors and the readership are the short publication timeframe and the world-wide distribution and exposure which enable both a wide and rapid dissemination of research output.

The series covers the theory, applications, and perspectives on the state of the art and future developments relevant to systems and networks, decision making, control, complex processes and related areas, as embedded in the fields of interdisciplinary and applied sciences, engineering, computer science, physics, economics, social, and life sciences, as well as the paradigms and methodologies behind them.

Indexed by SCOPUS, INSPEC, WTI Frankfurt eG, zbMATH, SCImago.

All books published in the series are submitted for consideration in Web of Science.

For proposals from Asia please contact Aninda Bose (aninda.bose@springer.com).

Carlos Montenegro · Álvaro Rocha ·
Juan Manuel Cueva Lovelle
Editors

Management, Tourism and Smart Technologies

ICMTT 2023 Volume 2

Springer

Editors
Carlos Montenegro
Universidad Distrital Francisco
José de Caldas
Bogota, Colombia

Juan Manuel Cueva Lovelle
Departamento de Informática
University of Oviedo
Oviedo, Spain

Álvaro Rocha
ISEG
Universidade de Lisboa
Lisbon, Portugal

ISSN 2367-3370 ISSN 2367-3389 (electronic)
Lecture Notes in Networks and Systems
ISBN 978-3-031-43732-8 ISBN 978-3-031-43733-5 (eBook)
https://doi.org/10.1007/978-3-031-43733-5

This Springer imprint is published by the registered company Springer Nature Switzerland AG
The registered company address is: Gewerbestrasse 11, 6330 Cham, Switzerland

Paper in this product is recyclable.

Preface

In this edition of the International Conference on Management, Tourism and Technologies—ICMTT 2023, papers were presented in the areas of: Managements, Tourism, Marketing Strategies in Management, Tourism and Technology, and Technology.

We would like to give special thanks to the Universidad Distrital Francisco José de Caldas, Fundación Universitaria Internacional de la Rioja, and Universidad de Cundinamarca, for hosting us, as well as to all the members and collaborators, since without them this dream would not have been possible.

We had more than 200 papers presented, we spent 3 consecutive days in 5 parallel rooms, and more than 400 people passed through the event and generated an academic space that allowed the exchange of experiences to advance in the era of knowledge, where we have so much data that the important thing is to transform it into knowledge. Another of the great experiences that the event left us is that virtuality is definitely a reality, since many of our works were presented in this way.

We still have many challenges, but a very important one and on which we are working is how to get that academic relationship that gives us the presence in these events we can also make up for with virtuality, and this reflection has helped us to understand what a visionary as Mark Zuckerberg CEO of Meta has envisioned in what he called the Metaverse and just put an oculus to understand that virtuality needs these visions to achieve the great challenge we have set ourselves: How will we relate to other people in academic, social, or other contexts through virtual scenarios?

I hope to see you all at the next edition of the International Conference on Management, Tourism and Technologies—ICMTT 2024, in Cusco, Peru, and as we will not stop doing virtual sessions, the challenge is that we all have some oculus to see ourselves in our Metaverse.

May 2023

Carlos Montenegro
Álvaro Rocha

Contents

Robotics

Strategic Management Innovation

Sustainability Management

Tourism Industry and Ecology

Tourism Management

Management of Supply Chain and Logistics

Impact of Logistics Training on Freight Transportation Companies in Bogotá

Hugo Almeiro Pachón Pedraza[1], Breed Yeet Alfonso Corredor[2](✉), Carlos Augusto Sanchez Martelo[1], and Martha Cecilia Cadena Chala[1]

[1] Universidad Manuela Beltrán, Bogotá, Colombia
{hugo.pachon,carlos.sanchez}@umb.edu.co,
martha.cadena@docentes.umb.edu.co
[2] Universidad Internacional de la Rioja, Logroño, Spain
breedyeet.alfonso@unir.edu.co

Abstract. Land cargo transportation companies have been growing, due to the increases in cargo volumes generated in the city, in the same way there are other relevant variables such as the expansion of population growth, the design and creation of new highways, Los Changes in the territorial planning plan, among others, have required these companies to have greater specialization in logistics processes, but this has been able to be improved, hand in hand with the use of technological and digital tools, which facilitate the process and to provide reliable and timely information to its customers. The foregoing is a demand from these companies to have personnel that demonstrate better skills, competencies and attitudes in the control and administration of internal logistics processes, which is reflected in greater efficiency in their daily operational and administrative performance. The purpose of this article is to show the impact of training and education in the logistics area in cargo transport companies in the city of Bogotá and the incidence of human talent in the industrial and logistics growth of these companies, due to the training academic that said personnel must have. This is achieved through the analysis and logistic performance of cargo transportation companies in Bogotá, and how their productivity improves significantly with the academic training of the personnel who work in them.

Keywords: Freight transport · skills · Logistics · Competitiveness

1 Introduction

Markets and the development of world trade are processes that have proven to be dynamic and seek to respond to the needs of consumers, who are increasingly demanding in terms of the products and services they require, and the national market is no stranger to this also. He is a good observer of the innovations that technology offers him, facilitating his access to knowledge. Although the Covid 19 pandemic has reduced the national economy, world economic growth was 5.9% during 2021 (World Bank 2021b), and this growth is mainly due to the strengthening of the labor market, which is the engine for reactivation of the national economy.

C. Montenegro et al. (Eds.): ICMTT 2023, LNNS 774, pp. 3–16, 2024.
https://doi.org/10.1007/978-3-031-43733-5_1

Similarly, according to reports from the World Bank in the report World Economic Outlook: Latin America and the Caribbean (World Bank 2023), it is expected that the dynamism and growth in demand will drop to 1.3% in 2023 and have a recovery and growth of 2.4% in 2024, based on the support that governments provide to companies that seek to advance in the search for strategies that lead them to stay in competitive markets and that provide total customer and consumer satisfaction.

In this context, companies should tend to establish commercial and Marketing strategies that cover the entire supply chain, therefore, the logistics processes of enlistment, loading, dispatch and distribution, among others, that are aligned with the defined objectives. In strategic planning, as Peter Drucker points out, logistics is the last frontier in organizations, which defines adequate competitiveness.

These strategic plans defined by the organization must include tactical and operational activities, through which it seeks to achieve corporate objectives, which become a fundamental element that includes human talent, who ultimately are the ones that allow achieving the long-awaited organizational development and the growth of the sector.

Consistent with the aforementioned, it is then established that the human talent used in the logistics field must be a fundamental tool that provides the necessary conditions to respond to this demand, which tends to provide contributions based on their skills and competencies, mainly in the management of digital technologies that the logistics sector has today within the current context, in order to guarantee the growth and evolution required for the adequate strengthening of companies in the sector.

At present, companies in the logistics sector, dedicated to the transport and distribution of land cargo in Bogotá, demand a high level of economic and commercial activities, for this reason, with the aim of deepening this, an analysis of the required logistics processes and likewise, the requirements of the skills and abilities of human talent in this sector.

In this order of ideas, we initially proceed to carry out a brief general conceptualization of logistics, later a review of logistics training in the country is made, then we proceed to define the structuring of the human talent requirements of logistics companies, to assess the needs inherent to specific training and training.

2 Materials and Methods

The research is framed under a qualitative approach, based on the methodological design called action research, with which an investigation of the problem was carried out to study the phenomenon in depth in the different phases of the investigation, also considering the documentation and initial experiences of the participants. This approach supports observation and analysis in a natural environment, which facilitates the perception of the study in its context, facilitating the interpretation of the information, Hernández, Fernández and Baptista (2020).

Data collection techniques are part of procedures and activities during the investigation that allow answering the research question. In this order of ideas, the instruments that were used were observation, open sources, and analysis of websites (Fig. 1).

The analyzes carried out based on the data collected are presented below:

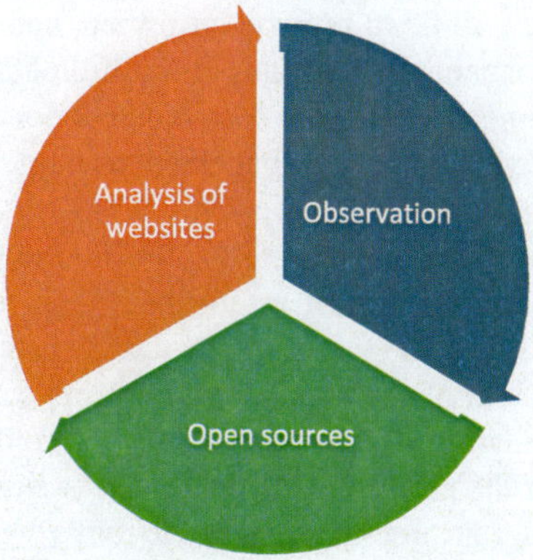

Fig. 1. Data collection instruments.

Logistics is present and has accompanied humanity in its development (Ospina Díaz and Sanabria Rangel 2017) since ancient cases, when the processes of storage, transport and distribution of merchandise were just being technified, likewise other authors define that logistics arose from military strategies, whose objective was to support the maneuvers of the armies.

Being logistics "an integrating discipline of the supply chain, no discipline is so complex or encompasses geographic positioning" (Bowersox, Closs and Cooper 2010), it is considered a fundamental element, capable of generating competitive advantages, which allows organizations access and permanent maintenance in global markets at the moment of improving the response times of their merchandise, in the face of the permanent demands of the market (Lambert, Cooper and Pagh 1998). Seen from the organizational level, it presents a relevant evolutionary process, in terms of the dynamics of globalization and market growth, the various implementations of technological tools that allow reliable and timely information, which have allowed logistics to cease to be simply an operational perspective that focuses on the administration and control of product flows, as expressed by Ballou: "it is all movement and storage that facilitates the flow of products from the point of purchase of materials to the point of consumption, as well as the information flows that are set in motion, in order to provide adequate levels of service to the consumer at a reasonable cost" (1999, p. 7), now turning to a totally strategic perspective, whose purpose is to add value to all operations of the organization.

Logistics then encompasses a comprehensive and flexible methodology, capable of adapting to changes resulting from the specific needs of the company, or the demand of the target market, thus allowing the articulation of various processes of the company (Mora García 2016).

The Council of Supply Chain Management Professionals (CSCMP) defines logistics as, "the part of supply chain management that plans, implements, and controls the flow and storage of goods, services, and related information between the point of origin and the point of consumption efficiently and effectively, to comply with customer requirements" (Vitasek 2013, p. 117). With the above, it is then defined that logistics is part of the supply chain and constitutes a process which integrates the activities of supply, production and transport, whose objective will be to provide customers and consumers with products and services in a timely manner, emphasizing the fulfillment of the logistics promise, defined site, opportune moment, reasonable cost, required conditions, understanding this as the optimization of information flow management as a relevant success factor for the entire process.

But logistics is not a process that is purely dedicated to the handling of materials, rather it goes much further, and is then established as an integral factor, as defined by Casanova and Cuatrecasas (2011): "the concept of logistics appears integral, which reaches the very management of the merchandise and its distribution, allowing better levels of competitiveness thanks to a better efficiency in the flow of products and information. In this sense, it should be clarified that logistics is extending its field not only to companies that operate tangible products, but also to those that provide services, considering that the economy relies more and more on this sector than on the industrial one" (p. 14). Which then implies that logistics is an integral process, which determines and establishes conditions of efficiency and competitiveness.

In Colombia, the National Planning Department (DNP) has developed the National Logistics Survey (ENL), whose objective is to generate data and logistics information from the business and industrial sector, which allows decision-making and the drawing up of public policies regarding logistics. Said survey then makes it possible to evaluate the entire national logistics context, under five premises: Logistics performance, Outsourcing, Foreign trade, Logistics services prospects and Regional Competitiveness.

For the year 2018, the results obtained in these surveys were, based on a business sample of 2,738, out of a total population of 848,986 (DNP 2018), which showed that 69.1% of the companies carry out measurements of their logistics cost, which in average of 13.5% of the total value of sales, which is made up as follows: 46.5% in storage costs, 35.2% in transportation costs, 11.1% in costs administration and customer service, and finally 7.2% for other miscellaneous costs.

Going back to the figures presented by the DNP, in its 2018 National Logistics Survey (ENL), it shows that 44.8% of the companies outsource their land freight transportation services and its distribution, referring to the logistics operation found that 22.4% of logistics companies have their own vehicle fleet, and that their urban routes are equivalent to 70.5%, while 29.5% obey national route routes. Based on these figures, it is confirmed that logistics companies, which involve ground cargo transportation operations, focus their efforts on purely operational operations, leaving management activities of a strategic nature because of their corporate objectives in the background.

The same survey produced other results referring to the measurement of times in the companies, it was found that 10% of these companies measure the times of storage, loading and distribution of the load, finding in this result figures of up to almost 3.8 h average in the waiting process for its reception, 3.3 h average, in the loading process, and 2.6 h average in the download process. The foregoing makes it clear that these "idle times", so to speak, have a direct impact on logistics costs, and if we articulate this to the percentages presented above, they clearly indicate that it is urgent that land freight transport companies, and companies dedicated to logistics operations in Colombia assume within their processes, measurement activities, which intrinsically entail opportunities for improvement in order to reduce the indicated times, which generates value to the operations described above.

For its part, Logyca (2017) carried out the Industrial Revolution Survey and it yielded results such as that 30% of the companies have implemented commercial digital tools (E-Commerce), tools related to the forecast of market demand, 20% do so. They have done in receipt, storage and dispatch processes, implementation of voice picking, pick to light and RDFI, and only 15% of companies apply digital tools in traceability processes (Higuera Gómez 2018, para 4).

Likewise, once again reviewing the results obtained in the National Logistics Survey (ENL 2015), the companies show great shortcomings in terms of road infrastructure and data capture in operations, for which they resort to the use of radio frequency to determine data that contributes to the forecast of demand, which impacts logistics costs and quality of service (DNP 2018).

Reviewing the data from the National Logistics Survey (ENL) a broad approach can be built, evidencing a high lack of an adequate and strategic logistics vision, which leaves out all the strategic and tactical aspects that directly affect the logistics operational processes of the companies. Companies in general, in the same way, the lack of use of some digital tools and technologies that allow adequate and timely control and monitoring of said processes, which should tend to guarantee the obtaining of data and evidence in decision-making processes. Decisions in favor of continuous improvement actions.

The Colombian government, based on the premise that the issues inherent to logistics that seek to obtain better results in terms of competitiveness against international global markets, as a result of the commercial expansion obtained with the signing of various free trade agreements, has formulated a National Competitiveness and Productivity Policy (COMPES 3527) focused on the modernization and improvement of competitive conditions, which include a section referring to logistics and transport infrastructure, and the National Logistics Policy (COMPES 3547) which is oriented to define a national and articulated logistics system (National Planning Department - DNP 2018b).

The Andean Development Corporation of the Development Bank of Latin America for Colombia, in its priority strategic guidelines, establishes training programs for logistical progress, where the impulse to the waterway axis of the Magdalena River is defined, using intermodal platforms that facilitate its navigability Similarly, promote a system of logistics platforms within the framework of a logistics plan for the region, which allows: "Evolve to become a subregional hub of the Andean region and Central America, from its position in the Colombian Caribbean" (Farromeque Quiroz 2016, p. 77). In this line, the training axis acquires a value from the educational approach.

The Development Bank of Latin America has defined the logistics profile of Latin America, and within this criterion, it establishes that, on average, companies in the region maintain between 18% and 35% of the value of the product in logistics costs, which is equivalent to between 16% and 26% of GDP. In this order of ideas, it can be seen that the costs for logistics concepts are very high, which has a negative impact on the supply chain, making these companies uncompetitive, mainly due to the topographic and geographical conditions of the country, which mainly implies long distances between collection centers and ports, and this in turn generates delays in the development of productivity or effectiveness of specialized logistics areas, as well as in companies in the transport sector (Ferromeque Quiroz 2016).

The performance of companies must be evaluated through various methods, to evaluate the strategies of these companies (Meade and Srkis 1998) define an adequate analytical framework to establish attributes of the Supply Chain, logistics and inter-business relations, (Perez-Franco 2010) establishes an adequate methodology to identify the strategies used by companies in the Supply Chain, evaluate and reformulate them, considers on this premise a diagnostic methodology for logistics in the Supply Chain, Ivan der Vosrst, van Kopten and Luning 2011) define an instrument that allows evaluating and optimizing quality (GCL), and finally, (Orjuela Castro and Adarme Jaimes 2017) propose tools for classifying media and logistic modes.

Based on all of the above, it is then defined that logistics is part of the supply chain, and it is assumed as a process that articulates the activities inherent to the supply, production, transportation and distribution processes, which implies complying with the Logistics promise, which involves activities such as delivering products and services to customers and consumers, in a timely manner, in the right place and at the right time, and under the agreed conditions, applying the concepts of effectiveness in the management of supply flows. Information, as a determining factor in continuous improvement.

3 Results

As evidenced in the paragraphs above, academic training and the development of human talent skills are understood as a fundamental element in the logistics field, understanding this aspect in a relevant and essential way so that transport companies can project themselves in the market and be at the forefront of international trends, which allow them to maintain a vision of competitiveness.

In the first place, the provisions of the National Qualifications Framework in Colombia – Transport and Logistics must be reviewed, which in 2017 defined that of the total number of people employed, that is, 1,585,363, 7.2%, which is equivalent to 74,951 people of these were focused on activities inherent to the logistics sector, among others such as storage, messaging, land, sea and river transport; data according to a report from the Public Employment Service (Mineducación and FITAC 2017).

On the other hand, the same study gives us data at the national level, where it is evident that 48.1% corresponds to supply and inventory control, 32.1% refers to messengers, messengers, porters and delivery men, 40.0% corresponds to drivers of trucks and heavy vehicles, and 3.6% is due to forklift operators, and finally 3.6% corresponds to various employees of postal services. These figures allow us to confirm then that most of the vacancies analyzed are around purely operational aspects of logistics functions. Which leads us to establish generalized guidelines inherent to the logistics operation.

The same study shows us that in terms of the academic training of the vacancies analyzed, 77% are at the primary and high school level, while 14% refer to the level of technical, technological, and professional training. It is worth noting that the study does not establish educational requirements at the master's, doctoral or postdoctoral level (Fig. 2).

Another reliable source of information at the national level is the National Logistics Observatory, which has a series of indicators of logistics activities, whose purpose is to

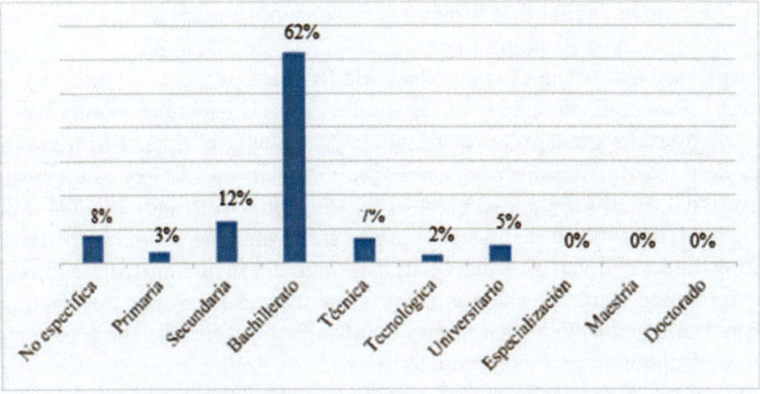

Fig. 2. Distribution of vacancies by educational level.

provide information about performance in different fields such as infrastructure, costs, and environmental impact. Within these studies, the operational scope is included, and this precisely involves indicators associated with the theme of logistics training, these indicators are Composition of the personnel in the sector and incorporation into the labor market of human capital that presents training in logistics issues.

The first indicator presents figures for the number of graduates per year in logistics specialties, and it is observed that in 2006 there were 387 graduates, and in 2014 there were 3,596 graduates, that is, there was an increase of 89%. The following indicator refers to the level of training of human capital in logistics issues, which established that for the year 2015 only 83.1% of graduates had studies at the baccalaureate, technical and technological level, of which more than 40% correspond to high school training, this shows a need for specialized training in logistics issues (Fig. 3).

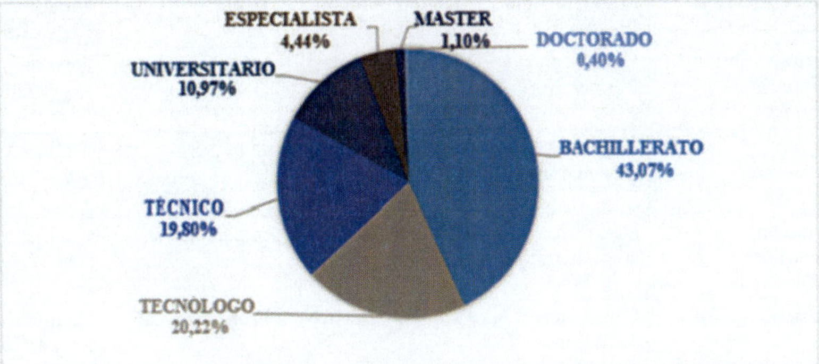

Fig. 3. Level of training of human capital dedicated to the performance of logistics.

Based on the National Productive Development Policy (CONPES 3866): "the main problem of the relationship between the labor factor and the increase in productivity, sophistication and diversification of the productive apparatus is that the country does not have the human capital that it requires to achieve substantial improvements in its productivity" (National Planning Department – DNP, 2018b, p. 85).

As expressed by Ospina Diaz and Sanabria Rangel 2017:

Taking into account that many of the studies on logistics in the country have shown that one of the relevant aspects for this field to develop is related to training in the logistics field, the increase in personnel specialized in the subject, the expansion of the educational offer, the improvement of existing programs and practical training in said programs, then it also seems necessary to start worrying about characterizing and monitoring the educational offer that exists in the country in relation to the subject (p. 22).

The foregoing ratifies what was already mentioned regarding the importance of training in the logistics field that allows companies to overcome some of the gaps in improving competitiveness and productivity.

Based on the above, information regarding a comparison of programs related to Logistics Administration in Colombia is presented below, taken from the source the National Information System for Higher Education (SNIES), there 20 undergraduate programs were found, Regarding geographic location, 11 of them are located in the city of Bogotá, which represents 55%, which represents an unfavorable aspect for those interested, since it is only evident that at the Manuela Beltrán University, the Administration program Logistics is taught virtually (Table 1).

Table 1. Table Professional logistics programs

Name IES	Program name	Academic level	Modality	Ministry recognition
Colegio Mayor de Nuestra Señora del Rosario	Logistics and production administration	Undergraduate	In person	High Quality Accreditation
Universidad Sergio Arboleda	Business Logistics	Undergraduate	In person	Qualified Registration
Uniempresarial	Marketing and Logistics	Undergraduate	In person	Qualified Registration
Fundación Universitaria Monserrate	Logistics engineering	Undergraduate	In person	Qualified Registration
Fundación Universitaria Compensar	Logistics administration	Undergraduate	In person	Qualified Registration
Corporación Universitaria Iberoamericana	Logistics administration	Undergraduate	In person	Qualified Registration
Universidad Santo Tomás	Logistics and operations engineering	Undergraduate	From distance	Qualified Registration
Universidad Manuela Béltran - UMB	Logistics administration	Undergraduate	Virtual	Qualified Registration

(*continued*)

Table 1. (*continued*)

Name IES	Program name	Academic level	Modality	Ministry recognition
Escuela de Logística	Logistics administration	Undergraduate	In person	Qualified Registration
Escuela Militar de Cadetes General José María Cordoba	Logistics administration	Undergraduate	In person	High Quality Accreditation
Fundación Universitaria Los Libertadores	Logistics administration	Undergraduate	In person	Qualified Registration

Referring to the level of academic training, it is evident that 34.98% of the programs correspond to specialization, followed by 34.08% of the programs that correspond to technologies, 12.56% are technical programs, and finally undergraduate and master's only 8.97%, to finally find the doctorate with 0.45% (Fig. 4).

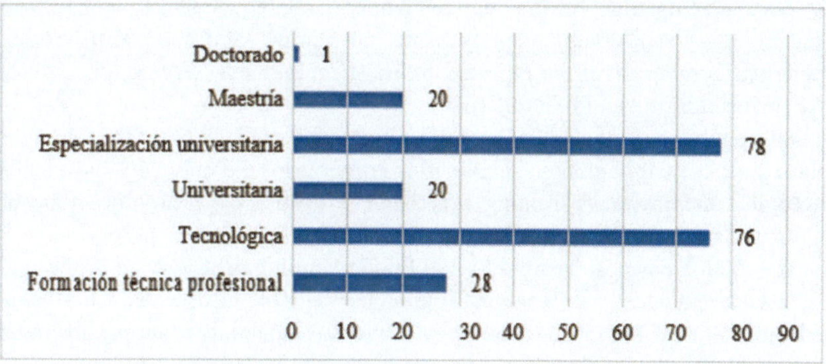

Fig. 4. Number of logistics training programs by training level.

On the other hand, it is evident that Bogotá maintains 24.66% of the logistics programs, and Cundinamarca has 2.36%, which indicates that the region accumulates 27.35% of the national total. 77.05% of these programs are face-to-face, in Cundinamarca all the programs are face-to-face, while in Bogotá 41 programs are face-to-face and 12 are virtual. It is also noted that in Bogotá 37.70% of these programs are technical and technological, 19.67% are undergraduate and 16.39% are master's and doctoral level (Fig. 5).

In summary, Bogotá - Cundinamarca maintains a better distribution regarding levels of academic training, despite this, the concentration of higher academic levels is low, which allows us to maintain the hypothesis that the logistics vision in Colombia tends to be operational and not tactical or strategic managerial. Said affirmation then can negatively affect the strategic logistics growth within the organizations, by not finding in the labor market human talent that has said competences.

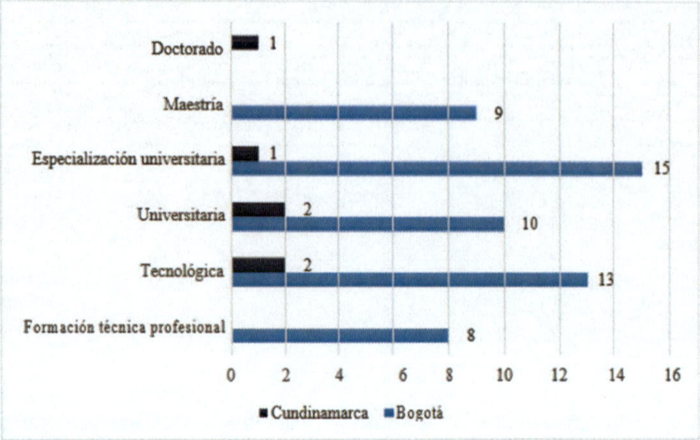

Fig. 5. Number of logistics training programs by level of training Bogotá - Cundinamarca.

According to data obtained in the National Qualifications Framework Colombia - Transport and Logistics, for the year 2017 only 7.2%, equivalent to 1,585,363, were focused on Storage and complementary issues, messaging, land, and maritime transport, compared to number of offers reported by the Employment Service, which reported 74,951 (Mineducación and FITAC 2017).

Regarding the issue of academic training, it was found that 77% of the jobs analyzed maintain basic primary and high school education levels, and only 14% have technical, technological and university training. It is then evident that the sector does not establish educational training at the master's or doctoral level.

Of the total vacancies reported by the Public Employment Service in the logistics sector, at a national level, 58.9% are in Cundinamarca - Bogotá, of which 5,050 vacancies are in Bogotá and 18,170 in Cundinamarca, dedicated especially to supply and inventory control.

Based on the previous study, a series of qualifications that will be required in the logistics and technological sector in the short and medium term is established, which refer to technological and planning knowledge, therefore, the business sector defines a series of knowledge and skills, as reflected in the following table (Table 2):

It is also evident that for the new vacant positions in the sector, professionals in the field of programming and developers are needed to provide technical support to digital platforms and technological tools, which are used in the various logistics processes, while tending to minimize them. Operational positions, as defined by Mineducación and TITAC 2017: "another aspect that will require technical personnel and reduce operators, a fact that now logistics goes further and requires more added work" (p. 50).

One of the shortcomings found in the study is the great gap that exists between government policies, private and public institutions responsible for academic training, compared to companies in the logistics sector and their demands for training human talent, which allows promoting skills and skills in human talent that allow us to be competitive.

Table 2. Table Knowledge and Skills

Subsector	Job title	Required skills and knowledge
Ground transportation	Director	Software management Information systems and platforms ICT Tools
	Operations Coordinator/Head warehouse section and distribution/dispatch coordinator	Georeferencing ICT Tools Technologies Software management
	Head of Operations transportation/transport operational coordinator in smart vehicles	Smart vehicles Routes Load types
	Transportation coordinator	Vehicle monitoring and maintenance (including electric vehicles) Distribution Environmental impact
	Trucker/Truck Driver	Data understanding and interpretation autonomous vehicles
	Data understanding and interpretation Autonomous vehicles	Load distribution Computer tools Database management Database analysis
Marine transport	Chief Engineer/Machinist/ First Mate Engineer	Electromechanical Reading comprehension Data interpretation Office tools
	Multimodal Transport Operations Coordinator (OTM)	Customs regulation update International regulations BASC
	Supervisor	Intelligent computer systems or programs Management of electronic documentation
	Operations Manager, Captain's Assistants and Port Operator	Software management
Storage	Vice President of Operations/Director/Manager	Handling of merchandise (regulations) Use of RFID technology Traceability technologies
	Warehouse Manager/Distribution Center Manager/Logistics Manager/Yard Manager/Logistics Operations Supervisor	Understanding and interpreting data operational logistics Understanding and interpretation of data, Operational logistics, Inventory management software
	Warehouse Manager (Supply Chain)	Use of RFID technology
	Logistics Manager	Information management (big data) Strategic and operational issues of the sector

(*continued*)

Table 2. (*continued*)

Subsector	Job title	Required skills and knowledge
	Operating Assistants	New tools and systems (Example: voice speaking) Use of new technologies in georeferencing Dynamic counting software GPS satellite tracking collection software Information analysis
	Operators	Operation of equipment such as forklifts and pallet trucks
Messenger service	Manager	Automated systems
	Courier office managers	Management of computer platforms Management of software and specialized information systems to handle large volumes of data
	Courier Manager/Distribution Center Manager	New software and routing platforms Big data y analytics
	Delivery courier	Drones Management of specialized inventory and storage systems

Companies in the logistics sector demand specific training and training in logistics-type skills, which requires including the management of digital technologies and technological tools, which must include professionals with specialization, master's, and doctorate degrees, capable of correctly managing the adversities and changes in the world that is increasingly demanding (Ospina Díaz and Sanabria Rangel 2017).

The logistics business sector is faced daily with responding to demands that are increasingly demanding, technological changes and the applicability of new digital tools, tend that companies must demand from their collaborators skills and competencies according to the needs of the changing market, in some specific topics.

With adequate academic training at all hierarchical levels of organizations, sustainability and economic growth can be promoted, both for small and large companies. The economic dimension consists of improving total value, optimizing processes, and reducing of operating costs, strategic and sustainable initiatives that enhance the financial stability and marketing of the organization.

Likewise, organizations may establish social sustainability policies, understanding this as all those actions that promote the development of people, communities and/or economic sectors in search of an improvement in living conditions, health, education, and social balance.

As established by Closs, Speier and Meacham (2011): The social dimension includes corporate social responsibility with ethical decision-making in areas such as employee relations, community participation and business management practices, which which is potentiated through education and the development of human talent (p. 235).

We can then define that in Bogotá - Cundinamarca they present a smaller gap than the rest of the country, in terms of training in logistics and supply chain and the latent needs of the market and business organizations, which allow them adequate growth and progress in these areas in accordance with world trends, which would allow them greater development in terms of costs, resources, operations, thus optimizing their processes, eliminating bottlenecks and reprocessing, optimal waste and recycling management, through the use of technological and digital tools allowing them further advancement and growth.

4 Conclusions

The financial results of the companies studied, is given by their competitiveness capacity, and the possibility of reacting to the changes that the market demands, understanding logistics competitiveness, as that differentiating factor that allows them to satisfy the needs of the market to which they are addressed your offer.

Although it is confirmed that in the country different proposals have been made to the business sector that invites them to develop training in their personnel on logistics conditions, most of the efforts of the companies have been implemented especially in the development of their technical and technological infrastructure, that allows them to enhance their capabilities in the operational factor, and although this aspect is very important, the human factor should not be left aside, as a differentiating element that contributes to continuous improvement and the financial and commercial growth of companies in the sector.

The various technological developments that the market offers companies in the logistics sector allow them to design operational and technological strategies that guarantee them to favor the growth and development of operations inherent to their logistics performance, which becomes a key differentiating factor in the search for continuous improvement and the competitiveness of organizations, for this reason it is considered highly relevant to have and maintain a highly competitive and comprehensive human team in operational, strategic and technological aspects, which allows organizations to maintain a strategic, tactical vision and operations that respond according to the needs of the market.

The governments of both Bogotá and the department of Cundinamarca have established various plans and programs that seek to strengthen the logistics infrastructure, the Connectivity Network that seeks to maintain adequate transport information for management, but the figures show that the results obtained up to at the moment they do not yield the best results. It is therefore necessary to define plans that articulate the government sector with the business sector in the region, seeking to promote greater logistics growth that involves human talent and provides better academic training programs in logistics issues, both operational and technological.

Bogotá maintains the highest concentration of people in Colombia; therefore, it is necessary to develop new paradigms, undertake some tasks that have already been previously programmed, in search of improving the logistics performance of the supply chain, with short, medium, and long plans. Term, which guarantee the development of logistics operations and specialized platforms, so that companies in the sector can, in collaboration with the government sector, overcome all those inefficiencies and lead the logistics sector to maintain international standards, guaranteeing stability and economic growth.

References

Restrepo Gómez, B.: Formación Investigativa e investigación Formativa: Acepciones y operacionalización de esta última (2004). http://hermesoft.esap.edu.co/esap/hermesoft/portal/home_1/rec/arc_3529.pdf. Accessed 15 Nov 2011

Medina Rivilla, A., Salvador Mata, F.: Didáctica General. Editorial UNED, Madrid (2009)

de Sánchez, M.A.: Desarrollo de habilidades del pensamiento. Procesos básicos del pensamiento, 2nd edn. Trillas, Ciudad de México (1991)

de Sánchez, M.A.: Desarrollo de habilidades del pensamiento. Discernimiento, automatización e Inteligencia práctica. Guía del Instructor. 5th edn. Reimpresión. Trillas, Ciudad de México (2004)

Restrepo, G.: Conceptos y aplicaciones de la investigación formativa y criterios para evaluar la investigación científica en sentido estricto. https://www.mineducacion.gov.co/CNA/1741/articles-186502_doc_academico5.pdf

Walker, D.F.: Methodological issues in educational research. A Project of the American Educational Research Association

Stake, R.E.: Investigación con estudio de casos, tercera edición. Ediciones Morata, Madrid (2005)

Schatzman, L., Strauss, A.: Field Research. Strategies for Natural Sociology. Prentice-Hall, Englewood Cliffs (1973)

Acceptance Sampling of Vegetables and Legumes in Units Based on the Military Standard (MIL-STD 105E)

Jacqueline del Pilar Villacís-Guerrero[1], Wendy Álvarez-Duque[1], José Roberto Bonilla-Villacís[2], and Fátima Avilés-Castillo[3](✉)

[1] Carrera de Ingeniería Industrial, Facultad de Ingeniería, Industria y Producción, Universidad Indoamérica, Quito 170103, Ecuador
{jacquelinevillacis,alvarezwendy}@indoamerica.edu.ec

[2] Industrial Engineering, College of Engineering and Applied Science, University of Wisconsin-Milwaukee, Milwaukee, WI 53211, USA
bonilla34@uwm.edu

[3] Centro de Investigaciones de Ciencias Humanas y de la Educación (CICHE), Universidad Indoamérica, Ambato 180103, Ecuador
faviles@indoamerica.edu.ec

Abstract. A low-quality batch brings with its adverse effects on the productivity and profitability of a product. The Military Standard (MIL-STD 105E) allows calculating the acceptance number for an attribute sampling plan under regular inspection with an acceptable quality level. This study is carried out in the national distribution center (CND) of one of the largest supermarket chains in Ecuador. There is a lack of quality control in the reception of essential products and processes carried out empirically. Ensure the quality and safety of vegetables and legumes that are received daily. Develop specific instructions for sampling vegetables and legumes in units based on MIL-STD 105E. When conducting a verification audit, an increase of 15.93% in compliance with the standard technical ARCSA-DE-067-2015-GGG was obtained. Food product reception, processing, and distribution must follow strict biosafety and quality protocols. This proposal provides business opportunities for improvement and standardized handling of documentation and processes, which other similar institutions can use.

Keywords: Military Standard · Safety Control · Sampling by Attributes

1 Introduction

Sampling for lot acceptance consists of periodically evaluating a portion of a lot, called a sample, of a given product. The result will determine whether the lot is suitable or should be discarded based on a criterion associated with these parameters. Over time, organizations are expected to invest in this method and other total quality management programs. The participation of industries and their suppliers will ensure high levels of product quality. With this action, the need for acceptance sampling will decrease, meeting customer satisfaction and ensuring benefits for the organization [1].

C. Montenegro et al. (Eds.): ICMTT 2023, LNNS 774, pp. 17–25, 2024.
https://doi.org/10.1007/978-3-031-43733-5_2

A low-quality batch implies economic losses and affects productivity and product profitability for any company. If these batches reach the final consumer, they would negatively affect customer satisfaction and the public image of the organization [2]. The sampling study, regardless of its type, has allowed companies to measure the quality of their products [3]. Investigations based on this methodology are tools used to obtain information; their implementation significantly improves the control of a company's production. Generally, investigations focused on sampling use quantitative techniques to collect information, which can be extrapolated to other areas since it comes from samples obtained from statistical calculations [4].

For attribute sampling, one of the most widespread systems in the world is the MIL-STD 105E, based on NTE INEN ISO 2859-1:2009. This system is also known sampling plan since it contains guidelines for carrying out individual sampling plans. The method consists of a set of tables with which the sample size is found for three different acceptance sampling plans for attributes. Over time, the standard has been adopted and generalized to other contexts, resulting in ISO 2859 [5]. The attribute sampling system has evolved from the original version of Military Standard 105E (1950) to the fifth version (1985). The latter is the most widespread and used by several organizations, as it is a system that shares characteristics with other methods. For the design of plans with this system, the acceptable quality level (AQL) is used, and the standard foresees twenty-six values for the AQL [5].

Using MIL-STD 105E, it is possible to develop new methodologies for attribute sampling or be a basis for developing new ones. This standard is applicable for several areas such as inspection of components, raw materials, operations, services, and materials in process, supplies in storage, maintenance operations, and administrative processes. The literature on acceptance sampling and inspection by attributes is scarce. Most evaluation and characterization documents used an earlier version of the standard (Mil-Std-105d); however, they could be updated for a few differences. This includes implementing some additional elements, editorial modifications, and reducing the number of lots in close inspection [6].

This work was done in one of the largest supermarket chains in Ecuador, dedicated to wholesale mass consumption products. It has a National Distribution Center (NDC), which receives food and staple products daily, supplying the entire supermarket chain. In 2018, a perishable products reception platform was set up and began operating. Quality assurance in the reception of these products, especially vegetables and legumes, is carried out solely based on the expertise of the operators. This process is done very empirically, generating delays in product dispatches, accumulation of suppliers in the receiving area, and rejection of complete lots. Consequently, products are delivered in poor condition, resulting in shortages and monetary losses due to the lack of products in the various branches nationwide.

2 Method

2.1 Methodology and Approach

According to the study's objective, this research is applied since it has identified a practical problem in the food supply industry. The proposed practices seek to solve the daily needs of this type of company. It is also of the exploratory type since a survey of information will be carried out based on the observation and analysis of the company's conditions. Although based on a methodology that has been validated and tested for decades, it is a process that involves empirical practices since it requires adaptation to its reality. According to the degree of manipulation of the variables, it is quasi-experimental research since it evaluates the cause-effect relationship of the problems in food reception without strict control of the variables. The approach used is quantitative since it uses mathematical techniques for data collection and evaluation. Similarly, descriptive statistics are used for the processing and presentation of results. Considering that the study applies to the entire organization, the whole payroll is indirect participants.

2.2 Ethical Guidelines for Research

This study involves the development of documentation for an organization, so no consent form or other similar documents were required.

2.3 Inspection Techniques and Levels

The standard has three ordinary inspection levels and four special levels. These levels depend on the complexity and importance of the product, i.e., the higher the level, the larger the sample size and the more discriminating the sampling plan. In addition, three sampling levels depend mainly on historical data from all the company's previous lots.

2.4 Types of Inspection

- Normal to stringent (strict): Apply rigorous inspection when two out of five or less than five consecutive lots are rejected during a routine inspection.
- Rigorous (strict) to normal: Normal inspection is reapplied when five lots are accepted consecutively.
- Normal to reduced: This type of inspection is applied when at least ten consecutive lots have been accepted, and therefore the products' quality is considered controlled or stable. The only condition is that when a rejected lot appears, the inspection returns to a routine inspection (Reduced to normal).
- Suspension of inspection: Sampling is suspended when a rigorous inspection is carried out, and five consecutive lots are rejected [9].

2.5 Inspection Levels

- I: This level is used for a few rejected products and requires less inspection than level II.
- II: Most commonly used, unless otherwise indicated.
- III: This level is used when many products can be rejected.

2.6 Special Levels

S1, S2, S3 and S4: It is applied when small sample sizes are required—taking risks of accepting poor quality.

2.7 Data Analysis Techniques

To better understand the internal processes of the organization, it was necessary to observe and talk directly with the workers. This way, the activities carried out were established, and immediate improvements could be defined. In addition, audits were carried out based on a standardized norm, where the main findings were determined to evaluate the company's performance.

2.8 Materials

Once the standard guidelines are detailed, we proceed to carry out the sampling based on the Military Standard (MIL-STD 105E) for the company described above. The flow chart shown in Fig. 1 details the activities carried out by the quality coordinator and the reception operator, respectively, to carry out the sampling.

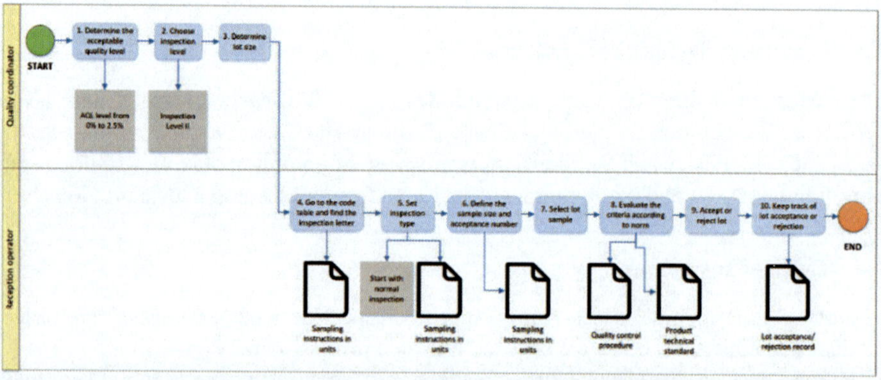

Fig. 1. Sampling flow chart for vegetables and legumes.

The sampling process requires identifying the key elements with which the methodology works. For the company in question are suppliers, inputs, processes, outputs, and customers. At the same time, a SIPOC diagram was developed, which complements the flow chart and summarizes the instructions to continue with the sampling. This diagram can be seen in the following link: https://github.com/faviles7/acceptance_sampling/ since it could not be included due to the length of the document.

2.9 Instructions for the Acceptance Sampling of Vegetables and Legumes in Units

Equipment and Utensils

- PPE (hard hat, steel toe shoes, reflective vest).
- Clothing provided by MSM Occupational Health and Safety area.
- Biosafety equipment (face mask, goggles or visors, gloves).
- Scale.

Materials

- Plastic trays.
- Plastic pallets.

2.10 Instructions for the Elaboration of the Sampling Based on MIL-STD 105E

- The sample should be randomly selected from the respective batch of products.
- The product lot and the inspection level give the sample size.
- Use the respective tables according to the type of inspection to be performed.
- Use the rules for changing inspection types when required.

2.11 Procedure

The procedure consists of several steps, which are described below: First, determine the acceptable quality level (AQL). The AQL level for product acceptance is 0% and 2.5%. Next, choose the inspection level (level I, II, II, or special). At the start of sampling, the most used inspection level is II unless another level (I, II, II) or (S1, S2, S3, S4) is indicated. Then, determine the lot size (Identify the units of each product according to the purchase order). Next, find the inspection letter in the code table according to the chosen level. Subsequently, establish the type of Inspection (Normal Severe, or Reduced). At the start of sampling, Normal inspection is performed unless otherwise indicated. Then, determine the sample size and acceptance number according to the inspection type: Normal Severe, or Reduced. Next, select the sample of the lot. Additionally, evaluate the criteria according to the technical standards of the product. Next, accept or reject the batch according to the number of defective products. Finally, keep a record of the acceptance and rejection of batches [6].

3 Results

3.1 Diagnostic Audit

The first step is inspecting the product reception process on a typical working day, without changes from the initial batch and in successive batches. Identifying the inspection levels is indispensable for sampling since they allow changing the number of products inspected without affecting the lot size and the sample size. As a starting point, the general assessment of compliance with the requirements of the technical standard ARCSA-DE-067–2015-GGG is performed. The results are shown in Table 1.

Table 1. Diagnostic audit results.

Requirements	Compliance	Non-compliance	N/A	Mandatory compliance
GMP Requirements	50	1	8	59
Equipment and Utensils	7	2	3	12
Hygienic Manufacturing Requirements	15	1	0	16
Raw Materials and Inputs	6	2	4	12
Production Operations	11	2	8	21
Packaging and Labeling	10	0	3	13
Storage, Distribution, Transportation, and Marketing	5	8	3	16
Quality Control	9	4	3	16
Total	113	20	32	165
Percentage	68,48%	12,12%	19,39%	100,0

3.2 Compliance Audit

The proposed methodology generated organizational changes from top management to operational staff. The best way to measure the effectiveness of this proposal is by performing a recheck through a verification audit of the ARCSA-DE-067-2015-GGG standard. The compliance calculation with the eight requirements that make up the standard is repeated to evaluate the influence of the application of this proposal. It should be considered that the standard was applied to the process of receiving vegetables and legumes. Most of it is based on the requirement of documents (mainly procedures), considering that some articles do not apply to this area or procedure.

The standard includes 165 points, of which 131 have been complied with, 32 are not applicable, and the rest are non-compliance. After applying the procedures manual, there is evidence of 79.39% compliance, rated on the descriptive scale as a satisfactory procedure. The 19.39% requirements do not apply to this organization, and only 1.21% represent nonconformities.

3.3 Comparison of Diagnostic and Final Results

Table 2 compares the percentages before and after the development of the procedure's manual. This analysis's overview determined that non-compliances were reduced from 20 to 2, i.e., 90%.

3.4 Standard Operating Procedures (SOP)

Considering the scope of application of the manual and the prioritization of corrective actions mentioned below. The complete description is available at the same link above.

- Five standardized operating procedures according to the needs of the manual.
- Five instructions for correctly executing activities involving contact with the vegetables and legumes received.

Table 2. Diagnostic audit and compliance audit

Requirements	Diagnostic audit	Compliance audit
GMP Requirements	98,04%	98,04%
Equipment and Utensils	77,78%	100,00%
Hygienic Manufacturing Requirements	93,75%	100,00%
Raw Materials and Inputs	75,00%	100,00%
Production Operations	84,62%	92,31%
Packaging and Labeling	100,00%	100,00%
Storage, Distribution, Transportation, and Marketing	38,46%	100,00%
Quality Control	69,23%	100,00%

- Twelve records for the control and follow-up of the procedures.
- Two reports for verification of compliance with procedures and other mandatory requirements.

3.5 Cost Analysis

Analysis of Fixed and Socialization Costs. To apply the manual, a quality technician must be hired to coordinate and direct its implementation. The work period includes the six months that the manual will be applied. In addition, the social security costs are calculated, giving a total of $5,872.80, as shown in Table 3. Similarly, the variable costs are described, where the participants of each meeting and the total cost per person-hour are considered. The annex is available at the same link.

Table 3. Fixed cost analysis

Position	Remuneration	Actual cost	N° months	Total
Quality technician	$ 700,00	$ 978,80	6	$ 5.872,80

4 Discussion and Conclusions

As indicated in [7], more effective procedures can replace empirical manual inspections. Statistical quality control is a technique that has been widely accepted in recent years, as mentioned by Madrigal [8]. This is due to the simplicity of its replicability and the standards used, which allow processes to be standardized. Lugo Cornejo et al. [2] describes that these techniques will allow improving the practices that industrial engineering currently presents, with which we fully agree. Acceptance sampling has a critical approval in the industrial field and has been used in other fields of knowledge. As indicated by Mendoza Rubiano and Zea García [1] it can be used to generate tools that allow students to learn better in subjects related to quality control. While in the industrial field, as in the paper by Rodriguez-Balsa [5] by using this technique, we have designed a sampling plan at a lower cost, despite the large number of products handled. The only work that maintains a similar methodology in the food industry is that of Khatiwada [9]; however, his analysis is different and, therefore, cannot be directly compared.

The main problem in our work was the lack of correct sampling for the types of products received. There was also a lack of quality control in the reception of products and a shortage of documentation. As a result, there was a misunderstanding of the specific activities to be carried out by the operators. All this directly impacts the quality of the products entering the company's NDC. In addition, a diagnostic audit was carried out based on a checklist of the ARCSA-DE-067-2015-GGG regulation. The initial result showed compliance of 68.48%.

Since there was no sampling process for the reception of vegetables and legumes, sampling by attributes was used for products in units following the MILITARY STANDARD (MIL-STD 105E), based on NTE INEN ISO 2859-1:2009. In addition, to mitigate quality failures in the reception of vegetables and legumes, a manual was developed containing the various procedures, instructions, records, and reports. For a better understanding, the SIPOC diagram is described as an appendix. With the development of this manual, five standard operating procedures, five instructions, 12 records, and two reports were generated. All this allowed us to comply with the standard's requirements, and when a new audit was carried out, the unfulfilled requirements were reduced by 90%.

The main limitation was the COVID-19 pandemic, which made it difficult to obtain information for the study. Another factor that should be highlighted is the lack of predisposition of some workers to new methodologies, which delayed the study. The authors of this study propose in future work that other methodologies based on similar standards be developed so that a comparison can be made. One of the initial objectives was to replicate this proposal in other companies in the food sector and be considered a reference.

References

1. Mendoza, A., Zea, Y.: Diseño de guías de aprendizaje para el tema lotes de aceptación en la asignatura de ingeniería de la calidad de la Universidad de Ibagué. Universidad de Ibagué (2019). https://repositorio.unibague.edu.co/jspui/handle/20.500.12313/1370
2. Lugo Cornejo, E., Guerrero Porras, M.G., Ponce Medina, B., Lugo Guerrero, B.R.: Técnicas de ingeniería industrial para el diagnóstico operativo de una empresa industrial. Rev. Tectzapic **3**(1) (2017). https://www.eumed.net/rev/tectzapic/2017/01/empresa-industrial.html
3. Chen, J., Tian, Z., Hang, W.: Optimal ordering and pricing policies in managing perishable products with quality deterioration. Int. J. Prod. Res. **59**(15), 4472–4494 (2021). https://doi.org/10.1080/00207543.2020.1766715
4. Kleeberg-Hidalgo, F., Ramos-Ramírez, J.-C.: Aplicación de las técnicas de muestreo en los negocios y la industria. Ing. Ind. (027), 11 (2009). https://doi.org/10.26439/ing.ind2009.n027.621
5. Rodríguez-Balza, M., Pérez-Ybarra, L., Quintana-Rivero, W.: Aproximación empírica para el diseño de planes de muestreo simple por atributos bajo inspección normal basados en las normas COVENIN 3133–1:2001 y MIL-STD-105E. Enfoque UTE **6**(2), 77–85 (2015)

6. Gutierrez, H., de la Vara Salazar, R.: Control Estadístico de la Calidad y Seis Sigma, 2nd edn., México (2004). https://www.uv.mx/personal/ermeneses/files/2018/05/6-control-estadistico-de-la-calidad-y-seis-sigma-gutierrez-2da.pdf
7. Solano, N.E.C., Bernal, L.J.M.: Método de evaluación de las reglas de cambio entre planes de inspección normal y reducida con base en tablas military standard 105e. Ingeniare **23**(1), 98–106 (2015). https://doi.org/10.4067/s0718-33052015000100012
8. Madrigal, R.: Control estadístico de la calidad – Un enfoque creativo, 1st edn., Ciudad de México (2021)
9. Khatiwada, R.P.: An alternative simplified attribute sampling plan for food product control. J. Food Sci. Technol. Nepal **5** (2009)

Check for updates

Assessing the Impact of Dry Ports to the Supply Chain Safety and Security

Jyri Vilko[1](✉), Antti Jakonen[2], Emmi Rantavuo[2], Oskari Lähdeaho[1], and Ville Henttu[2]

[1] Kouvola Unit, LUT University, Kauppalankatu 13, 45100 Kouvola, Finland
jyri.vilko@lut.fi

[2] XAMK University of Applied Sciences, Paraatikenttä 7, 45100 Kouvola, Finland

Abstract. The introduction of dry ports has changed the structure of the logistics systems in different countries. The impact of dry ports to the supply chain functions and flow of material have been studied in the scientific literature, however the role of the ports in the increased safety and security demand in the current turbulent international business environment have been sparsely discussed. The aim of this paper is to address the forementioned gap in the scientific literature by investigating the impact to supply chain safety and security of introducing new dry port to a multimodal maritime supply chain. The results of the study reveal how the safety and security of the supply chain differs with the introduction of dry port to a multimodal supply chain from different actors' perspective as well as the issues that remain unchanged. The results suggest that dry port can have a clear positive overall impact on the safety and security of the case supply chain.

Keywords: Supply chain management · logistics · dry ports · benefits and risks · safety and security · multimodal · seaport

1 Introduction

Dry port popularity in global transportation has increased in the previous decades to alleviate pressure from seaports. They are also used as substitutes for growing the size of existing seaports, which are usually located near city centers, as cargo flows increase. Thus, dry ports can be seen as necessary for efficient operations between seaports and the cities these seaports serve [1]. The expected benefits beyond those already mentioned are investment costs, cargo flow fluency, and operational efficiency, to name a few. There are previous studies that call for further studies in port safety and security [2, 3], which this research also contributes to in one hand by illustrating a multi-perceptual view of the most important safety and security issues in a multimodal supply chain in Northern Europe and on the other hand by illustrating the impact of a newly built dry port in this context.

The safety and security aspects of the multimodal supply chain have been increasing in importance in the last years with unexpected and high-impact disruptions. In addition, turbulence in global economies and politics (e.g. Covid-19 and the war in Ukraine, to

C. Montenegro et al. (Eds.): ICMTT 2023, LNNS 774, pp. 26–35, 2024.
https://doi.org/10.1007/978-3-031-43733-5_3

name the latest ones) has had an impact on supply chains, where some established supply chains have ceased to exist, but new routes and opportunities have come to replace those. Here the dry port arises as a logistics network node that can alleviate pressures and also offer substitute routes during supply chain disruptions. In this sense, dry port as a concept does not necessarily compete with existing ports but aims to support and complement them [4].

This qualitative case study aims to assess the effects of dry ports on supply chain safety and security aspects. More precisely, the study presents a case supply chain where a newly built dry port is opened, thus providing novel information on the changes it brings to the multi-modal supply chain. The case is built in two parts, where the first one focuses on the current safety and security concerns in a multi-modal maritime supply chain, especially in sea port-related activities. Secondly, the study investigates the impact of dry port to the identified safety and security factors. In doing this, the study contributes to supply chain risk management and logistics management literature. The rest of the article is structured as follows. The next section reviews relevant literature on dry ports as well as on overall port safety and security. Thereafter, methods and materials used in the empirical study are presented. This is followed by the results of the described empirical research. Lastly, the study is concluded with discussions relating the empirical findings to existing literature.

2 Literature Review

A dry port is an inland intermodal terminal, which has access to both rail and road transport networks. A dry port is connected to one or multiple seaports by rail access, and an optional proportion of freight between the dry port and seaport(s) is transported by rail transport mode [5, 6]. Furthermore, the freight is transported in intermodal loading units (such as containers) to speed up the modal change between road and rail and vice versa [7]. Only the final leg of the transport between the dry port and inland location is achieved by road transport.

Possible benefits of dry ports include [5–7]:

- Less congestion on the road network since less freight is being transported by road. One train can transport freight volumes of multiple road transport vehicles.
- Less congestion in the seaport area since direct rail connection is used instead of the road.
- Less congestion in the seaport city surrounding because fewer road transport vehicles are needed to transport the freight to and from the seaport.
- Trains (especially electric ones) emit fewer CO2 emissions, thus increasing the usage of trains between the seaport and dry port will decrease CO2 emissions.
- Trains create less noise.
- There are fewer accidents when trains are used instead of road transport vehicles.
- Rest time of road transport can be achieved easier with long distances because inland located dry ports can be used as resting points for road operators.

Port logistics in supply chains is a prerequisite for the import and export functions of Finland. Thus, the security and safety and the continuous development of them should be prioritized to a high level. A systemic approach to security and safety issues suggests that the communal safety of all port logistics operators will increase when in interaction with each other and secondly strengthen control of security risks [8]. Bye et al. [9] state that special attention to a systemic point of view and practical integration in safety and security related to seafaring should be implemented.

The awareness of human factors' effects is increasing which enables an open and flexible approach to new practices. High risks and dangers are recognized in port logistics. Processes and working stages are complex in the safety and security point of view. Common factors to risks are lack of an effective safety culture, extreme weather conditions, inadequate operating procedures, risk assessment, or operations management. Also lack of training and awareness, increased performance by faster ship turnarounds, increased port capacity, larger vessels, and bigger and faster port equipment are considered as contributing factors [10]. A great proportion of the safety deviations which lead to accidents in port operations are slips, trips, and falls, hits by moving or falling objects, or dust and noise-related injuries. High risk to port workers in container ships is noticed. Also, an increasing number of accidents involve port cranes or other mobile equipment [10].

The human factor can be seen as consistently involved in hazardous cargo accidents at port operations, but often there are also simultaneous managerial factors (e.g., organizational influence, execution issues, safety issues, bad operations, intellectual issues, and violations) at play when such accidents occur [11]. It seems that less serious accidents and near-misses go consistently without reporting, the latter possibly originating from missing consensus on what counts as a near-miss [12]. Port managers should consider the risks related to ships colliding when approaching or moored in their port [13].

Digitalization and the introduction of smart port systems also drives port operational and environmental safety [14]. Safety management systems in smart ports are established to manage safety-related risks, monitor the operations, further develop safety, and distribute safety related information via training and manuals [15]. The impact of weather conditions on maritime as well as port operations has been identified, and forecasting systems are being developed to ensure the fluency and safety of these operations [16].

According to Indonesian seafarers, safe and functional ports are weatherproof, with sufficient infrastructure for ships, have traffic control systems in place, services for ships, eco-friendly, and secure against external harm, such as thieves [17]. These seafarers deem being weatherproof as the most important factor, but also the other mentioned aspects should be developed [17].

Based on the above review of relevant literature on port cargo and working safety, we study the effects of the dry port through three perspectives. These are safety development, weather impact, and cargo safety. The research method is further explained in the following section.

3 Research Methods and Data

This qualitative study applied both primary and secondary data to investigate the impact of the dry port on the safety and security issues in a multimodal case supply chain in Northern Europe. In the research process, First, an integrated literature review was performed by searching the relevant scientific literature about supply chain risk management, maritime supply chains, and dry ports, after which the research questions were formed. Then, semi-structured interviews were performed with informants, who work among supply chains at seaports. The collected material was transcripted and open-coding, grouping, and sub-categorization were made as presented in a study by Elo & Kyngäs [18]. The analysis results were then introduced to other experts, who work among supply chains at or near dry ports. These experts working at dry ports were asked to rate the impact on the found items in analysis results if a dry port is added to the supply chain. The study process is presented in Fig. 1.

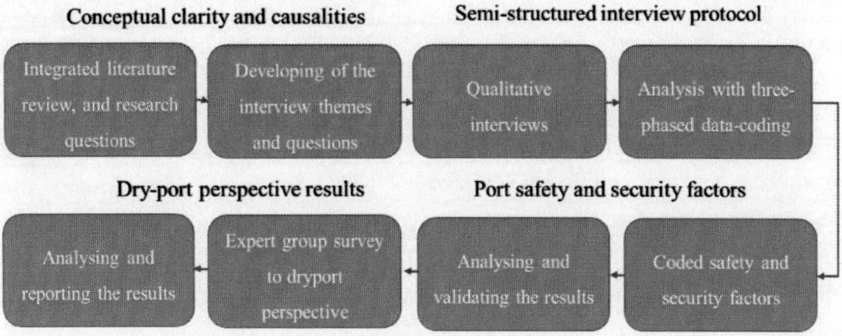

Fig. 1. Research process

The semi-structured thematic interviews (n = 22) were performed from November 2021 to May 2022 with experts and authorities who were working among supply chains at seaports in Finland. Target was to recruit a large scale of informants from the supply chain-related companies and authorities, which operate at the case seaport. The inclusion criteria were that the informants should have solid experience working in the supply chain and should have an interest in developing safety at seaports and the supply chain. The aim was to recruit informants from different levels in companies to gain information about seaport safety and security from different aspects.

The companies were contacted via email and asked to take part in the study resulting in recruitment of 14 informants from 12 supply chain-related companies. In addition, customs and fire & rescue services were contacted directly to take part in the study, as they are the essential supply chain-related authorities, which operate at the case seaport. As a result of the informant recruitment process, five informants from customs, three informants from fire & rescue services, and 14 informants from supply chain-related companies were recruited for the study interviews. The informants' profiles are presented in Table 1.

The interview contents were chosen in accordance with the research question and comprised the following themes about overall safety and security in seaports:

1) Lacks, needs and current state in safety and security at seaports
2) Effects of weather conditions on safety
3) The present state and development needs in cargo safety

Table 1. Informant information

Occupation	Degree	Experience (y)	Role in the supply chain
HSEQ Manager	Master's Degree	13	Supply Chain Auditor
Production Manager	Further Vocational Qualification	44	Port (Un)Loading planning and supervision
Docker/Chief Shop Steward	Further Vocational Qualification	16	Stevedor supervisor
Specialist in Railway and Occupational Safety	Basic Education Certificate	17	Railway safety and Health
Safety and Security Manager	Further Vocational Qualification	43	Port occupational safety and security
Stevedorer	Bachelor's Degree	20	Stevedoring
Stevedorer/Occupational Safety and Health Agent	Further Vocational Qualification	33	Stevedoring Health and Safety
Stevedorer	Further Vocational Qualification	19	Stevedoring
Development and Occupational Safety and Health Manager	Master's Degree	20	Port occupational safety development
Occupational Safety and Health Secretary	Basic Education Certificate	12	Labour union safety and health
Shipping Agent	Further Vocational Qualification	25	Maritime processes
Shipping Agent	Bachelor's Degree	32	Maritime processes
Labour Market Director	Master's Degree	30	Port operators' union representative
Foreman/Health and Safety Representative	Further Vocational Qualification	14	Stevedoring supervisor
Chief Executive Officer of Emergency Services Department	Bachelor's Degree	10	Supervision of Emergency services
Fire Protection Engineer	Bachelor's Degree	3	Accident prevention and risk management
Executive Fire Officer	Bachelor's Degree	20	Fire and Rescue team leader

(*continued*)

Table 1. (*continued*)

Occupation	Degree	Experience (y)	Role in the supply chain
Customs Inspector	Further Vocational Qualification	23	Customs officer
Customs Inspector	Undergraduate	24	Customs officer
Customs Inspector/Foreman	Further Vocational Qualification	39	Customs supervisor
Customs Inspector	Further Vocational Qualification	27	Customs officer
Customs Inspector	Undergraduate	35	Customs officer

The found sub-categorized safety and security factors were then introduced to the expert group, which consisted of four highly experienced professionals. These experts had at least over 20 years of experience working in the supply chain. Expert group members had a wide variety of experience in the field, and all had extensive knowledge about dry port operations and experience about seaport operations. The experts were hand-picked to get a holistic view of the supply chain in question, representing different organizations and different parts of the dry port supply chain. The composition of four industry experts was considered appropriate where two were representing the dry port organization, namely the construction and facilitation of the operations. The third expert was representing the rail connection between the port and dry port, and finally the fourth was representing the seaport operations. The role of the expert group was two-fold, firstly to further validate the found items and secondly, to evaluate the differences of the identified factors from the dry port perspective. More precisely, the experts were asked how they found items would be affected if a dry port is added to the multimodal supply chain. The results were coded on a seven-step scale, where −3/3 denotes high negative/positive impact, −2/2 denotes medium negative/positive impact, −1 denotes low negative/positive impact and finally, 0 denotes no impact.

4 Analysis and Results

The coded interview results revealed a total of 266 risk item development factors, these were classified according to three categories, namely (1) Risk development, monitoring, and reporting (102 items), (2) Cargo safety and security (68 items) and (3) Weather (96 items). The items discovered were from various positions in the supply chain, however, all were closely related to the port/dry port operations of the supply chain. The differing positions and responsibilities were illustrated in the results, and in terms of many items found the perspectives and justification for the factors were explained in detail.

The interviews revealed the current situation of safety and security culture in the port and focal companies. The informants identified several issues that would need improvements and development at various levels, namely at (1) the supply chain level and in transparency and visibility in the important issues, in (2) the organizational level where lack of reporting and handling of found grievances, in the (3) operational level and finally in the (4) individual level. A common factor in these different levels were the lacks in the safety culture. This could be observed for example in practices where the costs and efficiency were considered over human safety and security. Mostly, these issues were not reported as well which did not allow them to be visible for other members of the supply chain or executive level in the organizations. The safety and security culture and reporting-related issues could be noticed also in the number of identified items of Risk development, monitoring, and reporting. The informants from different levels allowed to gain a better overall picture of the lacks in the safety and security. The lower-level workers experienced that their concerns were not always taken into account appropriately. Surprisingly, from the perspective of the upper-level actors, the statistical monitoring of the accident frequencies were seen as a negative issue. The statistics were seen to cast an unfair shadow against the companies and thus in small logistics organizations, the accident statistics monitoring against a large amount of overall organizational working hours was seen as a dramatic consequence in safety and security figures. Resources and investments in Cargo handling safety were seen difficult as most of the equipment and machinery is expensive and some cases the organizations couldn't afford to keep all the equipment in proper working order and ensure the worker safety. One of the issues identified in the ways of operations seen as potentially dangerous was the forced piling of the containers. In the limited port area, the high piles of containers were seen to cause danger of falling containers.

Table 2. The most positive and negative items to safety and security

Safety and security items	Total ave.	St Dev
Shared responsibility for cargo safety	2	0.71
Locomotive safety automation	1.8	0.83
Rails constructed to needed locations	1.8	0.83
Before new construction, there should be discussions about the impact of construction/buildings on safety	1.8	0.43
Darkness and slipperiness during winter	1.8	1.09
Taking account extreme conditions in work planning	1.8	1.09
Consideration of extreme weather conditions	1.8	1.09
Lighting improvements in the dark and in the rain	1.8	1.09
Snow removal	1.8	1.09
Unloading the wagons in extreme conditions (indoors possibility)	1.8	1.09
Hazardous situation reporting	−0.8	1.48
Wrong safety instrument installing due to a lack of training	−0.8	1.3

(*continued*)

Table 2. (*continued*)

Safety and security items	Total ave.	St Dev
Customer demands affect cargo safety development	–0.8	0.83
Systematic safety reporting	–1	1.22
Individual dependent level and accuracy of safety reporting	–1	0.71
Accidents due to overestimating own strengths	–1	1.22
Truck overloading	–1	1.58
Overloading and poor tying	–1	1.58
Customers are not ready to pay for cargo safety	–1	1.22
Accidents caused by own actions are easily not reported	–1.3	1.3

When considering the dry port perspective the overall results indicated that the supply chain safety and security would benefit from adding a dry port. The expert group scores for the identified items varied highly. On average, the results indicated a minor improvement to the total safety and security issues (0.6) with responses deviating somewhat (standard deviation 1.1). The top ten most positive and negative scored items are shown in Table 2. The most positive items seemed to focus on the common safety and security responsibilities, which where the dry port was seen to add the collaboration between the parties, as it was lacking between the rigid port organizations. Furthermore, weather-related issues close to the sea were seen to have significant improvements as well. The most negative issues were seen to relate on the individual level of the workers, and to the safety culture as well as to the overall safety and security reporting when the number of actors in the supply chain increases. An interesting notion was that the planning of safety and security-related issues was seen to improve, however, the reporting of the occurred risk events was seen to worsen in some cases.

Most of the items were seen to have some kind of impact by the dry port, however, some items remained unchanged by the dry port. These were typically related to the cost of improving equipment and machinery, or to the lack of communication between the maritime vessels and other actors in the supply chain. Furthermore, the differences between managers' and workers' perceptions to safety and security issues remained unchanged and some individual-level responsibilities were not seen affected. Interestingly, the deviation between the responses of the expert group varied the least among those things that remained unchanged.

5 Concluding Discussion

This study aimed to discover the impact of a dry port on the safety and security issues in and multimodal supply chain. According to our results, the dry port seems to improve the overall safety and security levels in the case supply chain. The benefits of the dry port were seen as especially beneficial in the planning of the supply chain operations, thus allow sufficient time to conduct the operations as haste was considered a major safety and security issue in the port. A dry port allows flexibility to the operations of the port and improved planning of the operations. Furthermore, an inland dry port was not considered as vulnerable to the seaside extreme weather conditions, which decreased the amount of uncertainty due to weather and more safe and more secure way to operate. The additional space which caused some risks due to the high stacking of containers was also seen to be improved by using additional space in the dry port and thus allowing the port to operate with lower stacking of the containers.

Our study contributes to the current literature firstly by continuing studies on port safety and security (as called by Nguyen et al., [3]). Thereafter we strengthen the notion that dry ports support existing logistics operations in specific regions [4], improving operational efficiency [1]. Furthermore, on port safety, we find that weather is a central factor (similarly to [16, 17]). In addition, we recognize the human factor as a central part in causing safety issues (similarly as, [11, 12]) such as having a negative impact on hazardous situation reporting, systematic safety reporting, or for example, overloading. Our methodology allows a closer view of the effects of including dry port to logistics systems on cargo and worker safety and security. In this manner, we add nuance to the existing literature on dry ports and overall safety and security in the logistics industry.

The research has several implications, which help managers and scholars in following ways: Firstly, the study illustrates the less explored issue of dry ports and more precisely the impact of dry port to the safety and security aspects of supply chains. The found items illustrate the challenges that a multimodal face in focal operations near seaports, and how inland dry port impact those. While dry ports are known concept internationally, in some countries, they have not been introduced. Clearly, the dry ports have their advantages, especially in a turbulent and uncertain environments and supply chains. This study illustrates some of the benefits, but also takes into account the negative effects and issues that remain unchanged. This will help researchers and managers in the field in both future research and decision making in the field.

This study has limitations due to the limited number of participants. The generalizability of the results is also limited because the study is conducted in a certain geographical area in certain time, which can impact the results. Further studies should be conducted in order to confirm the results, with higher number of informants from different parts of the supply chain and in different geographical locations. Furthermore, we recommend that the risks involved in the seaport and dry port operations should be further investigated due to the findings and concerns expressed by the informants. Indeed, the research in safety and security can be considered a still insufficient to answer to all the challenges faced in the multimodal maritime supply chains.

References

1. Zain, R.M., Salleh, N.H.M., Zaideen, I.M.M., Menhat, M.N.S., Jeevan, J.: Dry ports: redefining the concept of seaport-city integrations. Transp. Eng. **8**, 100112 (2022). https://doi.org/10.1016/j.treng.2022.100112
2. Varese, E., Bux, C., Amicarelli, V., Lombardi, M.: Assessing dry ports' environmental sustainability. Environments **9**(9) (2022). MDPI. https://doi.org/10.3390/environments9090117

3. Nguyen, H.P., Pham, N.D.K., Bui, V.D.: Technical-environmental assessment of energy management systems in smart ports. Int. J. Renew. Energy Dev. **11**(4), 889–901 (2022). https://doi.org/10.14710/ijred.2022.46300
4. Jeevan, J., et al.: The symbiotic relationship between seaports and dry ports: an analysis of the ambidextrous functionalities of freight nodes and implications on regional development. J. Marit. Res. **19**(2), 44–53 (2022)
5. Roso, V.: The emergence and significance of dry ports: the case of the Port of Goteborg. World Rev. Intermodal Transp. Res. **2**(4), 296–310 (2009)
6. Roso, V., Lumsden, K.: A review of dry ports. Marit. Econ. Logist. **12**(2), 196–213 (2010)
7. Khaslavskaya, A., Roso, V.: Dry ports: research outcomes, trends, and future implications. Marit. Econ. Logist. **22**, 265–292 (2020)
8. Adjekum, D.K., Fernandez Tous, M.: Assessing the relationship between organizational management factors and a resilient safety culture in a collegiate aviation program with safety management systems (SMS). Saf. Sci. **131**, 104909 (2020). https://doi.org/10.1016/j.ssci.2020.104909
9. Bye, R.J., Holmen, I.M., Størkersen, K.V.: Safety in marine and maritime operations: uniting systems and practice. Saf. Sci. **139**, 105249 (2021). https://doi.org/10.1016/J.SSCI.2021.105249
10. Corrigan, S., Kay, A., Ryan, M., Ward, M.E., Brazil, B.: Human factors and safety culture: challenges and opportunities for the port environment. Saf. Sci. **119**, 252–265 (2019). https://doi.org/10.1016/J.SSCI.2018.03.008
11. Khan, R.U., Yin, J., Mustafa, F.S., Wang, S.: Analyzing human factor involvement in sustainable hazardous cargo port operations. Ocean Eng. **250**, 111028 (2022). https://doi.org/10.1016/j.oceaneng.2022.111028
12. Majumdar, A., Manole, I., Nalty, R.: Analysis of port accidents and calibration of Heinrich's pyramid. Transp. Res. Rec. **2676**(2), 476–489 (2022). https://doi.org/10.1177/03611981211044447
13. Vidmar, P., Perkovič, M., Gucma, L., Łazuga, K.: Risk assessment of moored and passing ships. Appl. Sci. **10**(19) (2020). https://doi.org/10.3390/app10196825
14. Paulauskas, V., et al.: Smart ports' influence on coastal sustainability. In: Transport Means - Proceedings of the International Conference, pp. 396–401, October 2021
15. Molavi, A., Lim, G.J., Race, B.: A framework for building a smart port and smart port index. Int. J. Sustain. Transp. **14**(9), 686–700 (2020). https://doi.org/10.1080/15568318.2019.1610919
16. Makris, C., et al.: Integrated modelling of sea-state forecasts for safe navigation and operational management in ports: application in the Mediterranean Sea. Appl. Math. Model. **89**, 1206–1234 (2021). https://doi.org/10.1016/j.apm.2020.08.015
17. Sulistiyono, A.B., Mutmainnah, W., Seputro, F.H., Furusho, M.: A proposal for criteria of safe and functional port: the perspective of seafarers. IOP Conf. Ser. Earth Environ. Sci. **557**(1) (2020). https://doi.org/10.1088/1755-1315/557/1/012046
18. Elo, S., Kyngäs, H.: The qualitative content analysis process. J. Adv. Nurs. **62**(1), 107–115 (2008)

Marketing Innovation

Associated Factors in the Shopping Experience that Influence Customer Satisfaction Mediated by Mood

J. Arce-Stojkic, S. Chicaña-Huanca, and O. Gutierrez-Aguilar(✉)

Catholic University of Santa Maria, Arequipa, Peru
{71327130,schicanah,ogutierrez}@ucsm.edu.pe

Abstract. The primary purpose of this research is to determine the factors associated with the shopping experience that influence customer satisfaction as mediated by the mood in the leading department stores in the city of Arequipa, Peru. The way the study was done was similar to non-experimental research. The questionnaire was given to a sample of 405 people (n = 20; = .918 = .914). The results obtained showed a positive relationship be-tween comfort factors and mood and purchase intention; mood towards customer satisfaction; innovation towards mood and customer satisfaction; purchase intention towards mood, perceived price, and customer satisfaction; perceived price towards mood, establishment innovation, and customer satisfaction. The related environmental factors, such as the comfort of the services in the shopping centres studied, are constituted as common complementary factors and would not have a statistically significant effect on customer satisfaction.

Keywords: Purchase experience · Convenience · Mood · Customer satisfaction · Purchase intention · Perceived price

1 Introduction

Different factors are associated with the purchase experience in the buyer's journey, such as eminently environmental factors, capable of generating comfort in department stores. However, these may vary according to the retail store types [1]. Aspects related to the establishment's innovation can positively influence the customer's moods; additionally, variables such as purchase intention and perceived price in satisfaction reflect judgments that capture the entire consumption experience [2]. In addition, customer satisfaction will significantly positively impact purchase intention [3]. Properly used ambient music can significantly strengthen satisfaction, promote approach behaviour, and increase the desire to navigate and explore; foreground music can significantly affect arousal emotion [4].

A sum of factors related to experiential experiences fosters favourable climates in the shopping experience in such a way that brands that can offer meaningful experiences to customers improve customer satisfaction, generate positive word of mouth and obtain

C. Montenegro et al. (Eds.): ICMTT 2023, LNNS 774, pp. 39–48, 2024.
https://doi.org/10.1007/978-3-031-43733-5_4

a competitive advantage over the competitors [5]. Companies must adapt their approach to the buyer journey to minimize pain points that can cause customers to abandon their store journey [6]. These insights help companies provide a personalized journey for their customers and improve customer satisfaction and loyalty.

The idea of comfort, explained by Daniels [7], Simmons [8] and Spake [9], occurs from a psychological perspective that represents a sense of tranquility that facilitates feelings of calm and carefree within an environment. Shopping in comfortable settings is perceived as more effective and/or efficient and tends to be reported as more emotionally rewarding. Therefore, if the objective of a retailer is to increase the purchase value, it is worth considering the convenience of the consumer [10].

Brand Equity is the most critical factor in customer satisfaction; consumers indicate that, in terms of satisfaction, Brand Equity is more important than Value Equity and Relation Equity, regardless of the type of retail industry. In any case, it is important to clarify that sustainable growth can be achieved in relation to these values [11]. Garaus & Wagner warn that digital signage in the waiting area of a retail store produces an increase in the general satisfaction of consumers in the store. For its part, digital signage reduces perceived waiting time and increases affective evaluations of waiting time [12].

On the other hand, innovation allows companies to increase their market share against their competitors and achieve objectives in a constantly changing external environment [13], in recent years, significant attention has been paid to marketing innovation, which is the implementation of new methods that involve significant changes such as packaging, product location, the price [14] and other types of innovation, such as technological [15]. In general, the most accepted interpretation of the nature of innovation is that proposed by Edwards-Schachter, which refers to the notion of «novelty» through the generation and successful implementation of ideas transformed into more or less «tangible» products and services, that is, technological and non-technological inventions [16].

Studies show that satisfaction is often studied in relation to service quality and perceived value; however, the idea of price, such as price fairness and perceived price, could expand on how to achieve higher levels of customer satisfaction [17]. The perceived price has an essential role in consumers' purchase behaviour towards brands as the main factor of the purchase objective [18], it should be noted that the attractiveness of the perceived price has a significant difference according to gender [19].

Numerous studies have shown that the intensity of sensory attributes affects purchase intention [20], in such a way that the purchase intention is influenced by sensory factors, non-sensory factors and emotional responses [21], as is the case with relaxing music, affects the consumer experience, which in turn affects the purchase intention [22], establishing a relationship of significant influence on the purchase intention [23].

Covid-19 has allowed people to internalize ICT, especially university students, generating latent emotional states to promote meaningful learning [24]. Moods are long-lasting emotional states that manifest themselves positively or negatively [25], therefore, the state of mind is a normal emotional state of a person, which tends to last a specific time [26]. Shoppers buy more when they have a powerful negative mood or emotion; in many cases, shopping helps them overcome their negative mood [27–29]. Taking the previous considerations, the following hypotheses are formulated:

- H1. There is a statistically significant effect between comfort and mood.
- H2. There is a statistically significant effect between comfort and customer satisfaction.
- H3. There is a statistically significant relationship between mood and customer satisfaction.
- H4. There is a statistically significant effect between establishment innovation and mood.
- H5. There is a statistically positive effect between store innovation and customer satisfaction.
- H6. There is a statistically significant relationship between purchase intention and mood.
- H7. There is a statistically significant effect between purchase intention and customer satisfaction.
- H8. There is a statistically significant effect between perceived price and mood.
- H9. There is a statistically significant effect between perceived price and customer satisfaction.

2 Methodology

For the study, online surveys were distributed to a sample of 405 people of both sexes in the city of Arequipa, Peru, between October and November of 2021. The sample studied corresponds mainly to employees of private companies (41.98%); independent workers (24.20%); government employees (13.09%); university students (9.38%); housewives (6.67%), unemployed (2.47%) and others (2.22%). Being. Women account for 62.46% of the population, while men account for 37.54%, with ages ranging from 18 to 65.

The applied instrument was specially designed for the present study and validated in the semantic dimension and specialized content by experts in advertising and marketing; in addition, statistical tests were carried out for reliability, the result of which was a Cronbach's (α) of 0.918 and McDonald's (ω) of 0.914. Subsequently, an Exploratory Factor Analysis (EFA) was performed, and the result for the Kaiser-Meyer-Olkin (KMO) test was 0.900 and Bartlett's test with a p-value of $<.001$.

In the EFA, the following criteria were used, manual number with six factors, an orthogonal rotation of varimax for independent variables, and output options with items greater than 0.50. To determine the robustness of the model, the Confirmatory Factor Analysis (CFA) was carried out, in which the additional adjustment indices were obtained that allow us to compare the theoretical matrix with the observed matrix from initial results, with its TLI being 0.953, which is a comparative fit indicator, and ideally a value greater than 0.950. Regarding the RMSEA, which is an indicator of total adjustment, the value obtained is 0.055, which is adequate. For the structuring and subsequent analysis of the data, Structural Equation Modeling was chosen, using partial least squares, based on variances (PLS-SEM) [30]. The development of the research followed at all times a systematic and rigorous process of data classification and analysis with the statistical software IBM SPSS Statistics (v. 27), Smart PLS (v. 3.3.9) and JASP (v.0.16).

3 Results

The results that are evident in Table 1 allow the evaluation of the measurement model for reflective constructs of common factor items, the individual reliability criterion of the indicator was used, accepting as desired threshold in external loads values equal to or greater than $\lambda \geq 0.700$, this indicates that the construct explains more than 50 percent of the variance of the indicator, thus providing an acceptable reliability of the item [31]. The values obtained exceed the desired acceptance threshold, so the items proposed in the proposed model express reliability traits.

Table 1. External Loads - Smart PLS.

	Comfort (CMD)	State of mind (EDA)	Establishment Innovation (INO)	Purchase intention (ITC)	Perceived Price (PRP)	Customer satisfaction (SAT)
CMD1	0.818					
CMD2	0.848					
CMD3	0.819					
EDA1		0.892				
EDA2		0.919				
EDA3		0.882				
EDA4		0.835				
INO1			0.816			
INO2			0.828			
INO3			0.877			
ITC1				0.850		
ITC2				0.857		
ITC3				0.877		
PRP1					0.823	
PRP2					0.928	
PRP3					0.858	
PRP4					0.905	
SAT1						0.893
SAT2						0.923
SAT3						0.881

For the reliability analysis of the construct or internal consistency, this was done through Cronbach's Alpha ≥ 0.7, whose values are between 0.771 and 0.905. For the Composite Reliability (ρc) ≥ 0.7 suggested by Dijkstra-Henseler (ρA) ≥ 0.7 [32], the results obtained exceed the minimum adjustment indices required and are between 0.868 and 0.934. Regarding the Convergent Validity criterion, the Average Variance Extracted (AVE) was used, the values must be ≥ 0.5, this indicates that the construct explains at least 50% of the variance of its elements, as shown in Table 2 the results are satisfactory because all its values exceed 0.500. As for using the construct reliability criterion, the coefficient (rho_A) is recommended to verify the reliability of the values obtained in the construction and design of the PLS-SEM [33], the results obtained in (rho_A) must have values of 0.7 or higher to demonstrate composite reliability, the results obtained are between 0.771 and 0.918.

Table 2. Assessment of the measurement model for reflective constructs.

	Cronbach's alpha	rho_A	Composite reliability	(AVE)
Comfort	0.771	0.771	0.868	0.686
Mood	0.905	0.908	0.934	0.779
Establishment innovation	0.793	0.798	0.879	0.707
Purchase intent	0.827	0.828	0.896	0.742
perceived price	0.902	0.918	0.932	0.773
Customer satisfaction	0.881	0.882	0.927	0.809

To verify the discriminant validity of the model, two criteria were used, which will allow knowing to what extent the construct is different from other constructs or measures different things. In Table 3, using the Fornell-Larcker criterion, there is discriminant validity [34] in the latent variables that are part of the structural model, this is due to the square root of AVE in each variable, its resulting values should be greater than the results of the correlation between the variables studied.

Table 3. Fornell-Larcker criterion

	Comfort	mood	Establishment innovation	Purchase intent	perceived price	Customer satisfaction
Comfort	0.828					
Mood	0.443	0.883				
Establishment innovation	0.451	0.504	0.841			
Purchase intent	0.38	0.398	0.355	0.862		
perceived price	0.243	0.207	0.363	0.569	0.879	
Customer satisfaction	0.427	0.68	0.573	0.508	0.374	0.899

Another criterion to know the Discriminant Validity in reflective models is the so-called Heterotrait-Monotrait (HTMT) [35] and is defined as the mean value of the item correlations between the constructs relative to the (geometric) mean of the average correlations for items measuring the same construct [31]. The acceptance threshold must be in the following proportions $\leq 0.85 \leq 0.9$, the results obtained suggest that the proposed structural model also meets the criterion of discriminant validity. See Table 4.

Table 4. Heterotrait Criterion - Monotrait –HTMT.

	Comfort	Mood	Establishment innovation	Purchase intent	Perceived price	Customer satisfaction
Comfort						
Mood	0.531					
Establishment innovation	0.576	0.595				
Purchase intent	0.475	0.458	0.435			
Perceived price	0.293	0.224	0.425	0.659		
Customer satisfaction	0.518	0.758	0.683	0.594	0.415	

Figure 1 is a graphical representation of the reliability and validity analysis of the model, based on the R^2 (coefficient of determination). In the proposed structural model, comfort; purchase intention; the perceived price and the innovation of the establishment assume the role of exogenous variables, while the mood assumes a mediation role between the endogenous variable customer satisfaction. The results obtained through

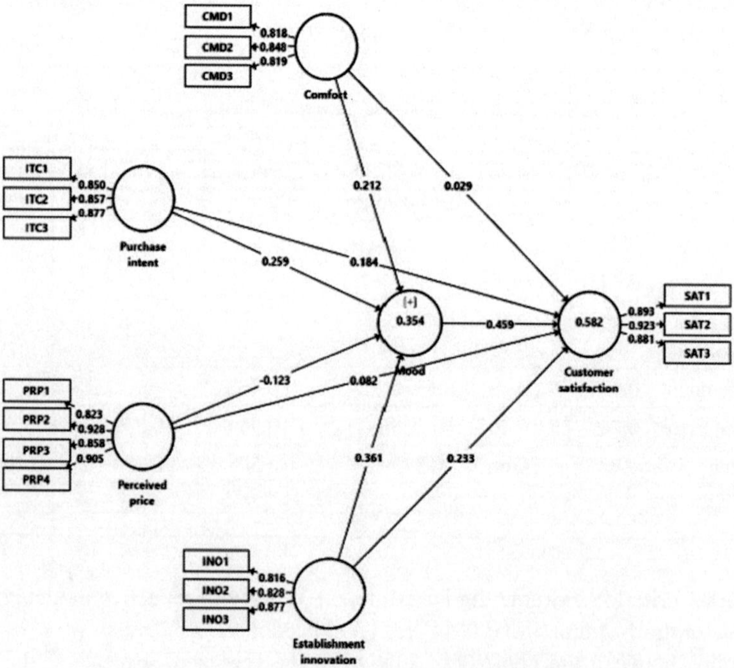

Fig. 1. Model R^2 - SmartPLS.

the results of R^2 show statistically significant effects of influence between one and the other variables in the model.

Table 5. Bootstrapping test results.

	Original sample (O)	Sample mean (M)	Standard deviation (STDEV)	Statistics t (|O/STDEV|)	P-Value
(H1) Comfort -> Mood	0.212	0.212	0.057	3,745	0.000
(H2) Comfort -> Customer satisfaction	0.029	0.029	0.045	0.647	0.259
(H3) State of mind -> Customer satisfaction	0.459	0.458	0.055	8.35	0.000
(H4) Innovation of the establishment -> State of mind	0.361	0.362	0.057	6,389	0.000
(H5) Establishment innovation -> Customer satisfaction	0.233	0.234	0.058	4,003	0.000
(H6) Purchase intention -> Mood	0.259	0.258	0.057	4.52	0.000
(H7) Purchase intention -> Customer satisfaction	0.184	0.183	0.05	3,677	0.000
(H8) Perceived price -> Mood	−0.123	−0.122	0.053	2.32	0.010
(H9) Perceived price -> Customer satisfaction	0.082	0.083	0.045	1,834	0.033

This study focused especially on the level of influence of the different factors as exogenous variables that could influence customer satisfaction, mediated by mood, the goodness of the structural model, has allowed the generation of other statistically reasonable relationships of influence and causality. Table 5 shows the results of the hypothesis tests - Bootstrapping, and how the model variables were hypothesized, therefore, considering the P Value ($p < 0.05$), the hypotheses H1, H3, H4, H5 are admitted. H6, H7, H8 and H9, and hypothesis H2 is rejected.

4 Conclusions

Customer satisfaction is a fundamental aspect of business success, so satisfied customers are more loyal and attract more customers through positive recommendations; therefore, in the long term, there is an excellent possibility of increasing profitability. Comfort is vital in achieving a satisfactory customer experience, so a comfortable and welcoming environment can make customers feel more relaxed and comfortable. Implementing actions related to the innovation of the commercial establishment, such as adequately used background music, can significantly strengthen satisfaction and facilitate behaviour and mood.

Price is also an essential factor that can influence customer satisfaction. Customers who feel they are getting good value for their money are more likely to be satisfied with their shopping experience. In addition, the price can also influence the mood of the customer. For example, if the customer perceives that he has obtained a reasonable offer, he may feel happier and more satisfied.

The study shows that purchase intention and mood positively influence each other; conversely, factors related to the environment and the comfort of services for the customer are common factors that do not statistically significantly affect customer satisfaction.

Therefore, companies must consider all these factors when designing their marketing strategy and improving the customer experience. By doing so, companies can increase customer satisfaction and loyalty, improving their profitability and long-term success.

References

1. Goić, M., Levenier, C., Montoya, R.: Drivers of customer satisfaction in the grocery retail industry: a longitudinal analysis across store formats. J. Retail. Consum. Serv. **60**, 102505 (2021)
2. Ahrholdt, D.C., Gudergan, S.P., Ringle, C.M.: Enhancing loyalty: when improving consumer satisfaction and delight matters. J. Bus. Res. **94**, 18–27 (2019)
3. Dash, G., Kiefer, K., Paul, J.: Marketing-to-Millennials: marketing 4.0, customer satisfaction and purchase intention. J. Bus. Res. **122**, 608–620 (2021)
4. Yi, F., Kang, J.: Effect of background and foreground music on satisfaction, behavior, and emotional responses in public spaces of shopping malls. Appl. Acoust. **145**, 408–419 (2019)
5. Becker, L., Jaakkola, E.: Customer experience: fundamental premises and implications for research. J. Acad. Mark. Sci. **48**(4), 630–648 (2020)
6. Tueanrat, Y., Papagiannidis, S., Alamanos, E.: A conceptual framework of the antecedents of customer journey satisfaction in omnichannel retailing. J. Retail. Consum. Serv. **61**, 102550 (2021). https://doi.org/10.1016/j.jretconser.2021.102550
7. Daniels, K.: Measures of five aspects of affective well-being at work. Hum. Relat. **53**(2), 275–294 (2000)
8. Simmons, R.G.: IO Comfort with the Self. Extending Self-Esteem Theory and Research: Sociological and Psychological Currents, p. 198 (2006)

9. Spake, D.F., Beatty, S.E., Brockman, B.K., Crutchfield, T.N.: Consumer comfort in service relationships: measurement and importance. J. Serv. Res. **5**(4), 316–332 (2003)
10. Ainsworth, J., Foster, J.: Comfort in brick and mortar shopping experiences: examining antecedents and consequences of comfortable retail experiences. J. Retail. Consum. Serv. **35**, 27–35 (2017). https://doi.org/10.1016/j.jretconser.2016.11.005
11. Kim, W., Kim, H., Hwang, J.: Sustainable growth for the self-employed in the retail industry based on customer equity, customer satisfaction, and loyalty. J. Retail. Consum. Serv. **53**, 101963 (2020). https://doi.org/10.1016/j.jretconser.2019.101963
12. Garaus, M., Wagner, U.: Let me entertain you – increasing overall store satisfaction through digital signage in retail waiting areas. J. Retail. Consum. Serv. **47**, 331–338 (2019). https://doi.org/10.1016/j.jretconser.2018.12.008
13. Azar, G., Ciabuschi, F.: Organizational innovation, technological innovation, and export performance: the effects of innovation radicalness and extensiveness. Int. Bus. Rev. **26**(2), 324–336 (2017)
14. OECD, E.: Oslo Manual 2018: Guidelines for Collecting, Reporting and Using Data on Innovation, The Measurement of Scientific, Technological and Innovation Activities. OECD Publishing, Paris, Luxembourg (2018)
15. Grimpe, C., Sofka, W., Bhargava, M., Chatterjee, R.: R&D, marketing innovation, and new product performance: a mixed methods study. J. Prod. Innov. Manag. **34**(3), 360–383 (2017)
16. Edwards-Schachter, M.: The nature and variety of innovation. Int. J. Innov. Stud. **2**(2), 65–79 (2018)
17. Cakici, A.C., Akgunduz, Y., Yildirim, O.: The impact of perceived price justice and satisfaction on loyalty: the mediating effect of revisit intention. Tour. Rev. (2019)
18. Calvo-Porral, C., Lévy-Mangin, J.-P.: Store brands' purchase intention: examining the role of perceived quality. Eur. Res. Manag. Bus. Econ. **23**(2), 90–95 (2017)
19. Büyükdağ, N., Soysal, A.N., Kitapci, O.: The effect of specific discount pattern in terms of price promotions on perceived price attractiveness and purchase intention: an experimental research. J. Retail. Consum. Serv. **55**, 102112 (2020). https://doi.org/10.1016/j.jretconser.2020.102112
20. Cerrato Rodriguez, W.A., Torrico, D.D., Osorio, L.F., Cardona, J., Prinyawiwatkul, W.: Taste perception and purchase intent of oil-in-water spreads: effects of oil types and salt (NaCl or KCl) concentrations. Int. J. Food Sci. Technol. **52**(10), 2138–2147 (2017)
21. Samant, S.S., Seo, H.-S.: Influences of sensory attribute intensity, emotional responses, and non-sensory factors on purchase intent toward mixed-vegetable juice products under informed tasting condition. Food Res. Int. **132**, 109095 (2020). https://doi.org/10.1016/j.foodres.2020.109095
22. Sharma, P., Agarwal, S.: A research to study the impact of lighting and music on customer purchase intention with special reference to Big Bazaar of Jaipur. Splint Int. J. Prof. **4**(2), 66 (2017)
23. Erlinda, R.P.: The effect of store atmosphere dimension toward purchase intention (study on Warunk Upnormal Malang consumer). Jurnal Ilmiah Mahasiswa FEB **8**(2) (2020)
24. Huanca, S.C., Aguilar, A.G., Perez, A.B.D., Aguilar, O.G.: The internalization of ICT and its influence on emotional states in university students. In: Iberian Conference on Information Systems and Technologies, CISTI 2021 (2021)
25. Karimi, S., Liu, Y.-L.: The differential impact of "mood" on consumers' decisions, a case of mobile payment adoption. Comput. Hum. Behav. **102**, 132–143 (2020)
26. Gao, Y., Rasouli, S., Timmermans, H., Wang, Y.: Effects of traveller's mood and personality on ratings of satisfaction with daily trip stages. Travel Behav. Soc. **7**, 1–11 (2017)
27. Dittmar, H., Drury, J.: Self-image–is it in the bag? A qualitative comparison between "ordinary" and "excessive" consumers. J. Econ. Psychol. **21**(2), 109–142 (2000)

28. Lee, J., Lee, Y.: Does online shopping make consumers feel better? Exploring online retail therapy effects on consumers' attitudes towards online shopping malls. Asia Pac. J. Mark. Logist. (2019)
29. Sneath, J.Z., Lacey, R., Kennett-Hensel, P.A.: Coping with a natural disaster: losses, emotions, and impulsive and compulsive buying. Mark. Lett. **20**(1), 45–60 (2009)
30. Henseler, J., Ringle, C.M., Sinkovics, R.R.: The use of partial least squares path modeling in international marketing. In: Sinkovics, R.R., Ghauri, P.N. (eds.) New Challenges to International Marketing, vol. 20, pp. 277–319. Advances in International Marketing. Emerald Group Publishing Limited (2009)
31. Hair Joseph, F., Risher Jeffrey, J., Sarstedt, M., Ringle Christian, M.: When to use and how to report the results of PLS-SEM. Eur. Bus. Rev. **31**(1), 2–24 (2019). https://doi.org/10.1108/EBR-11-2018-0203
32. Hair, J.F., et al.: Manual de partial least squares structural equation modeling (PLS-SEM). OmniaScience Scholar, España (2019)
33. Dijkstra, T.K., Henseler, J.: Consistent partial least squares path modeling. MIS Q. **39**(2) (2015)
34. Fornell, C., Larcker, D.F.: Evaluating structural equation models with unobservable variables and measurement error. J. Mark. Res. **18**(1), 39–50 (1981)
35. Henseler, J., Ringle, C.M., Sarstedt, M.: A new criterion for assessing discriminant validity in variance-based structural equation modeling. J. Acad. Mark. Sci. **43**(1), 115–135 (2015)

Factors that Influence the Quality of Customer Service in Telecommunications Companies

Luz Angelica Allpocc-Flores[1], Sarela Araseli Morales-Durand[1], Gisela Fernandez-Hurtado[1], and Franklin Cordova-Buiza[2,3](✉)

[1] Business Faculty, Universidad Privada del Norte, Lima, Peru
gisela.fernandez@upn.edu.pe
[2] Research and Innovation Department, Universidad Privada del Norte, Lima, Peru
franklin.cordova@upn.edu.pe
[3] Faculty of Business Sciences, Universidad Continental, Huancayo, Peru
fcordova@continental.edu.pe

Abstract. Customer service is the process of customer support before, during and after the purchase is made. The objective of this systematic review is to determine what is known about the factors that influence the quality of customer service in telecommunications companies in Latin America, in the scientific literature between 2011 and 2021. The PRISMA 2020 methodology was used, the search was performed in the databases Redalyc, Dialnet, Scielo and Scopus, applying inclusion criteria such as year of publication and countries of study. The result obtained shows that the countries of Mexico and Colombia have a high participation with 40% of articles, in turn, it was obtained that 18 articles are of mixed approach, where the most used tool was the questionnaire with 40% of participation and the type of study that prevails is descriptive with 36%. It is concluded that human capital, organizational climate, customer service time, accuracy of information, strategies and management tools have an influence at the time of providing customer service, so the correct operation and application of these factors should be verified.

Keywords: Quality · customer service · telecommunications

1 Introduction

In the last years, companies dedicated to the Telecommunications field have been fundamental pieces for the economic development of the country since, due to the appearance of COVID 19, they had to opt for remote work, thus increasing the number of clients. Therefore, nowadays, the quality of service provided to the customer is taken into consideration. Likewise, [1] point out that the perception of quality varies according to the type of customer; however, this has become a decisive factor when making a purchase. On the other hand, [2] mention that the concept of quality has two sides: the objective side and the subjective side, the former corresponds to the properties of the product and the latter refers to what the customer desires. In turn, [3] concludes that satisfaction is the customer's feeling of pleasure or disappointment after comparing the perceived performance in relation to their expectations.

C. Montenegro et al. (Eds.): ICMTT 2023, LNNS 774, pp. 49–59, 2024.
https://doi.org/10.1007/978-3-031-43733-5_5

In the national context according to [4], the companies with the largest market share still have a significant deficit, according to the standards established by the regulatory agencies in Peru. At the same time, it should be noted that there is no culture that allows knowing these annual indicators, so [5] considers that annual feedback should be carried out to verify compliance with the established objectives.

In the international context according to [6], Mexico was the country chosen to carry out the study on the structure of the telephone industry in Latin American countries, where they mention that they have obtained information provided by eight professional experts in the area of business strategy, commercial area and finance. This empirical analysis was carried out with 22 companies some of them are (Telecom, TIM, Oi, Nextel, Movistar, Comcel, Tigo, Telcel) from countries such as Argentina, Brazil, Chile, Colombia, Mexico; which were chosen to know which ones had the highest stock market capitalization, besides looking for differences in business performance, they came to apply 4 analyses: Cluster analysis, Discriminant analysis, Anova analysis, Analysis of variance and an exploration of the variables. Finally, according to [7] mentions that the final purpose is to identify the competitive advantages of the groups with the best performance and the strategies that have led to achieve it. This is how the importance of knowing the quality provided in customer service in telecommunications companies is evidenced. It also seeks to answer the question: What are the factors that influence the quality of customer service in telecommunications companies in Latin America, in the scientific literature between 2011 and 2021?

In this systematic review, the search for information related to the study variable was conducted. [8] points out that it is necessary to bet on innovation in customer experience, with good relationships and offers that adapt to customer needs. Therefore, it is important to understand the needs of customers, both current and future, this is one of the basic pillars in quality management as it aims to meet their requirements and expectations [9]. It is important to mention that currently there are quality strategies that help to eliminate defects and waste, so it is necessary to choose the appropriate strategy that meets the needs of the organization to obtain competitiveness and other benefits [10]. Likewise, [11] point out that after an evaluation of the queuing theory, 80% of non-conformity was presented, customers feel dissatisfied because they always find long lines, slow queues and among factors, customers have increasingly higher expectations and through improvement proposals and numerous studies they seek to meet the needs of the target audience.

On the other hand, [12] mention that the customer can establish communication with the company through telephone, fax, letter, email, being the most used means the telephone channel. Similarly, add that due to the increased use of the internet companies have implemented other means of communication such as social networks, applications, automated virtual assistants, among others. Complementing the above, [13] state that companies also consider the aspects of response times, proximity, service centers, product variety, among others, due to customer demand. Additionally, [14] affirms that according to the results obtained, there is a highly significant, positive and strong correlation of the service quality variable with customer satisfaction and customer loyalty. On the other hand, [15] obtained as a result that 80% of workers feel demotivated because their effort is not recognized through bonuses, incentives, vacations and others. Likewise, [16], point out the difference between the term's aptitude and attitude; whereby aptitude is understood as the set of skills and attitude as the state of mind. Therefore, the appropriate importance should be given to human capital in any organization.

The justification for this research is that in Latin American countries such as Colombia, Ecuador and Peru there is a technological growth of 300% per year that urge consumers to demand aspects that are not only governed by low prices, but also in the personalization of services in telecommunications companies [17]; Likewise, the diversity of services offered, such as telephony, cellular and television, makes customers demand greater responsibility from these companies; for this reason, companies now seek to position themselves in the consumer's mind to create differences in relation to the competition, which is another determining factor for quality to be increasingly perfected. The objective of the present study is to determine what is known about the factors that influence the quality of customer service in telecommunications companies, in a scientific literature between the years 2011–2021.

2 Methodology

This research uses a method of information gathering known as empirical literature search, which consists of the search, collection and analysis of scientific articles in recognized databases according to the study variable, in this case quality in customer service. In addition, according to [18] a systematic review is a review of a well-formulated question, for which systematic and explicit methods are applied to identify, select and rigorously evaluate the research. All this, in order to find reliable and truthful information. The information search strategy to obtain the suitable articles for the current research in the mentioned databases was the keyword method and was carried out using the synonyms of quality in customer service such as "quality", "customer service", "telecommunications", "customer experience" and "customer service". Another way used to search for information was to apply connectors such as "or" and "and" to obtain articles that are more related to the topic of study.

Within the exclusion criteria, it was of utmost importance to have the factors involved in the quality of service in telecommunications companies, discarding other types of reality at the same time the category of companies belonging to the telecommunications sector. On the other hand, the time span between 2011 and 2021, in order to have a close perception of the quality in force in the last ten years. In the inclusion criteria, companies such as telephony, television and internet were considered, discarding companies in areas such as hotels, education and manufacturing industries. Articles with IRMD structure were also considered, with elements such as introduction, results, methodology and discussion (Fig. 1).

3 Results

This chapter is of utmost importance, since a detailed analysis is made of the 37 articles selected from the different databases, where the data collected after searching for information on our research variable is consolidated. Further details on the databases examined are presented below (Fig. 2):

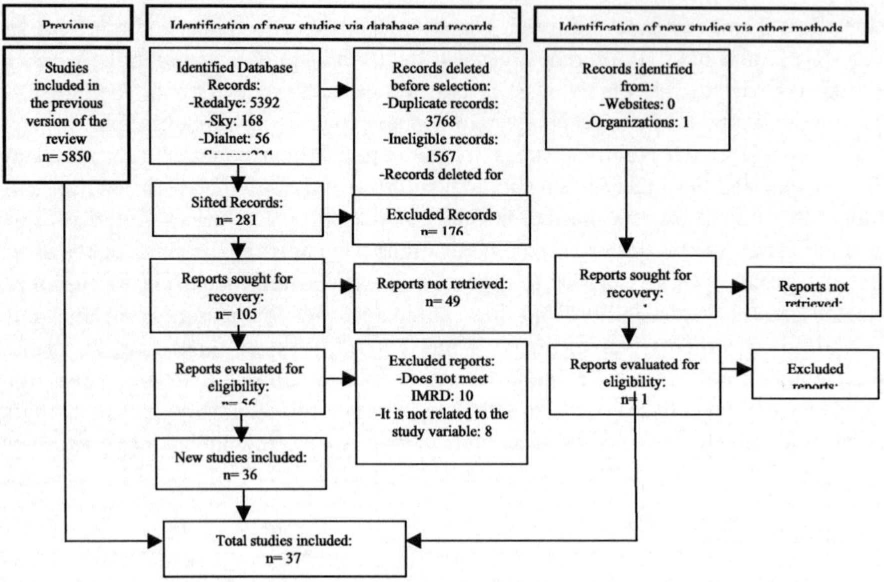

Fig. 1. PRISMA 2020 flowchart.

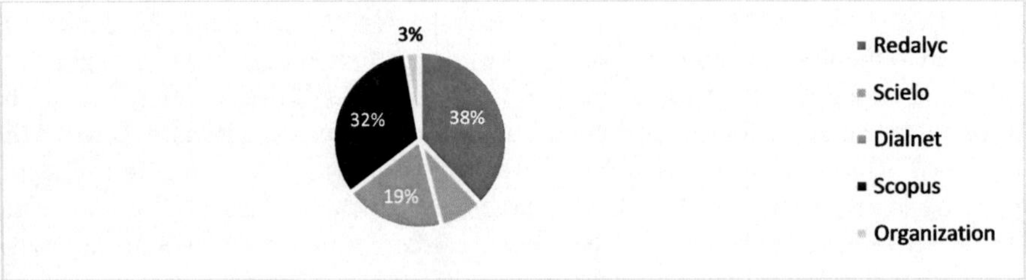

Fig. 2. Database.

The relation of articles destined for the development of the research presents a total of 37 selectively chosen sources, of which it can be observed that Redalyc leads with (38%) on the other hand, the one that has less participation is the Scielo database with (8%). In the identification of new studies through other methods, it is selected that in the identified records of the organizations there are 1 article with a 3% participation, being in this case a journal of the University of Mexico, and this leads to the reports sought for recovery, as well as the same number of reports evaluated for eligibility.

Table 1 shows the 31 names of the journals found where the selected articles were located, 3 articles from Accounting and Administration were found where article (23) refers to the attributes of service quality, followed by article (13) which refers to the value perceived by customers, and article (15) which is related to customer satisfaction.

Table 2 shows the percentage amount of the range of years chosen for this research between the years 2011–2021, and the time range with the highest participation is between 2017–2021 with 56%, being 2020 the year with the highest contribution with

Table 1. Name of journals found.

Magazine name	Articles
Accounting and Admnistration	(13, 15, 23)
Perspectives Magazine	(2, 4)
data industries	(6, 19)
University and society magazine	(24, 26)
National Journal of Management	(9)
Journal of Globalization, Competitiveness and Governance	(5)
Bilo	(7)
ScienceUAT	(21)
Latin American Controversies and Concurrences	(17)
Notebooks of Economy and Administration	(20)
School of Business Administration Magazine	(3)
Digital Publisher	(1)
Business Perspective Magazine	(18)
The financial quarter	(11)
Brazilian review magazine of business management	(25)
Eumedned Magazine	(12)
Venezuelan Magazine of Information, Technology and Knowledge	(10, 17)
Strategic Science Magazines	(22)
Magazine of Economy and Business of Markets of Esic	(8)
advances	(14)
industrial engineering	(16)
Utility Policy	(37, 32)
Telecommunications Policy	(36)
International Journal of Business Excellence	(35)
Procedia-Social and Behavioral Sciences	(34)
Journal of management and technological innovation	(33)
Interactive Marketing Magazine	(31)
Esic Market Economics and Business Journal	(30)
International Journal of Bank Marketing	(29)
International Journal of Electronic Customer Relationship Management	(28)
Mediterranean Journal of Social Sciences	(27)

22%, likewise, the articles of that year coincide in mentioning how telecommunications companies had to deal with the problem of the pandemic by COVID 19, so they decided

Table 2. Articles corresponding to each year.

Year	Article	Representation (%)
2011–2013	(6, 33, 36, 22, 5, 10, 17, 31, 34, 37)	26%
2014–2016	(3, 25, 11, 14, 15, 30, 23)	18%
2017–2021	(26, 29, 24, 2, 12, 19, 20, 32, 35, 7, 8, 9, 13, 16, 17, 27, 28, 1, 4, 18, 21)	56%

to look for other strategies to gain the loyalty of their target audience. On the other hand, it is observed that in the years 2014–2016 there is only 18% of studies present.

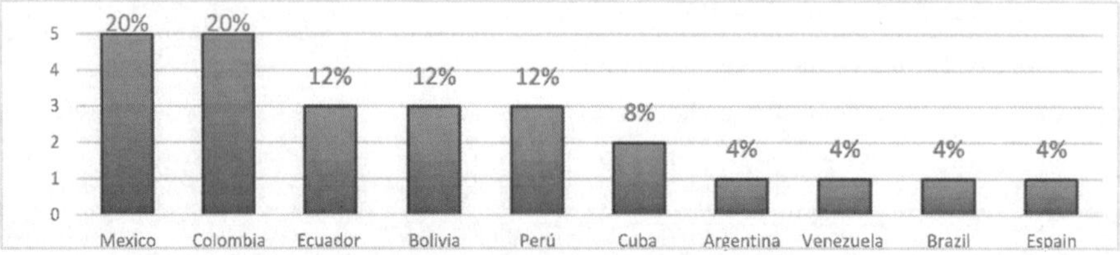

Fig. 3. Countries according to the articles collected.

With respect to Fig. 3, it is concluded that the countries of Mexico and Colombia represent 40% of participation in the selected articles due to their variety of studies, reliable information and affinity with the variables of analysis, in turn, the vast majority of these are focused on increasing the competitiveness of telecommunications companies through the quality of customer service, which is why several studies were conducted in Contac Center to then propose tools for improvement; On the other hand, the countries of Argentina, Venezuela, Brazil and Spain represent a total of 16% of participation, where each one is present with a research article. It should be noted that the country of Spain is included since this article contains information on the Latin American reality.

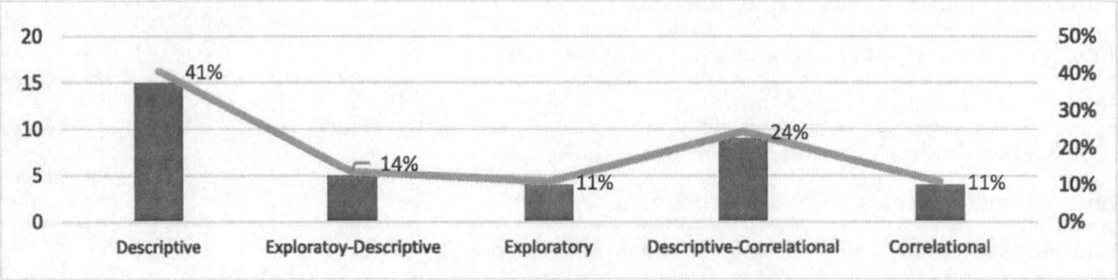

Fig. 4. Types of study.

In relation to Fig. 4, it is important to mention that the type of study with the highest participation is Descriptive with 41%, it should be noted that this type of study is used to analyze how it is and how a phenomenon and its components are manifested, considering the sample and instruments. In turn, the Descriptive-Correlational study only represents 24% of participation, mentioning that the type of correlational study seeks to evaluate the degree of relationship between two variables. Based on the above, it is concluded that the selected articles are prioritized in the Descriptive type of study for the present investigation.

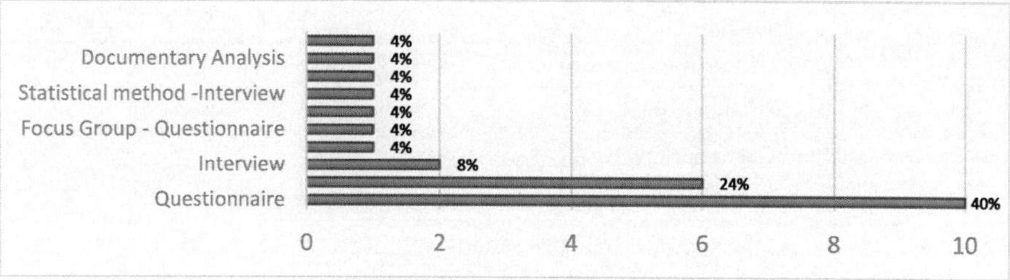

Fig. 5. Types of study.

The analysis in Fig. 5 shows that the most used tool was the questionnaire, representing 40% of participation, which is an indicator that most of the selected articles have a quantitative approach, which allows obtaining real and reliable information. In second place are the interview and questionnaire tools with 24% participation.

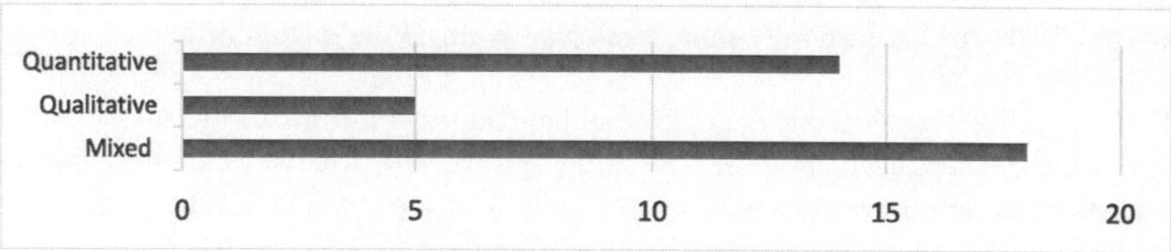

Fig. 6. Methodological approach.

In Fig. 6, first we analyze the techniques used and identifying that the questionnaires are quantitative and the qualitative ones are interviews, we note that the mixed approach is the one most used in the methodology of the articles found, with the participation of 18 articles, likewise the quantitative approach is present with 14 articles and then the qualitative approach with the least amount of only 5 articles, in this way the methodological approach of the 37 articles was developed.

4 Discussion

Table 3 shows the topics of similarity with the study variable, where the topic of customer satisfaction was the one that had the highest presence with 8 articles, according to [19, 20] define that customer satisfaction is the level of mood of a person after comparing the perceived performance with their expectations. Likewise, according to [21] mention

Table 3. Factors found in the articles.

Variable	Factors found	Articles
Attention quality	Work Culture	(17, 29)
	Service management	(16, 31, 32)
	care process	(10, 25)
	Customer satisfaction	(1, 6, 13, 14, 15, 19, 23, 25, 30)
	customer experience	(3, 8)
	Human capital	(20, 26, 28, 29)
	Customer loyalty	(12)
	quality strategies	(22, 18, 27,37)
	Management strategies and tools	(5, 7, 9, 25, 11, 21)
	Organizational climate	(24, 36)
	Corporate values	(2, 4, 33, 34, 35)

that in the Latin American mobile telephony industry competitive advantages can be achieved through the following variables: the number of subscribers, quota variable and customer loyalty. On the other hand, the customer experience through the telephone channel is not only determined by the level of service and productivity efficiency but also by the teleoperators. Some of the tools used are PS, CSAT, FCR and CES, these together with the application of other strategies evaluate the quality of interaction and customer experience [22].

The quality of service that exists in communications companies is linked to intrinsic factors in the company; that is to say, the philosophy or so-called work culture causes the human capital to consistently reflect a proactive function towards the client and at the same time raise the quality of service [23]; according to [24] the internal characteristics of the company is the ability to propose creative and innovative strategies thinking about market preferences; these strategies must be aligned to offer quality and differentiation due to the strong competition that exists in the sector. The organizational environment and climate cause an optimal performance in the collaborators with respect to the customer service systems; thus, the customer service processes are highly efficient [21, 25]. Likewise, ensuring the processes causes greater control over the quality actions directed towards customers, from the first contact and service offering to the follow-up and supervision of the same [26].

The customer experience is a factor that stands out in the decision to maintain appropriate standards in customer service, although it is true that customers increasingly live with new experiences that lead them to be more demanding, i.e., to expect more from the services received, for this reason many companies present differentiation strategies placing the customer at the center so that the treatment, cordiality and attention follow the same direction [22, 27]. On the other hand, there are objectives that direct efforts towards customer loyalty, this is one of the factors that has promoted the improvement of customer service processes, since, due to the level of competition, more and more users are looking for and comparing services, which makes them not permanent in the use of services [28, 29]. Loyalty causes the permanence in time of the company and its positioning and subsistence in the market [30].

Likewise, corporate values create appropriate environments and a climate that allows to interact appropriately with customers and to reinforce skills in the collaborators, such as: communication, team management, assertiveness, and so on [31, 32]. As part of the direction, planning and control of the processes, the corporate values are the basis for building strategies that do not move away from the company's philosophy, while the image of the organization is better cared for, the concept that customers handle the service will be as expected, and the level of preference over the competition will be significant to achieve greater positioning [33, 34].

5 Conclusions

It was possible to establish the factors that influence the quality of service of telecommunications companies; among them the internal elements of the company such as corporate values, organizational culture, service processes and mainly the human capital that is linked to a selection and constant supervision of the functions of the employees, since they are the basis of the relationship between company and customer; from this other factors arise such as the organizational climate, being this a factor that affects the way the market is served. Other factors are also shown, such as the experience that the customer wants to receive and how this pushes the competition to also try to meet the customer's demands to create customer loyalty.

Finally, the factors that influence the quality of customer service were determined, which are: human capital, organizational climate, service time, accuracy of information, management strategies and tools, customer experience, customer loyalty and corporate values. Therefore, it is important to have an adequate organizational climate so that the staff feels motivated in the performance of their duties, and it is also essential to have constant training on customer service procedures and the proper use of tools to streamline customer service and provide effective solutions.

References

1. Benzaquen, J.: Quality in Latin American companies: the Peruvian case. J. Glob. **7**(1), 41–59 (2013)
2. Espinoza, M., Lanza, J., Torrez, A.: Importance of an integrated management system: review of the current context applied to telecommunications companies in the case of Bolivia. Revista Nacional de Administración **11**(1), 61–68 (2020)
3. Moench, E., López, O.: The formation of a work culture. The managerial-performative ethos among the telephone agents of Call Centers. Revista Controversias y Concurrencias Latinoamericanas **11**(20), 157–172 (2020)
4. Arévalo, R., Gabriel, Y.: Methodological proposal to increase competitiveness in contact centers and telephone solutions for companies in the telecommunications sector through the development of the workforce management process. Revista Escuela de Administración de Negocios **1**(76), 92–126 (2014)

5. García, M., Bajo, N., Martínez, F.: Competitive advantages of mobile telephony companies in Latin America. El Trimester Econ. Mag. **82**(235), 89–116 (2015)
6. Grilo, A., Lopes, P.: Characteristics of the work and its implications in the levels of satisfaction of call center employees: a study in a large telecommunications company. Braz. J. Bus. Manag. **16**(53), 658–676 (2014)
7. Henao, L.: Service quality and perceived value as antecedents of customer satisfaction of telecommunications companies in Colombia. Account. Adm. Mag. **65**(3), 1–23 (2020)
8. Ibarra, G., Casas, E.: Application of the Servperf model in Telcel service centers, Hermosillo: a measurement of service quality. Account. Adm. **60**(1), 229–260 (2015)
9. Linares, J., Vilalta, J., Garza, R.: Queuing theory applied to a commercial telecommunications office. Revista Ingeniería Industrial **41**(2), 1–15 (2020)
10. Almeida, R., Laredo, F.: Brand loyalty of mobile phone companies in the students of the Bolivian Catholic University "San Pablo" Cochabamba Academic Unit. Perspect. Mag. **1**(43), 79–106 (2019)
11. Colin, M.: Call center service level: customer experience model from benchmarking and multivariate analysis. Esic's J. Mark. Econ. Bus. **51**(3), 497–526 (2020)
12. Aguirre, J., Pillaca, C., Quispe, R.: Relationship between service quality and mobile service customer retention in telephone operators. Digit. Publ. **6**(2), 164–175 (2021)
13. Montes, H.: Evolution of the concept of quality and contribution to regenerative development from business strategy. Perspect. Bus. Mag. **8**(2), 48–64 (2021)
14. Moreno, E.: Internet service quality and customer satisfaction. Ind. Data Mag. **22**(2), 105–110 (2019)
15. Granda, G., Moya, D.: Loyalty of users of the national telecommunications corporation, Plaza Colonia branch, Guayaquil, based on the theory of service marketing. J. Eumed **1**(12), 11–21 (2019)
16. Camargo, G., Troncoso, A.: Incidence in the sales of a telecommunications company in the city of Barranquilla. Bilo **2**(1), 1–5 (2020)
17. Galbán, O., Clemenza, C., Araujo, R.: Quality of service in the telecommunications sector competitive element in television enterprises by subscription. Venezuelan Inf. Mag. Technol. Knowl. **10**(2), 61–82 (2013)
18. González, J., Buñuel, J., Aparicio, M.: Guideline checklists for systematic reviews and meta-analyses: PRISMA statement. https://evidenciasenpediatria.es/articulo/5902/listas-guia-de-checklist-of-systematic-reviews-and-meta-analysis-prism-declaration. Accessed 10 Oct 2022
19. Riccio, M., Astudillo, B., Vega, M.: Analysis of perception of the quality of customer service in a telecommunications agency. Econ. Adm. Noteb. **6**(3), 130–147 (2019)
20. Silva, J., Macías, B., Tello, E., Delgado, J.: The relationship between service quality, customer satisfaction and customer loyalty: a case study of a commercial company in Mexico. Revista CienciaUAT **15**(2), 85–101 (2021)
21. Azero, V., Almeida, R.: Impact of service quality and customer loyalty satisfaction of mobile phone companies. Perspect. Mag. **1**(48), 37–56 (2021)
22. Hernández, Y., Lazo, G.: Tool to measure the level of customer satisfaction at ETECSA Pinar del Río. Adv. Sci. Mag. **17**(3), 296–306 (2015)
23. Smith, E.: The impact of quality strategy on organizational performance. Revista Ciencias Estratégicas **24**(2), 15–31 (2016)
24. Vera, J.: Mobile telephony service quality attributes for Mexican customers and their impact on satisfaction and brand loyalty. Account. Adm. **58**(3), 39–63 (2012)
25. Vera, N., Suarez, A.: Incidence of the organizational climate in work performance, customer service: telecommunications corporation of the Canton of La Libertad. Univ. Soc. Mag. **10**(1), 180–186 (2018)
26. Rendón, C., Vásquez, A., Benjumea, M., Valencia, A.: Proposed model for measuring customer satisfaction with telecommunications services. Mediterr. J. Soc. Sci. **8**(2), 15 (2017)

27. Kaur, B.: Role of e-service quality in articulating e-WOM: a serial mediation approach. Int. J. Electron. Cust. Relatsh. Manag. **12**(4), 309–325 (2020)
28. Özkan, P., Süer, S., Keser, K., Kocakoç, D.: The effect of service quality and customer satisfaction on customer loyalty: the mediation of perceived value of services, corporate image, and corporate reputation. Int. J. Bank Mark. **38**(2), 384–405 (2020)
29. Díaz, G.: The influence of satisfaction on customer retention in mobile phone market. J. Retail. Consum. Serv. **36**, 75–85 (2017)
30. Aponte, J.: Determiners of consumer trust towards electronic commerce: an application to Puerto Rico. Esic Mark. Econ. Bus. J. **46**(1), 125–147 (2015)
31. Aksoy, L., Buoye, A., Aksoy, P., Larivière, B., Keiningham, T.: A cross-national investigation of the satisfaction and loyalty link for mobile telecommunications services in eight countries. Interact. Mark. J. **27**(1), 74–82 (2013)
32. Resende, M., Cardoso, V.: Mapping service quality in electricity distribution: an exploratory study of Brazil. Utility Policy **5**(6), 41–52 (2019)
33. Paulrajan, R., Rajkumar, H.: Quality of service and customer preference of mobile cellular service providers. J. Manag. Technol. Innov. **6**(1), 38–45 (2011)
34. Arıkan, E., Güner, S.: The impact of corporate social responsibility, service quality and customer-company identification on customers. Procedia Soc. Behav. Sci. **9**(9), 304–313 (2013)

Behavior of the Peruvian Consumer Towards American Casual Dining Restaurant Chains

Lucero Nicole Huerta-Tantalean[1], Nataly Alexis Carrasco-Carrasco[1], Maria Eugenia Castillo-Silva[1], Franklin Cordova-Buiza[2,3](✉), and Wilver Auccahuasi[4]

[1] Business Faculty, Universidad Privada del Norte, Lima, Peru
[2] Research, Innovation and Social Responsibility Department, Universidad Privada del Norte, Lima, Peru
franklin.cordova@upn.edu.pe
[3] Faculty of Business Sciences, Universidad Continental, Huancayo, Peru
fcordova@continental.edu.pe
[4] Faculty of Business and Systems Engineering, Universidad Cientifica del Sur, Lima, Peru
wauccahuasi@cientifica.edu.pe

Abstract. Consumer behavior is changing rapidly and with this companies have to be constantly updating and improving their products and services for customer satisfaction and loyalty. The present investigation has the objective of identifying the most predominant factors in the behavior of the Peruvian consumer towards American Casual Dining Restaurant Chains. A descriptive and non-experimental design investigation was carried out. Likewise, the selected sample was 390 consumers, to whom a survey was applied. The results show that a cultural level does not recognize being influenced in their purchasing behavior; at a social level, they distinguish that the comments or testimonies given by their families and/or friends are relevant; at a personal level, more than half of salary income and economic situation influence consumption; finally, psychologically, they allude to the comfort, location and quality of the restaurant. It is concluded that cultural, social, personal and psychological factors have a strong influence on consumer behavior and should be considered by restaurants in order to grow the business.

Keywords: Consumer Behavior · Cultural · Social · Personal · Psychological

1 Introduction

Critical, conscious and decisive are now the characteristics of the new consumer before the moment of purchase in the market; characteristics that businesses or companies know well [1]. More and more companies are concerned with guaranteeing the benefit of the consumer, understanding and adapting to their evolution, their priorities, desires and interests [2].

Globally, consumer behavior has changed in recent years and more so now after Covid-19, for example, consumer behavior in Europe, specifically in Spain, has undergone a significant change compared to the previous year, 2020. According to García [3],

C. Montenegro et al. (Eds.): ICMTT 2023, LNNS 774, pp. 60–70, 2024.
https://doi.org/10.1007/978-3-031-43733-5_6

Empathy.co in its II Annual Survey of Spanish Consumers, resulted in a 12% increase in online purchases compared to the previous year. In addition, regarding their purchasing preferences, they indicate a decrease in food and an increase in spending on fashion, clothing and sports equipment. The attitude of consumers towards online purchases and the responsibility of choosing a product that protects the concept of personal data have become a remarkable learning in the last 12 months, explained Lara Menéndez, Product Director of Empathy.co [3].

Internationally, the United States consumer has changed his perception of purchase; currently, they choose their products carefully, worrying about the price, the benefits and the difference value that the product promises to provide [4]. Likewise, the American consumer prefers variety, is very selective in his interests and tastes, which is why he is very open to acquiring foreign products [5].

At the national level, the Peruvian consumer has also changed their needs, consumption habits and expectations regarding the different products or services offered, since consumers are clear about what they prefer and what they avoid. There are five trends that characterize the consumer of 2021, determined by the Association of Media Agencies based on the event held by the Interactive Advertising Bureau (IAB Peru), these trends are preferences in online purchases, emphasis on brand activism, experiences from home, savings and entrepreneurship, and greater awareness of food [6].

Regarding Peruvian gastronomy, it has been evolving over the years, the restaurant business grew by 4.60%, unlike previous years [7]. Peruvian consumers seek to live unique experiences in relation to discounts, promotions, personalized attention and quality of service. According to research conducted by NielsenIQ [8], 7 out of 10 people say they focus on ingredients in both the drinks and foods they eat.

Chili's and Fridays are restaurants that offer a variety of dishes and drinks, located in different districts of Peru. Likewise, they are leading restaurant chains that seek to provide experience and quality to their guests, spreading gastronomy through their peculiar dishes and salads. These companies are characterized by their identity and commitment, while their main means of dissemination is through Word Of Mouth (WOM) or word of mouth, one of the oldest allies of marketing, but one of the most truthful in the influence of the market [9]. However, even the most requested restaurants by Peruvian consumers need to be constant with the service, seek loyalty and preference above all things, it is necessary to know the changing behavior of their customers, in addition to providing them with adequate services to guarantee their stay with a Satisfactory treatment and service [10]. Considering the above, the following question is formulated: What are the most predominant factors in the behavior of the Peruvian consumer towards American Casual Dining Restaurant Chains?

There are many theories that address consumer buying behavior. The dimensions are defined from each factor and indicators that affect the consumer's purchasing behavior. Thus, there are cultural factors, social factors, personal factors and finally psychological factors.

Cultural factors are the basis of the different tastes, desires and preferences, these being the main ones that originate the purchasing behavior; finding culture, subculture and social classes as the main ones [11]. Also, social factors are influenced by small groups, family and the roles they play in society. Meanwhile, personal factors correspond to the decisions made according to the characteristics of each individual, age, stages of life, trades, economy, lifestyle, personality, among others. While psychological factors are beliefs, attitudes and perceptions, directly affecting consumer purchasing decisions either positively or negatively [11].

Angnes and Mello [12] obtained a list of 27 attributes that were identified as influential in the choice and evaluation of services in restaurants, among the most relevant attributes are the search attributes, characterized by their tangibility in service. Likewise, Iofrida et al. [13] mention that the most important general attributes for customers are the quality and type of food, comfort and cleanliness, quality of service and value for money; and demographic variables such as gender, age, education and income can also influence the choice of restaurants. For Ryu et al. [14], the general image of the restaurant influences the perceived value as well as the general image and the perceived value have a great impact on customer satisfaction. Bedinelli et al. [15] determined that the price is an endogenous variable and the environment, service, food and cleanliness are exogenous variables in the consumer's perception of value of services in a restaurant.

Carvache [16] mentioned that the main motivations that influence the purchase are friends, families and others to a lesser quantity: the quality of food, facilities, quality of service as well as price. Additionally, Chou et al. [16] mention that "For green restaurant consumers, higher perceived quality will cause higher evaluation of the marketing mix and consumer disposition" (p. 8).

Castro [18] determined that the marketing factor has the greatest relevance in the consumer's purchasing behavior in the restaurant, followed by the culture factor, psychological and finally the person factor. On the other hand, Li et al. [19] mentioned that marketing exposure through different purchase channels does not always produce a long-term behavioral change, but rather a short-term one. In addition, Truong and Truong [20] analyzed the variables that influence purchasing decisions during the COVID-19 pandemic and mention that purchasing behaviors can be affected by demographic variables (age, gender, race, marital status and income), health fears (COVID-19 infection and prevention) and fears about financial conditions (job loss, economic impact payment, difficulty with expenses and use of cash); which shows that consumers respond with emotional actions in the face of risk.

Therefore, in light of all of the above, the research aims to identify the most predominant factors in the behavior of the Peruvian consumer towards American Casual Dining Restaurant Chains.

The present study had no hypothesis since it only seeks to describe the purchasing behavior of a group of people, in this regard, through the aforementioned factors of Chili's and Fridays casual dining restaurants. The importance of the study lies in the finding of information, the purchasing behavior of customers, the evaluation and analysis for subsequent decision making. In addition, the work may be used by enterprises similar or different from the field that have a vision of success. Finally, it will allow expanding knowledge through subsequent investigations that wish to continue with the development of the investigation, and/or make a change in the study methodology.

2 Methodology

The type of research was descriptive with a quantitative approach, this approach is represented by numerical measurement at the same time it is analyzed with statistical methods, in order to establish the conclusions of the study as behavior patterns and the approval of theories [21]. The design was non-experimental - cross-sectional, non-experimental since an investigation was carried out without manipulating the variable, observing the phenomena as they occur in their natural context, for later analysis [21]. Likewise, cross-sectional, because the data collection was carried out in a certain time, adapting the instrument in a single moment and space, in order to describe the variable, analyzing its incidence and interrelation [21].

The population is constituted by all the clients of Lima, capital of Peru, who attend the Chili's and Fridays casual dining restaurants. We worked based on an infinite population, since according to INEI data based on the inclusion and exclusion criteria, statistically the number exceeds 10,000 clients throughout Lima, that is, the exact totality of units and/or elements that make it up is unknown [22]. Likewise, the study had a type of probabilistic sampling - simple random, due to the use of statistical data, with a reliability of 95% and a margin of error of 5%. Finally, through this calculation, based on the infinite population formula, a representative sample of 385 clients was obtained. The study unit was made up of a client from Lima, who, based on their psychographic segmentation, were fans of casual food classified as Bar & Grill. In addition, among the inclusion and exclusion criteria, Lima customers who have ever used one or both casual dining restaurants, that is, Chili's and/or Fridays, in any district of Lima where said restaurants are located, were included. On the other hand, clients under 18 years of age were excluded since they were not considered within the economically active population (EAP).

Based on a quantitative approach, the technique used was the survey, this is a research technique that seeks to investigate a specific sample of the population through questions formulated directly or indirectly, for data collection [23]. The instrument used was adapted from Tongo [24] where he analyzed the predominance of consumer purchasing behavior factors at the Huancahuasi tourist restaurant in the Pachacamac district; as data collection was a questionnaire. The questionnaire was applied according to the representative sample of the population, to 390 clients from Lima who go to the casual dining restaurants Chili's and Fridays. Validity was performed through expert judgment and reliability was obtained through Cronbach's Alpha, where the highest value of Alpha is 1 and in general 0.65 is considered as an acceptable value, while the lowest value is 0; in this sense, the values provide a reliability of 0.789, demonstrating that the instrument is reliable.

2.1 Data Collection

The collection of data and information was carried out by first validating the instrument, consequently, the questionnaire was designed using the Google Forms tool. Said questionnaire was made up of 24 items with response alternatives based on a Likert scale, ranging from a range of (1) totally disagree to (5) totally agree, in addition, general questions used as a filter were formulated, regarding the age, in relation to consumption in one or both of the casual dining restaurants, that is, Chili's and/or Fridays, and based on the district. Finally, the link provided by this tool was sent to a group of relatives, friends and acquaintances. The data analysis was processed with descriptive statistical methods, therefore, shown through statisticians and frequency tables regarding the study variable, that is, the consumer's purchasing behavior. These data were presented, both in statistical tables and in figures, showing the information collected in a summarized and detailed way. Afterwards, the analysis, interpretation and presentation of the data was carried out, for which the programs Microsoft Office Excel 2016 and Microsoft Office Word 2016 were used.

The research was reliable, promoting ethical values, from the beginning of this to the end of the study, underlining its authenticity, it was also carried out protecting the identity of the people surveyed, responsibly using the data obtained solely for academic research purposes. On the other hand, very important criteria were taken into account, such as examining information from the various databases that supported the development of the research, it should be noted that this process was carried out without altering the content exposed in each study found, respecting the corresponding reference manual.

3 Results and Discussion

Next, the results obtained from the characteristics of the sample from the collection carried out through the questionnaire are presented. Most of the respondents were between 26 and 33 years old (43.3%), followed by 22.1% who represent the population between 18 and 25 years old, in third place, there are 80 people who represent 20.5% who are 42 years old or older, and finally with 14.1% those between 34 and 41 years old. Also, most of the interviewees are residents of North Lima (49.7%), followed by 22.4% from East Lima, in third place, there are 70 people represented by 17.9% belonging to South Lima, finally, Lima Center with 10%.

3.1 Cultural Factors

Regarding this dimension, the population is a little polarized, since 39% consider that they disagree with the premise that their cultural beliefs influence their purchase decision in the restaurants studied, 22.3% believe they agree, followed by 26.9% who have ambiguity, on the other hand there are those who strongly believe that it does not influence with 8.2%, finally 14 people represented by 3.6% consider that it does have an influence in its entirety.

Also, it was determined if the dishes offered are part of the customs of the respondents, where 45.6% disagree, followed by 31% who have a neutral position, in turn 16.7% of the sample who agree, while 17 people totally disagree, that is to say that the dishes offered in the restaurants investigated are not part of their customs and finally 2.3% believe that these dishes are part of their customs.

In adition, it was possible to recognize that 36.4% of the respondents consider that the society in which they live is a factor for consuming in these places, followed by an impartial population with 31.8%, in disagreement 106 people were counted represented by 27.2%, while 3.6% think they totally agree, while 1% say they totally disagree.

On the other hand, 55.4% of the respondents agree that their social class facilitates their access to consume in the restaurants investigated, while 25.6% have a neutral position, 11.5% totally agree with this premise, 29 people, represented by 7.5%, disagree and finally, in a null way, there are those who totally disagree.

3.2 Social Factors

In this dimension, 6 items were proposed starting with the first that refers to the influence of friends in the act of purchase, it is observed that 43.6% equivalent to 170 people agree with this information, on the other hand, 30.6% present an ambiguous position regarding this premise, while 21% are not influenced by their group of friends. In the same way, including the group of friends as a factor in the consumption of the investigated restaurants, it was possible to see that 78.2% represented by 305 people believe that it is a factor to be able to attend, while 12.8% have a neutral position. At the same time, it was wanted to know if the respondents make their purchase from a family recommendation, 56.2% agreed with this information, before an impartial position, 28.7% is shown, which is equivalent to 112 people, while 11% disagree. Then it was learned that 80% of the sample consumes in restaurants with their family, who represent more than half, in turn 38 people have an ambiguous position and finally only 3.1% do not consume with any family member.

As for consuming in restaurants gives the respondents a certain status, the majority of these represented by 39% agreed that they disagreed with this premise, while 29.7% said they agreed, however, 27.4% equivalent 107 people present a neutral position, being indifferent to this information. Finally, 199 people consider that their status would not be affected if they stop attending said restaurants, this is equivalent to 51% of the sample, followed by 25.6% who are in total disagreement, this indicates that the status of people would not be affected if they stopped going to the mentioned restaurants.

3.3 Personal Factors

There are 6 proposed items corresponding to this dimension starting with whether the age of the respondents determines what they consume in the proposed restaurants, 40.5% considered that they agreed, while 36.7% had a neutral position, with 18.7% represented by 73 people who disagree at the other extreme. Next, for the item of job occupation as a factor for consumption, it is observed that 161 people mention that they disagree, while 109, represented by 27.9%, consider that they agree, while 25.4% have an impartial position regarding this premise.

As for the third item that refers to whether salary income influences consumption, it can be seen that 59.7% agree with this information, while 13.8% have an impartial position, finally 11.9% represented by 46 people are in disagreement. Also, 279 people respond that their economic situation allows them to access the services of said restaurants, this is equivalent to 71.5%, while 21.3% totally agree and finally only two people considered that their economic situation does not affect finding totally disagree.

On the other hand, considering whether you go to restaurants voluntarily at lunchtime, almost more than half of those surveyed have an impartial position, that is, 53.3%, which is equivalent to 208 people, the other figure that predominates is a 22.3% who agree, while 21.8% have an opposite position.

Finally, the sixth item refers to the accessibility of the price of products and services, in this item the sample considers with 74.3% that the prices are accessible to them, however, 22.1% represented by 86 people have an impartial position to the price.

3.4 Psychological Factors

In this dimension, 8 items were proposed. First, regarding to the fact that the atmosphere of the restaurants was a factor to consume comfortably 66.4% of the respondents consider that it is an important factor and they agree, at the same time, 109 people totally agree with this premise, while one person disagrees. In addition, taking into account the location of the restaurants, it can be seen that for the sample it is a predominant factor, since 56.4% and 33.3% consider that this factor is important to be able to attend, agreeing and totally agreeing, respectively.

It is known that quality is an important factor for any product or service, which is why 84.4% consider that the products and services offered in restaurants have a high quality standard. Meanwhile, 43.1% of those surveyed present an impartial or neutral position regarding the premise that they know in detail the services offered by Chili's and/or Fridays, while 116 people do not know in detail what is offered compared to 24.4% who agrees with the opposite opinion. Likewise, 165 people are neutral to the premise that beliefs are an important cultural factor to consume in the restaurants studied, this is equivalent to 42.3%, followed by 32.3% who disagree and 21.3% represented by 83 people who are agree with the item.

It was decided to know if the restaurants transmit values to the diners in order to share them and more than half with 62.5% of the sample is considered impartial, this being the most predominant quantity, while 96 people consider to be in agreement with the premise. On the other hand, 67.2% mentioned agreeing that they felt comfortable consuming restaurant products, while only 2 people, represented by 0.5%, disagreed with this information. Finally, in the eighth item, we wanted to know if tastes and preferences were a purchase factor, where 56.7% and 37.9% agreed and totally agreed, respectively, followed by 5.1% who showed a neutral position, while only 1 person considered disagreeing with this information.

Finally, of the various factors to describe consumer purchasing behavior, four were identified: cultural, social, personal, and psychological. The results show in general that Chili's and Fridays are one of the most recognized and successful restaurants in the Peruvian market, located mostly in shopping centers, in addition to a large part, in Lima North (49.7%) and Lima East (22.4%). Being young adults between the age range of 26 to 33 years.

Similarly, specific outcomes were identified. Most of the respondents, more than 50%, at a cultural level, do not recognize being influenced in their purchasing behavior, however, socially they distinguish that the comments or testimonials given by their families and/or friends are relevant. Meanwhile, on a personal level, they refer to work and economic situation, finally, psychologically, they allude to restaurant environments.

As for the predominant factors in consumer purchasing behavior in the restaurants studied, at a general level, the ones that have the greatest predominance, with more than

65%, are psychological factors. That is, the characteristics with reference to the environment, quality, values, tastes and preferences are essential to determine the purchase of a product. These results coincide with Tongo [24] and Meléndez [25] who also recognize the psychological factor as the one with the greatest impact for consumers of the Huancahuasi tourist restaurant and the Shalom pastry shop, respectively. On the other hand, Castro [18] differs despite also studying a food item, since as part of his results he identifies the marketing factor as the most relevant, with an average of 3.76, followed by the cultural factor. In addition, the difference in factors or indicators studied for the finding of the studies can be commented on.

Regarding the social factor and/or participation of friends and relatives, it was observed that more than 43.6% of the respondents refer to being influenced by the different comments or testimonials provided by their social circle for the decision to purchase a certain product. This allows us to infer that the participation and good performance of care and/or service to any diner is key to establishing permanent ties. Similarly, Carvache [16] in his study coincides with the analysis of the gastronomic consumption behavior of Samborondón, because in his results family and friends were recognized as the main purchase motivations, which corresponded to more than 80%. However, in the study by Schnettler et al. [26] about the behavior of consumers in the main cities of southern Chile, the results differ despite recognizing the behavior of moments between friends as frequency of consumption, since a minimum percentage of the totality is observed. Likewise, this can be evidenced by the different population and sample in the study, for example, in the aforementioned study, the southern zone of Chile is specifically mentioned and in an indistinct category.

About the indicators that determine the behavior of consumers at a psychological level, it was identified that the environment, the spaces within the restaurant (appearance, texture, colors) are related to the choice of a certain place, since more than 80% of the respondents indicate that they agree and totally agree, followed by the quality that they can perceive regarding the attention or supplies inside the restaurant. Angnes and Mello [12] coincide with these results, since in their research they found the evaluation of the service provided to be influential in selection. Similarly, Carvache [16] identifies the quality provided to a lesser extent, but just as relevant, in this aspect, he groups it in different terms: food, facilities, service and price.

4 Conclusions

The conclusions determine that the purchase behavior of the Lima consumer who goes to casual dining restaurants such as Chili's and Fridays, are young adult consumers between the age range of 26 to 33 years, highlighting among its main clients Lima Norte (49.7%) and Lima Este (22.4%) located within shopping centers. In addition, these present a very active participation in consumption, specifically influenced mostly by psychological factors as factors of greater predominance.

It was identified that, in the consumer's purchasing behavior regarding the cultural factor, the highest percentage of respondents represented by 39%, indicate not being influenced by cultural beliefs in the purchase process, in addition, 45% indicated not recognizing that the dishes and/or products are part of their customs. Another point to note is that the behavior of consumers, mostly with 36.4%, is impacted by the comments or references they receive from society.

Regarding the social profile, the purchasing behavior of consumers is distinguished by the influence of the group of friends and recommendation of family members, equivalent to 43.6% and 56.2% respectively. Therefore, it can be inferred that people close to any consumer encourage their active participation in purchase as long as they have a positive testimony or appreciation. In addition, in this dimension, the perceptions of consumers against the type of "status" were evaluated, the majority with 39%, commented that they felt in disagreement with the premise.

Regarding the personal profile, several items were identified to know the purchasing behavior of consumers in this dimension. It was observed that the work occupation does not respond to being influential in the consumption of the respondents, 41.3% affirm to be in disagreement, on the other hand, the economic situation such as monthly income is decisive for the continuity of consumption among the respondents with 59.7%.

Finally, the psychological factor against the purchasing behavior of consumers who go to these restaurants, consumers are mostly influenced by the environment of the place, 66.4% consider it a relevant factor when being in a food establishment , as another important item, the location of the restaurants is observed, 56.4% indicate it is important to be in a safe place where you can spend pleasant moments, followed by a high quality standard with 84.4% of the respondents. On this point, the restaurants surveyed were mostly located in shopping centers.

During the investigation, limitations were identified, these are mainly concentrated in the collection of information and application of the questionnaire, in addition, the lack of willingness, time, the new virtual modality. The limitations respond to the already known confinement and preventive measures taken as a result of COVID 19. On the other hand, the large number of consumers and premises of the restaurants investigated throughout Lima made it difficult to find the respondents.

The proposals for future research of the work, from a practical point of view, will allow expanding knowledge and improving decision-making, promoting critical thinking and developing strategies to achieve success. Likewise, the research may be used as a guide and/or analysis for other companies in the same or different fields.

From the theoretical point of view, the reliability of the theory described was demonstrated, the identification of the most predominant factors that influence consumers, however, it is recommended to look for other theories that allow the combination or mixture of indicators in order to improve the result of the investigation. Finally, from the methodological point of view, the reliability and viability of the evaluation instruments have been demonstrated and can be used for future research, however, it is recommended to implement the development of a mixed research, as far as possible, since it includes the interview, which will allow a much more specific approach.

References

1. Amitabh, A.: Factors affecting consumer buying behaviour towards gold jewellery. Int. J. Recent Technol. Eng. **8**(3), 8088–8093 (2019)

2. Auf, M., Meddour, H., Saoula, O., Majid, A.: Consumer buying behaviour: the roles of price, motivation, perceived culture importance, and religious orientation. J. Bus. Retail Manag. Res. **12**(4), 177–186 (2018)
3. García, V.: Este es el comportamiento de los consumidores en España tras la Covid-19. Revistabyte 20 April 2021. https://revistabyte.es/actualidad-it/consumidores-espana/. Accessed 16 May 2021
4. Siminelli, C.: Consumer behaviours and attitudes towards a circular economy: knowledge and culture as determinants in a four-market analysis. Econ. Policy Energy Environ. (1), 135–169 (2017)
5. Xavier, T., Ferreira, M., Bizarrias, F.: The relation between financial attitude, consumer purchasing behavior and debt background. Revista Brasileira De Marketing REMark **18**(4), 220–243 (2019)
6. El Comercio: Cinco características que marcarán al consumidor peruano en 2021, El Comercio, 3 January 2021. https://elcomercio.pe/economia/personal/consumidor-peruano-cinco-tendencias-que-marcaran-en-2021-nndc-noticia/. Accessed 20 June 2021
7. INEI: Negocios de restaurantes aumentó 4,60% en diciembre de 2019, INEI, 11 March 2020. https://www.inei.gob.pe/prensa/noticias/negocios-de-restaurantes-aumento-460-en-diciembre-de-2019-12132/. Accessed 30 June 2021
8. NielsenIQ: El 49% de los peruanos sigue dietas bajas en grasa, ubicándose en el segundo lugar de Latinoamérica. NielsenIQ, 5 October 2016. https://nielseniq.com/global/es/insights/report/2016/el-49-por-ciento-de-los-peruanos-sigue-dietas-bajas-en-grasa/#:~:text=Perú. ocupa el segundo lugar, y en carbohidratos (23%25). Accessed 05 July 2021
9. Rai, R., Tripathi, S.: Consumer buying psychology and brand perception: influence of word of mouth communication. J. Content Commun. Commun. **12**, 159–168 (2020)
10. Rondán-Cataluña, F., Arenas-Gaitán, J., Ramírez-Correa, P.: Travel buying behavior in social network site users: to buy online vs. offline. J. Theor. Appl. Electron. Commer. Res. **10**(1), 49–62 (2015)
11. Kotler, P., Armstrong, G.: Marketing, Versión para Latinoamérica, 11th edn. Pearson Educación, México (2007)
12. Angnes, D., Mello, C.: Atributos de escolha em serviços de restaurantes: Um estudo exploratório. Revista Brasileira de Pesquisa em Turismo **7**(2), 317–336 (2013)
13. Iofrida, N., et al.: Italians' behavior when dining out: main drivers for restaurant selection and customers segmentation. Int. J. Gastron. Food Sci. **28** (2022)
14. Ryu, K., Han, H., Kim, T.H.: The relationships among overall quick-casual restaurant image, perceived value, customer satisfaction, and behavioral intentions. Int. J. Hosp. Manag. **27**(3), 459–469 (2008)
15. Bedinelli, G., da Silva, D., Nogueira, F., Neves, M.: Percepção de valor dos consumidores de serviços de restaurantes: um estudo com modelagem de equações estruturais. Revista brasileira de marketing **11**(3), 27–52 (2012)
16. Carvache, S.: Análisis del comportamiento del consumidor de restaurantes de gastronomía típica del segmento BA de Samborondón. Universidad Católica de Santiago de Guayaquil (2016)
17. Chou, S.-F., Horng, J.-S., Sam Liu, C.-H., Lin, J.-Y.: Identifying the critical factors of customer behavior: an integration perspective of marketing strategy and components of attitudes. J. Retail. Consum. Serv. **55** (2020)
18. Castro, J.: Factores que influyen en el comportamiento de compra del consumidor en el Restaurante El Cristal de Trujillo del distrito de Trujillo, año 2016. Universidad César Vallejo (2016)
19. Li, X., Dahana, W.D., Li, T., Yuan, J.: Behavioral changes of multichannel customers: their persistence and influencing factors. J. Retail. Consum. Serv. **58** (2021)

20. Truong, D., Truong, M.D.: How do customers change their purchasing behaviors during the COVID-19 pandemic? J. Retail. Consum. Serv. **67** (2022)
21. Hernández, R., Fernández, C., Baptista, M.: Metodología de la investigación, 6th edn. McGRAW-HILL/Interamericana Editores, México (2014)
22. Arias-Odón, F.: El Proyecto De Investigación: Introducción a la metodología científica, 5th edn. Editorial Episteme, Caracas (2006)
23. Trespalacios, J., Vázquez, R., Bello, L.: Investigación de mercados: Métodos de recogida y análisis de la información para la toma de decisiones en marketing. Paraninfo Cengage Learning, Madrid (2005)
24. Tongo, J.: Comportamiento de compra del consumidor del restaurante turístico Huancahuasi del distrito de Pachacámac, año 2017. Universidad César Vallejo (2017)
25. Meléndez, G.: Factores que determinan la decisión de compra del consumidor en la pastelería Shalom E.I.R.L en la ciudad de Trujillo, año 2016. Universidad César Vallejo (2016)
26. Schnettler, B., et al.: Tipologías de consumidores según el estilo de vida en relación a la alimentación: un estudio exploratorio en el sur de Chile. Revista Chilena de Nutrición **39**(4), 165–172 (2012)

Consumer Interaction in the Digital Environment: A Systematic Review

Brigitte Alondra Aguayo-Villodas[1], Sara Esther Reyes-Gomez[1], Franklin Cordova-Buiza[2,3](✉), and Wilver Auccahuasi[4]

[1] Bussiness Faculty, Universidad Privada del Norte, Lima, Peru
{n00041250,n00171054}@upn.edu.pe

[2] Research and Innovation Department, Universidad Privada del Norte, Lima, Peru
franklin.cordova@upn.edu.pe

[3] Faculty of Business Sciences, Universidad Continental, Huancayo, Peru
fcordova@continental.edu.pe

[4] Faculty of Business and Systems Engineering, Universidad Cientifica del Sur, Lima, Peru
wauccahuasi@cientifica.edu.pe

Abstract. Since the confinement worldwide, consumer interaction in the digital environment has evolved significantly. The companies have adapted to the new means of communication to reach the target audience. The research was carried out with the objective of describing what has been written in the scientific literature on the impact of the pandemic on the post-pandemic environment of the digital consumer, before and during COVID-19. The information was collected in databases: Scopus, Redalyc, Scielo and Dialnet; using specific routes such as digital consumer, digital era and pandemic, filtering and categorizing the research collected in relation to the digital consumer based on technology and the digital world, the digital consumer based on consumption, business and commerce and the digital consumer on use of information and communication technologies (ICT). The collection of information used inclusion and exclusion criteria, only scientific articles published in the last 5 years, it was obtained that 68% were published in 2021, 58% of the articles are written in Spanish, 31% of the articles are from Mexico and 42% were published in the Redalyc database. The results highlight that consumer interaction in the digital environment has increased in the scientific literature between 2017 and 2022 due to the impact of COVID-19 and adaptation to digitalization. Finally, it is concluded that new technologies, the digital world, social networks and electronic commerce have become fundamental tools to satisfy the primary needs of the human being.

Keywords: Digital consumer · Digital era · Pandemic

1 Introduction

The digital era basically comprises a time interval in which there is a connection between technology and the human being, which totally revolutionized the traditional ways of communication with the help of essential tools such as "Information and Communication

C. Montenegro et al. (Eds.): ICMTT 2023, LNNS 774, pp. 71–80, 2024.
https://doi.org/10.1007/978-3-031-43733-5_7

Technologies" or also called ICT, opening free expression through the digital participation of society [1]. The evolution of the digital era induces the constant adaptation of people who interact with technology, this is where digital consumers are born. As [2] explains, the new digital consumer presents advantages regarding the speed of access to information and it's simultaneous connection with the whole world, added to this the constant innovation of technology and its absolute freedom as a consumer.

Digital marketing generates a bond with the user to find out their needs, providing a 24-h connection and generating new habits and behaviors; In addition, it adds that digital marketing gives to the user the freedom to comment on the product or service, with the option of sharing the content, which is linked to measurement tools to find out about online activity and connect with the public of interest [3].

At the beginning of the year 2020, the world went through one of the biggest changes, both socially and emotionally, which triggered new ways of thinking, restructuring the human being, looking for new ways to satisfy their interaction with the outside world, forcing them to interact with others in the digital world, the human had to adapt the ways of interaction in recreational, study and work environments, one of the biggest changes that COVID-19 brought with it, was the increased need to be part of a group social, which resulted in the birth of new digital communities, generating stronger relationships between people [4].

Neuromarketing analyzes consumer behavior by applying neuroscience techniques such as Encephalogram, Functional Magnetic Resonance, Magnetoencephalography, Electromyography, Eye tracking based on the study in the emotional aspect; The new emerging technologies have generated the continuous development of neuromarketing due to its influence on the consumer's purchasing decision through the information of the attributes of the product or service. Based on the above, Neuromarketing is considered a modern discipline with emotions as its main component, developing methodologies to understand the functioning of the brain in the face of Marketing strategies that surround the consumer [5, 6].

The question that sustains the research process is the following: What has been written about consumer interaction in the digital environment in the scientific literature between 2017 and 2022? Therefore, the objective of this research is to describe what has been written in the scientific literature on the impact of the pandemic on the post-pandemic environment of the digital consumer, before and during COVID-19.

2 Methodology

The systematic review arises from a specific structured question, which it is analyzed and summarized according to the sources of information, it can treat, diagnose or predict according to the included and evaluated studies, it is characterized by its transparency when collecting, selecting and evaluating according to the evidence in relation to the subject matter [7]. For the preparation of the systematic review, scientific articles on topics related to the impact of the pandemic on the post-pandemic environment of the digital consumer, before and during COVID-19, were used. The collection of information focused on using mainly databases such as Scielo.org (Scientific Electronic Library Online), Scopus, Redalyc.org and Dialnet, in which they explain the impact of the pandemic on the population, specifically on the digital consumer. Post pandemic environment. The following table shows the specific routes applied (Table 1):

Table 1. Specific routes used in the Database.

Database	specific routes
Scopus	"digital consumer" and "digital era" or pandemic
Redalyc	"digital consumer" and "digital era" or pandemic
Scielo	"digital consumer" and "digital era" or pandemic
Dialnet	"digital consumer" and "digital era" or pandemic

All those open access articles, national and international scientific articles, articles published in the last 5 years, articles published in English or Spanish, articles published in the SCOPUS database, were taken as inclusion criteria. REDALYC, SCIELO and Dialnet, articles related to the topic of the post-pandemic digital consumer focused on the digital age, articles related to the area of business, social sciences and humanities. Next, all those articles with restricted access, articles published more than 5 years old, articles written in languages other than English and Spanish, articles published that do not belong to the Scopus, Redalyc databases, were used as exclusion criteria, Scielo and Dialnet, articles whose content is not related to the post-pandemic digital consumer, articles outside the area of business, social sciences and humanities (Table 2).

Table 2. Specific routes used in the Database.

Selection details	No. of articles
Potential studies identified in the databases	1684
Studies not considered scientific articles	10
Articles not considered "open access"	3
Items out of 5 years	88
Articles in languages not considered English or Spanish	322
Items outside the boarded area	1002
Articles not directly related to the central topic	240
Total Items Included	19

This section provides a descriptive analysis, by classifying the results obtained from the search in databases, with respect about articles that met the aforementioned keywords. The result obtained from this search was 1684 articles, which are distributed as follows: 9 articles in Scopus, 1530 articles in Redalyc.org, 76 articles from Scielo.org and 69 articles from Dialnet. After applying the inclusion and exclusion criteria, it was determined that only 19 articles (1%) are the most appropriate due to the filters applied and mainly have a direct relationship with the theme addressed in this systematic review. Initially, the amount of 1684 writings were obtained, of which only 19 of them met the characteristics of interest.

3 Results

In the selection process, a quick reading of each article that passed the filters established by the researchers was carried out, to identify the articles most related to the theme and research objective of this systematic review. The main problem identified during the review was the broad concept of the digital consumer in different areas, for which many articles were obtained as a result. At the beginning, 260 scientific articles were obtained based on the applied filters, however 249 articles were not related to the main theme and the area to be investigated in this systematic review. After reading the abstracts of the most attractive articles, validating their origin, and investigating the main topic, 19 scientific articles were chosen (Table 3).

Table 3. Scientific articles for the systematic review.

Author	Year of publication	Article title	Country
Kowalczuk, I., Stangierska, D., Gębski, J., Tul-Krzyszczuk, A. & Zmudczyńska, E	2021	Digital consumers in the food services market	Swiss
Rodríguez-Martínez, G. & Arango Lozano, C	2021	Use of the internet and social networks in the framework of the Covid-19 contingency in Colombia: analysis in the youth population considering their socio-economic level	Colombia
Pasquel-López, C. & Valerio-Ureña, G	2021	Emotional contagion on social networks: the case of COVID-19 on Facebook	Mexico
Montesdeoca Calderón, MG, Zambrano Alcívar, KG, Lemoine Quintero, F. Á., Villacis Zambrano, LM, Mendoza García, KA, & Zamora Cusme, YA	2022	Evolution from traditional marketing to digital marketing as a training process	Ecuador
Morales, M	2021	Translation: covid-19, new technologies, productivity and business' emergency plan	USA
Chavarria-Zambrano, KL, & Medina-Chicaiza, RP	2022	Implementation of Growth Hacking in small and medium-sized companies. Case: Ambato – Ecuador	Ecuador
López-Vázquez, A. S., Garcia-Grajales, J., & Buenrostro-Silva, A	2021	The internet in times of SARS-Cov-2 (COVID-19) in Mexico	Mexico

(*continued*)

Table 3. (*continued*)

Author	Year of publication	Article title	Country
Meza Jimene, M. & Reyes Ramos, M	2021	Changes in lifestyles during confinement by COVID-19	Mexico
Menéndez-Mera, MH, & Arteaga-García, ML	2021	Social Networks: Innovative resource for entrepreneurs	Ecuador
Castro, A., Zartha, J., Solleiro, J., Montes, J., Vargas, E., Palacio, J., & Hoyos, J	2018	Innovation management model proposal for a sales company to the final consumer	Mexico
Kim, D., Ammeter, A	2017	Shifts in Online Consumer Behavior: A Preliminary Investigation of the Net Generation	USES
Belling, C., Enslin, C., & Bozas, A	2021	A proposed media channel framework for integrated brand communication planning	South Africa
Sources, I	2019	Legal recognition of the digital trade in personal data	Australia
Cervantes Hernandez, R. & Chaparro Medina, PM	2021	Transformations in communication and sociability habits through the increased use of digital partner networks in times of pandemic	Mexico
Garcia-Madurga, MA, Grillo-Mendez, AJ, & Morte-Nadal, T	2021	The adaptation of companies to the COVID reality	Mexico
Torres Lajo, M., Torres Valladares, M., & Tinoco Gómez, O	2021	Emotional Control, Concern for Covid-19 and Consumer Behavior in the Lima Population	Peru
Vazquez-Martinez UJ, Morales-Mediano J. & Leal-Rodriguez AL	2021	The impact of the COVID-19 crisis on consumer purchasing motivation and behavior	Spain
Bermeo-Solorzano, N., Ortega Barre, D., & Villavicencio-Rodas, M	2021	Retail consumer purchasing behavior regarding the use of digital tools in a pandemic	Ecuador
Arana Arrieta, E., Mimenza Castillo, L. & Narbaiza Amillategi, B	2020	Pandemic, audiovisual consumption and future trends in communication	Spain

All those open access articles, open access scientific articles, national and international scientific articles, articles published in the last 5 years, articles published in English or Spanish, articles published in the SCOPUS database, were taken as inclusion criteria. REDALYC, SCIELO and Dialnet, articles related to the topic of the post-pandemic digital consumer focused on the digital age, articles related to the area of business, social sciences and humanities. Then, all those articles with restricted access, articles published more than 5 years old, articles written in languages other than English and Spanish, articles published that do not belong to the Scopus, Redalyc databases, were used as exclusion criteria..org, Scielo.org and Dialnet, articles whose content is not related to the post-pandemic digital consumer, articles outside the area of business, social sciences and humanities.

4 Discussion

(Fig. 1)

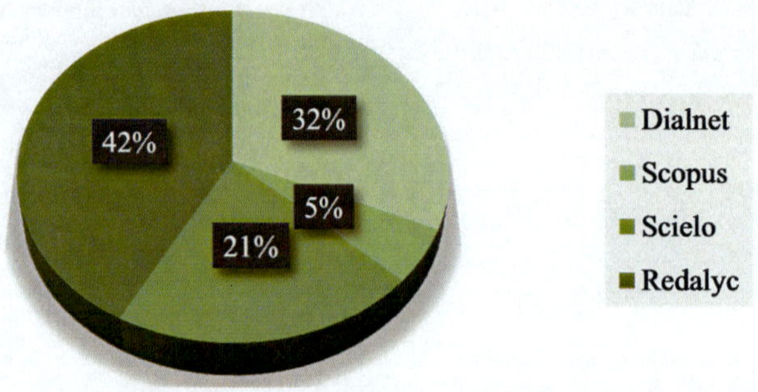

Fig. 1. Databases of leaked scientific articles.

The first figure shows the weight in percentage scale of the origin of each scientific article collected, this graph shows that 42% of the articles come from Redalyc, followed by 32% originating in Dialnet, 21% from Scielo and 5% of Scopus. A strong tendency of choice was observed on the part of the articles found in Redalyc, which potentially influenced the elaboration of the results found (Fig. 2).

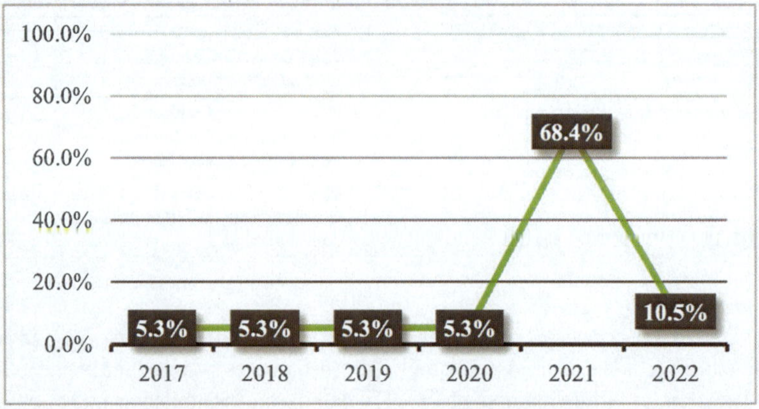

Fig. 2. Years of publication with respect to the scientific articles collected.

The second figure shows the year of publication of each chosen article, 5% from 2017 was chosen, followed by 5% from 2018, 5% from 2019, 69% from 2021 and 11% from year 2022. These selected articles obtained more information from the year 2021, due to the appearance of COVID-19, where more information was collected due to the new topics of interest that arise based on the problem of the context (Fig. 3).

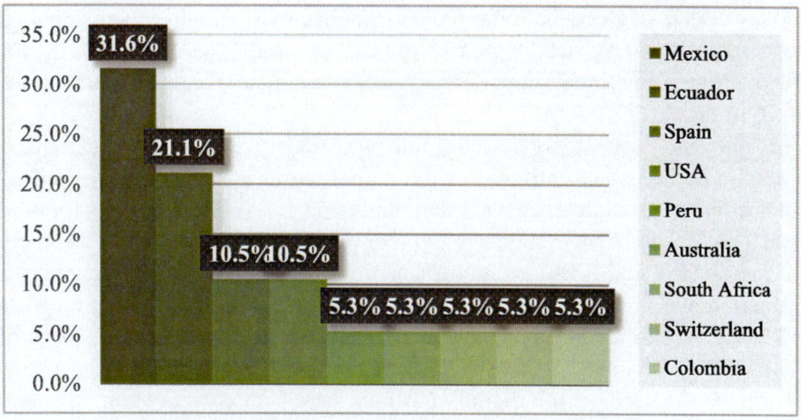

Fig. 3. Country of origin of the scientific articles collected.

On the scale from highest to lowest, with respect to the origin of the selected items, there are Mexico, Ecuador, Spain, the United States, Peru, Australia, South Africa, Switzerland and Colombia, followed by weights of 32%, 21%, 11%, 11%, 5%, 5%, 5%, 5% and 5% respectively. Which a greater preference for articles of Mexican and Ecuadorian origin can be observed (Fig. 4).

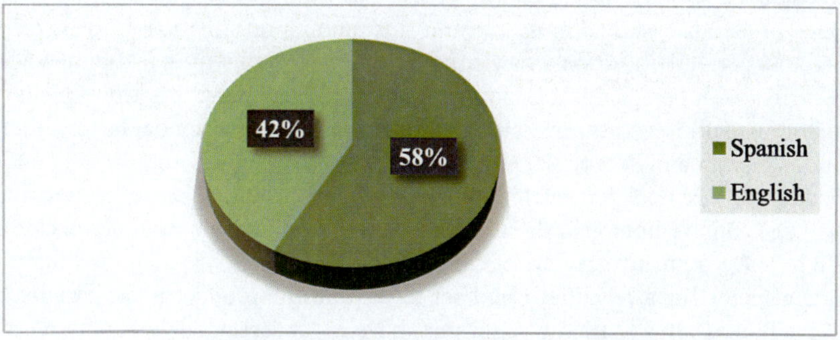

Fig. 4. Language of the collected scientific articles.

Since this systematic review takes a Latin American context, there is a greater preference for articles in Spanish, with a percentage of 58%, followed by 42% written in English.

Of the 19 selected articles, 4 articles related to the digital consumer focused on their communication interaction on social networks. The internet became a fundamental tool in every home as a result of the pandemic, being the smartphone the most used device to interact in the digital world, sometimes specific devices are used, this may vary depending on the activity you want to perform, in order to access the digital world in a certain way, with Instagram, YouTube and Facebook being the most used social networks to find news such as events of a public nature, multimedia files and entertaining content [8].

Social networks play a very important role in emergency situations, maintaining human relationships, without the need to be face to face, motivating people to inform and stay informed, provide emotional support to their environment and create closer ties between digital communities [9].

Over the years, consumers have become more sensitive to what the exterior offers them, this is how a greater affinity for the digital world arises, currently people have a greater affinity for expressing what they think and feel, collaborating with others to promote a comfortable environment with other people and brands [10]. The quarantine helped the takeoff and popularization of audiovisual content among people, staying entertained is one of the key factors that motivates interaction in the digital world, as well as being accessible and personalized to the tastes and preferences of the audience. [eleven].

11 articles focused on the digital consumer based on consumption, business and commerce were found. The pandemic brought with it the closure of commercial premises and the order to stay at home, this triggered the impulse of consumption through different digital media, which in turn encourages the growth of the use of the home delivery service, specifically in the food sector, companies have promoted the use of digital tools as a new means to take orders, the only impediment that people have to access this activity is not having internet on an electronic device [12]. The constant need for evolution of human is something that companies must learn to identify, since the key to their survival is to be productive, so staying at the forefront of technology benefits the relationship with customers, partners and collaborators, reduces costs and time [13]. It was recognized that the digital consumer has two aspects as a guide before choosing a product, the chosen brand must transmit trust and its information must be credible, both factors correctly position a new company in the digital world, taking into account it's security. With personal information at the time of payment, it is worth mentioning that among the digital platforms with the biggest number of potential buyers, there are mainly Instagram and Facebook [14].

Today, it's essential for small companies to have a digital presence, specifically, a web page, this is how growth strategies can be applied, for example encouraging people to leave a positive review of the product, because It was identified that a very important factor for a potential customer to acquire the product is the experience of another who has already tried it, after that, have an understandable purchase process, no more than 3 steps and finally encourage a repurchase. Through attractive promotions [15]. The participation and prominence given to customers with respect to the brand plays a very important role, since through feedback the objectives can be aligned in a more personalized way, in addition to the fact that the customer likes to be taken into account for comments regarding opportunities for improvement and with the help of social networks, is encouraged in a very natural way [16].

Information about digital natives and the phases during the online purchase are highlighted. In the information phase, digital consumers search for a wide variety of products. In the agreement phase, the shopping experience of other digital consumers and the quick and sincere attention of the staff is valued, they are willing to start a conversation via call or message. In the resolution phase, they assess the fulfillment of their rights, the ease of returns and the registration of claims of the website, it is the final purchase decision of the digital consumer. Finally, in the environment phase, the online purchase process and their experience during this journey are vital background information for the purchase decision [17]. Companies are constantly adding different communication channels, consumers are willing to try it, but do not have the necessary information about the benefit or objective of the channels. Therefore, it is required that the information for digital consumers be through relevant channels because it influences the purchase decision [18].

Companies during COVID-19 have had to adapt to the new characteristics of the consumer by expanding their variety of products and services, taking into account economic problems, supply, demand and uncertainty; technology is also essential to ensure efficiency. in online commerce processes [19]. The pandemic influenced the behavior of the digital consumer, however, it has progressively moderated due to the disappearance of the restrictions implemented at the beginning. Likewise, the consumer is influenced by emotion during the purchase decision, the items with the greatest increase in online purchases were restaurant delivery and the acquisition of technological products [20]. COVID-19 has influenced the purchase of digital consumers in the different categories, prioritizing the category of essential products, affecting the balance with other categories [21, 22].

5 Conclusion

It is concluded that, regarding consumer interaction in the digital environment, it has increased in the scientific literature in recent years and it is highlighted that social networks are used to search for news, audiovisual content increases interaction, companies must transmit trust and information Real in the digital world, consumers saw the need to make use of electronic commerce, companies saw the opportunity to boost their digital presence with the help of relevant content and consumer big data for new digital strategies.

References

1. García-Umaña, A., Ulloa, M., Córdoba, E.: La era digital y la deshumanización a efectos de las TIC. Reidocrea **9**, 11–20 (2020)
2. Flores Cornejo, F.O., Valdiviezo Vásquez, H.: El nuevo consumidor digital. Be lawyer (2020)
3. Viteri Luque, F.E., Herrera Lozano, L.A., Bazurto Quiroz, A.F.: Importancia de las Técnicas del Marketing Digital. Recimundo **2**(1), 764–783 (2018)
4. De Molina Benavides, L., Ramos, E.: Habitar en la era digital. Modelos colaborativos y su respuesta en tiempos de crisis. Rita Rev. Indexada de Textos Académicos **14**(4), 94–101 (2020)
5. Salas Canales, H.J.: Neuromarketing: Explorando la mente del consumidor. Rev. Científica de la UCSA **5**(2), 36–44 (2018)
6. Suárez Lugo, N.: Neuromarketing a debate. Rev. Cubana Salud Pública **46**(3), e1645 (2020)
7. Moreno, B., Muñoz, M., Cuellar, J., Domancic, S., Villanueva, J.: Revisiones sistemáticas: definición y nociones básicas. Rev. clínica de periodoncia, implantología y rehabilitación oral **11**(3), 184–186 (2018)

8. Rodríguez-Martínez, G., Arango Lozano, C.: Uso de internet y redes sociales en el marco de la contingencia Covid-19 en Colombia: análisis en población juvenil considerando su nivel socio-económico. Texto Livre: Linguagem e Tecnologia, **15**, 1–21 (2022)
9. Pasquel-López, C., Valerio-Ureña, G.: Contagio emocional en las redes sociales: el caso de COVID-19 en Facebook. Texto Livre: Linguagem e Tecnologia **14**(1), 1–14 (2021)
10. Montesdeoca Calderón, M., Zambrano Alcívar, K., Lemoine Quintero, F., Villacis Zambrano, L., Mendoza García, K., Zamora Cusme, Y.: Evolution from traditional marketing to digital marketing as a training process. J. Bus. Entrepreneurial Stud. **6**(2), 73–84 (2022)
11. Arrieta, E.A., Castillo, L.M., Amillategui, B.N.: Pandemia, consumo audiovisual y tendencias de futuro en comunicación. Rev. de Comunicación y Salud **10**(2), 149–183 (2020)
12. Kowalczuk, I., Stangierska, D., Gebski, J., Tul-Krzyszczuk, A., Zmudczynska, E.: Digital consumers in the foodservices market. Sustainability **13**, 7403 (2021)
13. Morales, M.: Covid-19, new technologies, productivity and business emergency plan. Telos **23**(3), 764–773 (2021)
14. Menéndez-Mera, M., Arteaga-García, M.: Redes Sociales: Recurso innovador para emprendedores. Rev. Arbitrada Interdisciplinaria Koinonía **6**(12), 566–578 (2021)
15. Chavarría-Zambrano, K., Medina-Chicaiza, R.: Implementación del growth hacking en las pequeñas y medianas empresas. Caso: Ambato – Ecuador. Rev. Arbitrada Interdisciplinaria Koinonía, **7**(13), 8–41 (2022)
16. Castro, A., et al.: Propuesta de modelo de gestión de innovación para una empresa de ventas al consumidor final. Rev. Lasallista de Investigación **15**(1), 75–89 (2018)
17. Kim, D., Ammeter, A.: Shifts in online consumer behavior: a preliminary investigation of the net generation. J. Theor. Appl. Electron. Commer. Res. **13**(1), 1–25 (2018)
18. Belling, C., Enslin, C., Bozas, A.: A proposed media channel framework for integrated brand communication planning. Communitas **26**, 20–32 (2021)
19. García-Madurga, M.A., Grilló-Méndez, A.J., Morte-Nadal, T.: La adaptación de las empresas a la realidad COVID: una revisión sistemática. Retos Rev. de Ciencias de la Administración y Economía **11**(21), 55–70 (2021)
20. Torres Lajo, M., Torres Valladares, M., Tinoco Gómez, O.: Control Emocional, Preocupación por la Covid-19 y Comportamiento del Consumidor en la Población Limeña. Rev. De Investigación En Psicología **24**(2), 63–80 (2021)
21. Vazquez-Martinez, U.J., Morales-Mediano, J., Leal-Rodriguez, A.L.: The impact of the COVID-19 crisis on consumer purchasing motivation and behavior. Europ. Res. Manag. Bus. Econ. **27**(3) (2021)
22. Bermeo-Solorzano, N., OrtegaBarre, D., Villavicencio-Rodas, M.: Comportamiento de compra del consumidor retail frente al uso de herramientas digitales en pandemia. 593 Digit. Publisher CEIT **6**(6), 212–221 (2021)

City Brand as a Promotional Strategy: A Perception of the Tourist in Peru

Maria Elena Silvia Gamarra-Ramirez[1], Woendy Peralta-Ugaz[1], Franklin Cordova-Buiza[2,3](✉), and Wilver Auccahuasi[4]

[1] Business Faculty, Universidad Privada del Norte, Lima, Peru
n00123014@upn.pe

[2] Research, Innovation and Social Responsibility Department, Universidad Privada del Norte, Lima, Peru
franklin.cordova@upn.edu.pe

[3] Faculty of Business Sciences, Universidad Continental, Huancayo, Peru
fcordova@continental.edu.pe

[4] Faculty of Business and Systems Engineering, Universidad Cientifica del Sur, Lima, Peru
wauccahuasi@cientifica.edu.pe

Abstract. The city brand is a set of attributes, values and strategies to create an identity that allows differentiating one city from another. The city of Chiquián is located on the Cordillera de los Andes, in the center of Peru, it has various natural and cultural tourist attractions that would allow a greater development of tourist activity; therefore, this research aims to determine the perception of tourists about the Chiquián city brand as a promotional strategy. The methodology is quantitative, cross-sectional and non-experimental; the sampling was probabilistic, by the simple random method, 384 people who visited the city participated; In the data collection, the survey technique was used and the questionnaire was used as an instrument. The main results highlight that the city is considered welcoming and the residents as hospitable, conservative and generous; but many claim that it offers low-quality products and services; Tourists agree that the city be identified by a brand that reveals all the attributes it possesses, since by promoting the city as a tourist destination it will contribute to the economic growth of Chiquián and improve the quality of life of the inhabitants. It is concluded that the perception of tourists proposes to create a city brand for Chiquián that helps to be known as an attractive tourist destination, since the city has tourist potential despite the poor management of the municipal government.

Keywords: Branding · tourism · brand attitude · brand value · Chiquian · Ancash

1 Introduction

Currently, branding focused on destinations or territories is a form of growth for cities, since with proper management of the identity of the territory it becomes a brand [1–3]. The city brand concept is based on its cultural characteristics, economic conditions,

This paper is of interest to the participants of the conference as there are few studies on city branding and the application of branding focused on tourist destinations. This tool would help cities to better position themselves and differentiate themselves from nearby competitors.

C. Montenegro et al. (Eds.): ICMTT 2023, LNNS 774, pp. 81–89, 2024.
https://doi.org/10.1007/978-3-031-43733-5_8

industrial distribution, geographical and living conditions, government capacity, among other factors [4]. For this reason, it is important from the beginning to exercise good brand creation and management to generate value through it in order to attract tourists and investors, making the city a more competitive environment [5, 6].

One of the benchmarks in the city brand is New York, in the 70s a great marketing campaign was promoted to attract tourists from all over the world, it sought to modify the image it had due to a high level of crime rate and considered violent, for a completely different one; This remains best known to this day with the slogan "I love New York" [7].

The awareness of the brand can be evaluated taking into account the factors: characteristics of the city, the preference of tourists, the commitment and positioning among the competition [8]. Therefore, it is necessary to know the perception of the brand, for which it is necessary to know the quality of the brand, the differentiation, the budget and its availability for different audiences [9]. There are four essential elements: brand identity, meaning, response and relationship with loyalty [10]. Currently, Mexico City, which has a deep-rooted country brand, since it has found in territory marketing a way to develop various country brand strategies, enhancing its image internationally; this has made Mexico become a city with great potential [11, 12].

The city brand as a reference in Peru began with the city of Trujillo, on the north coast of Peru, in which a City Marketing and brand creation program was proposed, this was carried out through research that allowed promoting and reinforcing the image of Trujillo, it was determined to highlight its cultural heritage and thus encourage tourism; proposed a logo showing the Chimú culture (an important pre-Inca civilization), the sailor dance as the flag dance of Peru and positioning itself with the slogan "the city of eternal spring" [13–15].

The city of Chiquián is located in the Cordillera de los Andes of Peru, province of Bolognesi, in the mountains of the department of Ancash at 3,374 m above sea level; This area has a great view of the snowy Yerupaiá that has a diverse flora and fauna that has been the inspiration for important literary compositions; On the other hand, it also has little-known archaeological sites dating back eight thousand years before the arrival of the Spanish in Peru. It was created at the time of Independence, after the 1970 earthquake, many families emigrated to other cities, in the The year 1993 began with the improvement of basic services in Chiquián, in the following years the promotion of adventure tourism was successfully developed, however, currently the largest number of visits are made on August 30, the day of the celebration of Santa Rosa de Lima [16].

Due to the above, the Chiquián city brand is necessary, which will allow proper brand management and make known the added value it offers, reinforcing it with experience and differentiating it from other tourist destinations in the region [17, 18]. The inhabitants of Chiquián and some frequent tourists know the tourist attractions that this city has, however, it is still unknown to some people in the region and even in the country, therefore, the following research question is formulated: What is the perception of tourists about the city brand as a promotional strategy for the city of Chiquián in 2021?

The inhabitants of Chiquián and some frequent tourists know the tourist attractions that this city has, however, it is still unknown to some people in the region and even in the country, for this reason the following research question is formulated: What is the perception of the tourists about the Chiquián city brand as a promotional strategy?

The trademark refers to a name, word, sign or peculiar sketch with which the products or services of a merchant are identified; In addition, it is pointed out that the brand influences the daily life of users and is of vital importance to determine the purchase, which is why branding is essential in any organization and a certain amount of money must be designated in advertising, positioning of brand, representative management of the organization, and attract public interest; On the other hand, it is important to emphasize that the brand has the objective of positioning itself in the mind of the consumer, so that they remember images and sensations related to the brand [23].

According to [1] the dimensions associated with the city brand are the attitude of the brand and the value of the city brand, which will be the study dimensions of this research work. On the one hand, the attitude of the brand is based on the nature of the brand, business creativity, shopping, transportation, cultural activities and social ties, each item can be adjusted according to field conditions; On the other hand, the value of the city brand is related to brand loyalty, unique image, and perceived quality.

There are various antecedents on the study of the city brand, [19] presents a study referring to the marketing of cities and socially responsible urban management applied to 400 people. Subsequently, [20] studied about the knowledge of the brand in the city of Ipiales in 377 people, their results show that the term city brand is unknown and they consider it irrelevant, which they describe as negative, for this reason an awareness should be carried out. About the pros of creating a brand for the city and population, 100 tourists were also interviewed, to find out the reason for their stay in Ipiales, the purpose was rest and pleasure, followed by visits to relatives. For their part, [21] carried out an investigation on the city brand to 362 tourists from the city of Salcedo, it is concluded that tourists prefer architectural and colonial places, the tourist perception is favorable regarding the basic services of the city, in addition there is a significant impact between tourists and cultural traditions. Finally, [22] identify the strengths of the Bahia city brand in Brazil, among them the people, clean and orderly city, tranquility and good tourist services.

2 Methodology

The design of this research is quantitative, descriptive, non-experimental and transversal [23]. An infinite population was chosen that is made up of tourists, of legal age, who have visited the city of Chiquián at least once in the last 5 years. The probabilistic sampling was used by the simple random sampling method, where the formula with the infinite population was applied. To carry out the equation, the following data were chosen: the confidence level of 1.96, the margin of error is 0.05, the probability of failure 0.5, the probability of success 0.5, resulting in a sample of 384 tourists.

The technique was the survey and the instrument a questionnaire that consisted of 17 questions, with dichotomous and multiple choice variables, which was extracted from the scientific article of [24] published in an indexed journal; It was validated by expert judgment criteria and its reliability was determined using Cronbach's Alpha, obtaining a result of 0.762, demonstrating that the instrument is reliable.

Between July and October 2021, the information was collected with a questionnaire in Google forms, sending the link through social networks such as WhatsApp, Facebook and Instagram, also sharing in newsgroups about the Chiquian city, cultural pages and with the collaboration of the Chiquián Sub-prefect through his travel and tourism agency. Finally, the data analysis was carried out in the SPSS program, from which tables and figures were obtained that are shown in the results.

3 Results

The results reveal the main discoveries found through the application of the survey addressed to 384 tourists from the city of Chiquián. The city brand variable was described using the dimensions of city brand attitude and city brand value, taking into account the following data:

Table 1. Profile of the Respondents.

Age	No. Tourists		Male		Female	
	F	%	F	%	F	%
18–25	56	fifteen%	29	8%	27	7%
26–32	134	35%	73	19%	61	16%
33–39	78	twenty%	Four. Five	12%	33	9%
40–47	42	eleven%	24	6%	19	5%
48–55	41	eleven%	twenty	5%	twenty-one	5%
56 and over	33	9%	22	6%	eleven	3%
Total	384	100%	213	55%	171	Four. Five%

Table 1 presents the results of the sociodemographic aspects of the respondents; where the mean and median age is 36 years, on the other hand, 35% of tourists belong to the age group of 26 to 32 years, while 20%, in the range of 33 to 39 years, followed by 15% of tourists in the age range of 18 to 25 years, in addition, 11% of the age range of 40 to 47 years has a similar proportion of 48 to 55 years. On the other hand, 45% of tourists are women and the other 55% men.

Figure 1 shows that 56% of tourists perceive the city of Chiquián as welcoming, while 30% perceive it as calm, 24% as supportive and 22% think that the city offers good quality products.

Figure 2 shows that 31% of tourists perceive the inhabitants of Chiquián as hospitable, while 28% perceive them as conservative, 25% as generous and 9% think that the inhabitants are supportive and 7% who are entrepreneurs. On the other hand, 0% of the environmentalist option is observed, which is probably due to the fact that they do not have sustainable plans with the environment.

Table 2 shows that 97% of tourists indicate that the city of Chiquián has tourism potential, while 3% of tourists think that it does not have tourism potential.

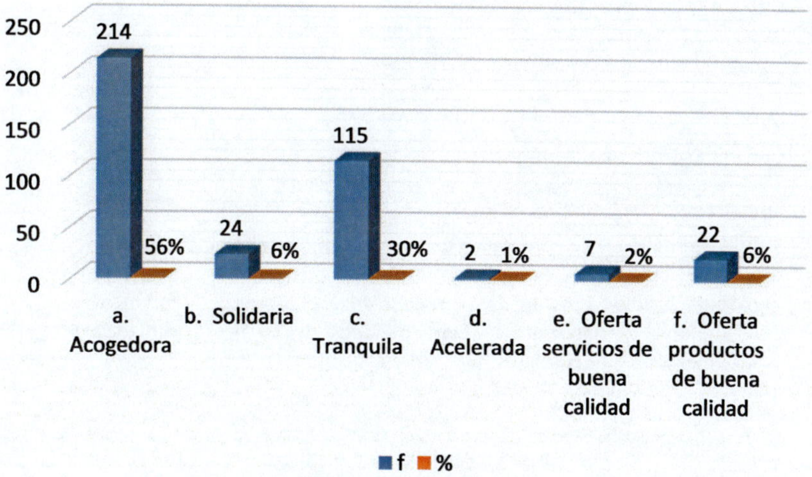

Fig. 1. Profile on the city.

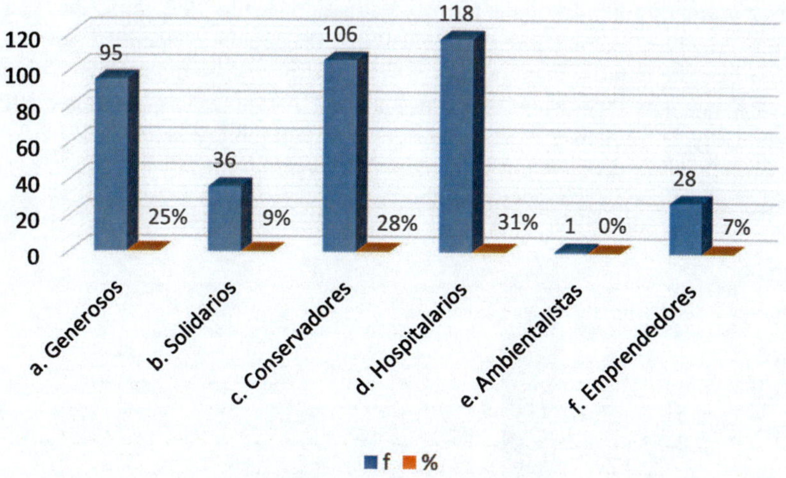

Fig. 2. Profile of the inhabitants.

Table 2. Tourism potential.

Condition	Frequency	Percentage
With tourist potential	372	97
No tourist potential	12	3
Total	384	100

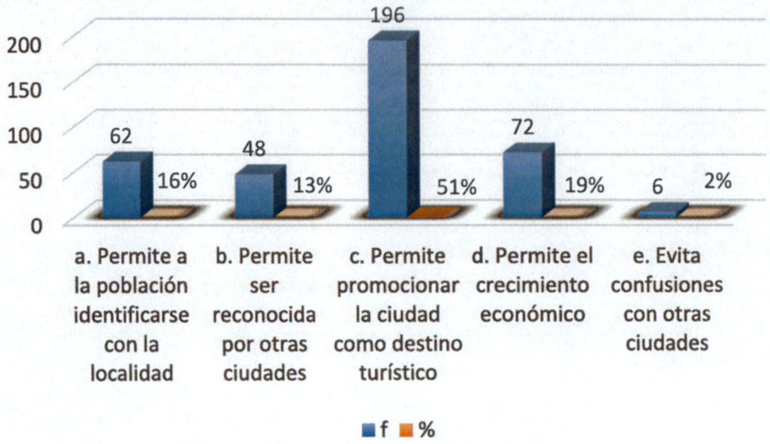

Fig. 3. Perception of the logo.

Figure 3 shows that 51% of tourists perceive that a brand is important because it allows promoting the city as a tourist destination, while 19% think that it allows economic growth, 16% think that the brand allows the population identify with the town and 13% indicate that the brand allows it to be recognized by other cities. At another point, 98% rate it as very important to have a brand that represents them nationally and internationally as tourism potential and only 2% think otherwise. On the other hand, 72% affirm that they do not know any brand that identifies the city of Chiquián.

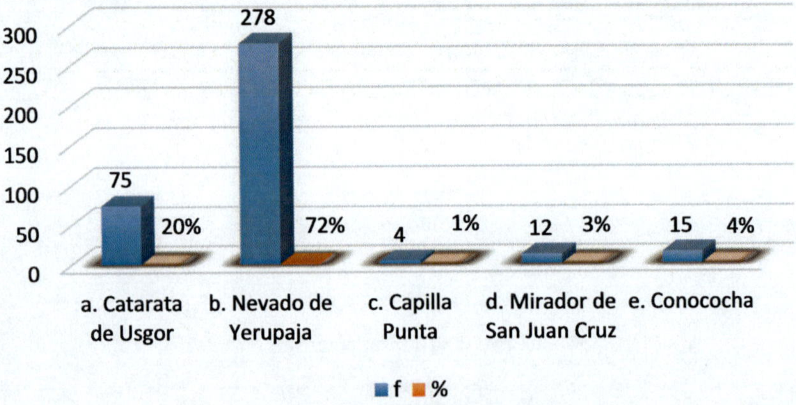

Fig. 4. Symbol of the city.

Figure 4 shows that 73% of tourists perceive that the symbol of the city of Chiquián is the Nevado de Yerupajá, while 20% indicate that it is the Usgor Waterfall, 4% opt for the Conococha lagoon and 3% think it is the Mirador de San Juan Cruz.

4 Discussion

According to the surveys carried out, 97% of tourists affirm that the city of Chiquian has tourist potential, for this reason, 98% of those surveyed support that it be represented by a brand, they also consider that the symbol of the city is the snow-capped Yerupajá since it represents the city, finally, 83% chose the slogan "Mirror of Heaven" which currently represents this city.

Tourists find the city of Chiquián with great tourist potential, describing it as cozy and calm; This result coincides with the results obtained in the investigation that It reflects that visitors request hospitality and that security together with tranquility projects confidence in tourists [24]. On the other hand, tourists perceive that low-quality services are offered, which can be a weakness to promote the city, since it would need to improve these aspects and generate services that respond to the demand of tourists who visit it, this coincides with the research in the city of Salcedo [21].

In addition, according to the perception of tourists, 45% affirm that the District Municipality of Chiquián carries out an inadequate management of tourism in the city, this is due to the fact that the city does not plan tourism, nor is it prepared to receive visitors due to lack of products. Complete and sustainable tourism; This result coincides with those who studied the brand of the city of Cayambe, in Ecuador, concluding that there is a mismanagement of tourism, and that most of those surveyed would agree to see a promotional spot related to its history, culture, tourist sites and gastronomy [25].

Regarding the value dimension of the city brand, 69% of tourists have not heard about the city brand, while 31% of tourists are aware of these terms, this result is related to the data obtained by in their research on the city brand in Ipiales where most people are unaware of the term city brand and awareness of the benefits of developing a brand is recommended [20].

Likewise, 40% of tourists know the different areas of Chiquián and its natural resources and 97% agree that the city has great tourist potential, however, it is essential to establish the unique characteristics of the city and determine its tourist attraction. Throughout the year, this result coincides with those who indicate that functional urban areas in Poland identify tourism as a primary factor for the economy and that it should be planned based on its natural resources, followed by infrastructure, unique characteristics and be attractive throughout the year [26].

5 Conclusions

Regarding the perception of tourists about the Chiquián city brand as a promotional strategy, it was determined that the city is considered welcoming and the residents as hospitable, conservative and generous. Almost everyone agrees that the city of Chiquián has the potential to become a tourist destination, despite the fact that 45% of tourists disagree with the management of tourism by the Municipality. For this, strategies must be developed to improve the products and services offered to tourists' hand in hand with the municipality of Chiquián.

According to the perception of tourists, the attitude of the brand is positive, they describe the city as a cozy and quiet place, but many affirm that it is a city that offers low-quality products and services. On the other hand, the perception of the city in relation to the inhabitants is positive, tourists agree that the city be identified by a brand that reveals all the attributes it possesses, since by promoting the city as a tourist destination it will contribute with the economic growth of Chiquián and will improve the quality of life of the inhabitants.

Finally, the perception of tourists suggests creating a promotional strategy based on the Chiquián city brand, using the snowy Yerupajá as a symbol, which helps it to be known as an attractive tourist destination for tourists, since this city has tourist potential; The first thing would be to sensitize the inhabitants of the city, then formulate a tourism plan and then develop the city brand.

References

1. Miftahuddin, A., Hermanto, B., Raharja, S.J., Chan, A.: City brand attractiveness on tourism using rasch model approach. Int. J. Supply Chain Manag. **9**(2), 150–156 (2020)
2. Ye, Z., Jeon, H.: Chinese city brands and semiotic image scales: a tourism perspective. Soc. Semiot. **33**(1), 1–24 (2020)
3. Yuwo, H., Purwanegara, M.S.: Does city brand equity have any impacts in tourism world? a case study of bandung city - Indonesia as a tourism destination. In: Plangger, K. (eds) Thriving in a New World Economy. Developments in Marketing Science: Proceedings of the Academy of Marketing Science, pp. 84–87. Springer, Cham (2016). https://doi.org/10.1007/978-3-319-24148-7_25
4. Torres, M., González, Y., Manzano., O.: City brand as a strategy for urban competitiveness in intermediate cities. Spaces Mag. **41**(36), 171–185 (2020)
5. Matlovičová, K., Tirpáková, E., Mocák, P.: City brand image: semiotic perspective a case study of prague. Folia Geog. **61**(1), 120–142 (2019)
6. Jokela, S.: Transformative city branding and the evolution of the entrepreneurial city: the case of Brand new helsinki. Urban Stud. **57**(10) (2020)
7. Ruiz, E., De la Cruz, E., Aramendia., G.: Territory brand and city brand, utility in the field of tourism. the case of Malaga. Int. J. Sci. Manag. Tour. **3**(2), 155–174 (2017)
8. Benedek, J.: Measuring place brand equity: a review of current models. J. Media Res.-Rev. de Studii Media **10**(27), 42–57 (2017)
9. Górska-Warsewicz, H.: Factors determining city brand equity—a systematic literature review. Sustainability, **12**(19) (2020)
10. Haarhoff, R., De Klerk, B.: Destination South Africa: analysis of destination awareness and image by international visitors. Geo J. Tour. Geosites **24**(1), 201–211 (2019)
11. Echeverri, L., Niculcar, C., Herrera, C., Ossorio, J.: Country brand development and tourism: the case study of Mexico. Stud. Perspect. Tourism **22**(6), 1121–1139 (2013)
12. Won, J., Lee, J.Y., Jun, J.W.: Influences of SNS (social network service) uses and musical consumption on city branding: a focus on broadway, New York and the West end, London. Sustainability (Switzerland) **12**(9) (2020)
13. Regalado, O., Berolatti, C., Martínez, R., Riesgo, G.: Competitive Identity and Brand Development for the City of Arequipa, 1st edn. Esan University. Lima, Peru (2012)
14. Pasquinelli, C., Trunfio, M., Rossi, S.: Beyond the authenticity–standardization paradox in international gastronomy retailing: twisting the hosting city brand with the place of origin. Br. Food J. **123**(13), 561–578 (2021)
15. Sevin, H.E.: Understanding cities through city brands: city branding as a social and semantic network. Cities **38**, 47–56 (2014)
16. Alvarado Balarezo A.- Dear Chiquian. https://naloalvaradochiquian.blogspot.com/2019/06/club-ancash-presentacion-del-libro.html. Accessed 21 Aug 2022

17. Priporas, C., Stylos, N., Kamenidou, I.: Image, city brand personality and generation Z residents' life satisfaction under economic crisis: Predictors of city-related social media engagement. J. Bus. Res. **119**, 453–463 (2020)
18. Tran, V., Nguyen, N., Tran, P., Tran, T., Huynh, T.: Brand equity in a tourism destination: a case study of domestic tourists in hoi an city, vietnam. Tourism Rev. **74**(3), 704–720 (2019)
19. Jara, R.: The marketing of cities and socially responsible urban management, the case of the city of Chiclayo 2016–2035. Acc. Bus. **11**(21), 93–112 (2016)
20. Arteaga, A., Pianda, E., Sandoval, X.: City branding as a market strategy to promote tourism in the city of Ipiales. Trends **20**(1), 22–52 (2019)
21. Manobamba, F., Garcés, L., Guerrero, C.: City brand as a promotional development strategy for the city of Salcedo. Digit. Publisher **5**(1), 174–186 (2019)
22. Carbache, C., Delgado, Y., Villacis, L.: Influences of experiential marketing to position the city brand in Bahía de Caráquez. Investig. Bus. Mag. **13**(22), 33–42 (2020)
23. Hernández, R., Fernández, C., Bautista, M.: Research Methodology, 6th edn. McGraw-Hill, Mexico DF (2014)
24. Castro, J., Sánchez, D.: City brand: nature and scope from the Anholt hexagon. Ecociencia Sci. Mag. **7**(3), 1–29 (2019)
25. Calderón, Z., Toapanta, D.: I love you Cayambe: city brand campaign. (15), 145–160 (2020)
26. Glińska, E., Matwiejczyk, A., Barkun, Y.: Tourism as an aspect of city branding in functional urban areas. WSEAS Trans. Bus. Econ. 301–312 (2021)

Robotics

Lucas-Bot: A Robotic Assistant to Carry Out Relaxing Pre-therapy Activities for Children with Disabilities

José Panamá-Mazhenda, Karina Panamá-Mazhenda, Sofía Bravo-Buri, Vladimir Robles-Bykbaev(✉), and Efrén Lema-Condo

GI-IATa, Cátedra UNESCO Tecnologías apoyo para la Inclusión Educativa, Universidad Politécnica Salesiana, Cuenca, Ecuador
{jpanamam,jpanamam1}@est.ups.edu.ec, {vrobles,elema}@ups.edu.ec

Abstract. According to the latest estimates of the National Council for Equality of Disabilities (CONADIS), in Ecuador, there are a total of 215,156 people with motor disabilities. Along these lines, the rehabilitation or therapy processes are one of the main challenges physiotherapists have due to the different conditions and limitations that are in the middle. Therefore, incorporating technological tools in this area is very convenient so that experts in this area can have a support tool. In this work, a robotic assistant called "Lucas-Bot" is presented, capable of making movements with its arms and legs providing relaxation exercises as a previous step to therapies with children with motor disabilities. With this, it is sought that children have a greater predisposition to work with experts in the development of physiotherapies. The robotic assistant has different routines adapted under the criteria of experts, in order to determine its viability in real use. In orr to validate the robotic assistant, Kendall's Coefficient of Concordance W was used with 4 experts in the area, obtaining an almost perfect level of consensus and a *p-value* of 4.5e−06.

Keywords: Robotic assistant · physiotherapy · children · disability · inter-rater agreement

1 Introduction

Nowadays, we evince the efforts to introduce technological tools that provide support for different disabilities with the main objective of improving the quality of life of people with disabilities to develop themselves independently within society. Motor disability is one of the limitations that creases the total or even partial mobility of the members of the body, it also presents a loss of communication with the environment due to the lack of mobility in the extremities, which generates a great problem in the daily life of people who suffer from it [6]. In this regard, with the advancement of technology, robotic assistants have been developed to aid the rehabilitation of people with motor disabilities and to provide the necessary support to professionals in communication with their patients.

C. Montenegro et al. (Eds.): ICMTT 2023, LNNS 774, pp. 93–102, 2024.
https://doi.org/10.1007/978-3-031-43733-5_9

The use of robotic assistants is of great relevance and importance in existing practices by experts for children with motor disabilities or special educational needs, even allowing them to create and implement new practices, models, or methodologies for rehabilitation. This provides the opportunity to manage processes based on specific indicators [11]. Robotic assistants form a dynamic system through contact with children to perform a task, simulating a cooperative and motivational scenario that makes the sessions velop more efficiently [2].

In Ecuador, there is a total of 471,205 people with disabilities, of which 215,156 people have motor disabilities [5]. One of the main challenges presents itself during therapies, due to the different conditions and needs that users present, one of which is to overcome communication barriers to include and instruct patients during interventions. This is due to the lack of motivational resources or didactic means that maintain a level of concentration and attention in the users. To begin with, the methodologies for rehabilitation require a lot of effort, not to mention that in some cases, these take a lot of time to prepare, especially to find a way to get the attention of the children to do the rehabilitation, not to mention that the repetition of physical exercises can be tedious and tiring, generating a lack of interest during therapy. Therefore, the use of robotic assistants would greatly facilitate the preparation time of the professional in charge, looking for the children to interact with the robotic assistant more dynamically. It is important to mention that the trained personnel oversee the patient's recovery, and also of improving the self-esteem of children with disabilities [4].

The use of this technology could be an alternative solution for the execution of different functionalities, improving the experience of people, their physical state, and behavior during therapy sessions, involving their emotional state, and promoting comprehensive rehabilitation [6]. For this reason, it is important to emphasize that the sign of these robots must consider several factors of the targeted group of users and the professionals in charge of the interventions in relation to the context of use and the fulfillment of the objectives to carry out a significant rehabilitation adapted to the patient's needs. Thus establishing a level of commitment, attention, and emotional state during the interaction between humans and robots [4].

This article presents a robotic assistant that allows the execution of several routines composed of relaxation exercises before the development of therapies. In this case, the robotic assistant will work as a therapy partner to teach warm-up exercises to children with motor disabilities, in which the child will have to interact with the robot by imitating the movements that it executes, all this process is carried out with an expert in charge so all sessions are supervised.

2 Related Work

Digital transformation involves complex development in any organization due to various factors that can be technological, organizational, and cultural. In classrooms and educational centers, the incorporation of new technologies or ICTs

(Information and Communication Technologies) has a radical impact on the development of new proposals for integrating pedagogical models into teaching and learning as innovative responses. However, there are certain obstacles that prevent the incorporation of ICTs in these processes, such as the lack of ICT skills, low motivation, inadequate teacher training, lack of infrastructure, limited access to equipment, among others. Therefore, it is important to address these digital gaps by advancing the articulation of educational policies that facilitate the use of these technologies [1]. For example, robotics encompasses its use in different areas as tools for support or process automation. In the educational field, robots or robotics as such are used to impart technical knowledge, or a more attractive proposal is the use of them as intervention companions to impart knowledge and encourage active participation in a more enriching learning environment. In this line, [10] it describes the design, creation, and control process of a physiotherapy support robot for shoulder rehabilitation. The robot uses a 5DOF mechanism (2 active and 3 passive) adjustable to different body sizes and EMG sensors as a control subsystem to be able to adjust the robot's torque depending on the movement order sent by the user. It also uses a microcontroller to record movements, session time, signals, and unexpected events, among other data, which can be used to improve the calibration process to the patient's needs. Finally, to determine how efficient the use of the robot is, the authors experimented with a patient for 15 days, affirming its efficiency through the results in the increase of the patient's arm mobility from 10 to 75°. Other important proposals are the following:

- DAREBOB is a double-arm robot that provides support in the rehabilitation of upper limbs, mainly focused on the shoulder and elbow joints. This robot can be adjusted to different arm lengths, admitting a larger range of movement (ROM) that allows a greater variety of movements during therapies, for this, they used the Denavit-Hartenburg method to establish coordinates of 5 degrees of freedom in the arm in a three-dimensional space of the upper extremity. The DAREBOB system is made up of two UR5 robots with 6 degrees of freedom limited to the kinematics of the upper extremity of the human being by a master/slave system, whose objective is to avoid joint injuries in the patient in cooperative control of degrees of freedom [9].
- ZORA is a therapeutic robot for the rehabilitation of children with severe physical disabilities based on existing NAO hardware that integrates unique and accessible software adapted to the care area. ZORA interacts with users through different scenarios where it performs activities such as dancing, movement exercises, and games that can be programmed by the user as new functionalities. The results of the use of ZORA reflected that it can be used with children with a minimum level of specific skills and that it could be useful in other areas, highlighting the level of motivation that the robot provides in the activities, which is very convenient [7].
- Similarly, in the NAO robot was used as a social assistant in physiotherapy sessions for children with cerebral palsy. The robot works as a mediator during physiotherapy, it behaves like a partner that is present throughout the session,

motivating the development of exercises through audible phrases and movements that it performs with its arms and head, these actions are programmed using the Choregraphe Suite software and Python SDK - NAO software. The therapies that were performed in conjunction with the robot were based on the principle of promoting goal-directed functional performance. The therapeutic intervention in conjunction with the robot was composed of 16 sessions of around 45 min each, the results of the use of the robot reflected that it can positively benefit the therapeutic processes, meeting the proposed objectives [3].

3 System Architecture

Figure 1 shows the structure of the Lucas-Bot robotic assistant whose objective is to perform relaxation exercises before the physiotherapy intervention. Lucas-Bot performs the exercises through interaction with a mobile device. The brain of the robot is the ESP32 microcontroller that integrates Bluetooth communication to be able to receive the signals sent from a mobile application. This microcontroller commands the different actuators (arm and leg motors, OLED display, and the PCA9685 PWM module). The robotic assistant is static (does not walk) and has 3 joints in both arms and legs. To carry out the movements of the extremities, it uses 12 micro-servos controlled by PWM with 12 bits of resolution and a frequency of 1.6 kHz. In this way, the ESP32 uses an I2C communication channel to control the servos through the PCA9685. Based on this, the frequency of the PWM signal can be manipulated to establish the work cycle and thus modify its rising and falling flank and execute the different exercises that will be required in the session prior to therapy.

The OLED screen shows the eyes of the robotic assistant through the ESP32 and I2C communication that controls the data to display the images. This device works with a matrix of rows and columns that can be programmed to display the desired image, in our case the emotions of the robot.

The robotic assistant power system has a 7.2 V lithium battery to power the ESP32 module and the PCA9685 module. However, these modules only receive 5 V, so we make use of a voltage regulator that allows us to convert the direct current voltage of our battery from 7.2 V to 5 V through the integrated chip MP3207, not to mention, we can control the output of voltage with a potentiometer associated with it. Finally, to prolong the use of the battery, a PUSH BUTTON LED was incorporated to control the on and off, of the robotic assistant, the green color being the lighting of the LED as the ignition mode. The robot has an autonomy of 5 h of uninterrupted operation.

3.1 Physical Therapy Routines Assisted by the Robot

Lucas-Bot has three levels of routines, the first and second level is made up of four exercises, and the third level of five exercises, in all levels the level of repetitions that the robot performs varies; in level 1, 5 repetitions are performed, in level 2,

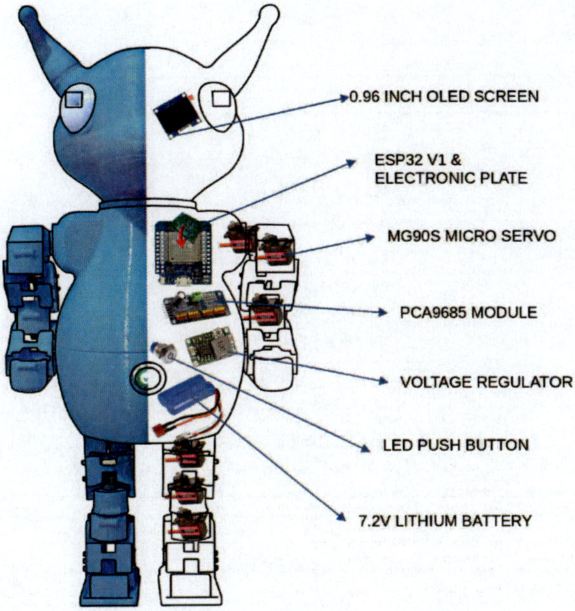

Fig. 1. Structure and main electronic components of the Lucas-Bot robotic assistant.

10 repetitions and at level 3, 15 repetitions. To access the exercises, a connection must first be established with the mobile device where the application is located. The application was developed in Unity and is only manipulated by the expert in charge of the therapy.

Table 1 shows the exercises by level routines that can be developed with the support of the Lucas-Bot robotic assistant.

The exercises listed in Table 1 are relaxation exercises before the development of therapy, which aims to prepare the body for the activities to be carried out and prevent any type of injury. In addition, it is very important to mention that Lucas-Bot can be used for children with motor disabilities of different degrees of disability and depends solely on the execution plan of the expert in charge. Not to mention that all the exercises proposed were reviewed and approved by the same experts, even in the control of execution time, meaning it adapts to the needs of the user.

4 Pilot Experiment and Preliminary Results

To carry out an initial validation process of the main functionalities of the robotic assistant, a form was applied to 4 experts from the field of physiotherapy. The profiles of the experts are described in Table 2.

As can be seen, the experts (all physiotherapists) have extensive experience in working with and caring for people with motor disabilities (mean of 16.5

Table 1. Physical therapy routines that can be performed by children with the support of the Lucas-Bot robotic assistant.

Level	Repetitions	Exercises
1	5	Raise arms forward. Raise left foot. Raise right foot. March on your ground.
2	10	Crouch with arms out in front. Open feet on both sides. March on your ground. Perform squats.
3	15	Raise both arms forward. Raise left and right foot alternating. March on your ground. Crouch with arms out in front. Perform squats.

Table 2. Profiles of the experts who participated in the validation process of the robotic assistant functionalities.

Expert	Gender	Age	Years of professional experience with children with motor disabilities	Years of professional experience in ICT
1	Male	47	15	7
2	Male	48	26	5
3	Female	48	20	5
4	Male	55	5	5

years, Standard Deviation of 8.88). The age range of the experts goes from 47 to 55 years (mean of 49.5, Standard Deviation of 3.69), with which it can be seen that the profiles are homogeneous. To determine if there was consensus among the experts, Kendall's Coefficient of Concordance W [11] was used. With this coefficient, an analysis of the results of the forms that were organized in two large blocks was carried out. In the first, 11 questions of a demographic nature were included (names and surnames, gender, age, ethnicity, marital status, years of experience in ICT work, years of experience in caring for people with motor disabilities, and current activity in which performs), while in the second block, 34 questions were raised to determine various aspects of the robotic assistant. The groups of questions are detailed below according to the criteria they focus on:

- Aspects of the robot's functionality: support in relaxation activities before therapy, incorporation of appropriate routines for subsequent initiation of therapy, exercises provided by the robot (raising arms, raising feet, walking on the ground, bending down with arms outstretched when front, foot opening,

squats), the relevance of the number of levels of the routines and difficulty to execute the routines described above. Likewise, the relevance of the number of repetitions in each level of the routines is also considered.
- Robot appearance: 3D structure colors, general appearance, the most suitable color of the structure, and dimensions.
- Incorporation of the robot in physiotherapy practices: feasibility of using the robot during the physiotherapy intervention and relevance of using the robot in its current state for physiotherapy.

To this end, several demonstration sessions were held where the experts were able to interact with the robotic assistant and observe how children with disabilities reacted when they interacted with it. To determine if there was consensus among the experts, an analysis of iterrater agreement was analyzed. For this, Kendall's Coefficient of Concordance W was calculated [12]. The calculation process was carried out using the statistical software R (version 4.1.2) and the following hypotheses were raised concerning the agreement between the experts, considering the pair of hypotheses for each criterion:

- H_0: There is no consensus among experts for the proposed criteria.
- H_a: There is consensus among experts for the proposed criteria.

As can be seen below, Kendall's Coefficient of Concordance W, we worked with the interpretation proposed by [8]:

- $0.00 \leq w < 0.20$ - Slight agreement
- $0.20 \leq w < 0.40$ - Fair agreement
- $0.40 \leq w < 0.60$ - Moderate agreement
- $0.60 \leq w < 0.80$ - Substantial agreement
- $w \geq 0.80$ - Almost perfect agreement

According to Kendall's Coefficient that was calculated, we can appreciate that there is an "Almost perfect agreement" among the 4 experts since a value of 0.825 was obtained with a *p-value* of $4.5e{-}06$ (highly significant, with a level of confidence of 95%).

Given that the *p-value* obtained is less than 5%, the level of consensus obtained is considered to be valid and significant. Consequently, the null hypothesis is rejected.

Similarly, Fig. 2 shows the preference results that the experts had when analyzing the 18 questions. The questions related to the level of difficulty for the execution of the routines (16 in total) were not considered to carry out the calculation, since this factor depends on the profile of each child, their level of disability, and their clinical history. As seen in Fig. 2, the level of consensus, there are differences for the questions related to the size of the robot (Q05) and the colors of the structure (Q03 and Q04).

On the other hand, Fig. 3, shows the experts in the area interacting with the robotic assistant during the validation process of its main functionalities.

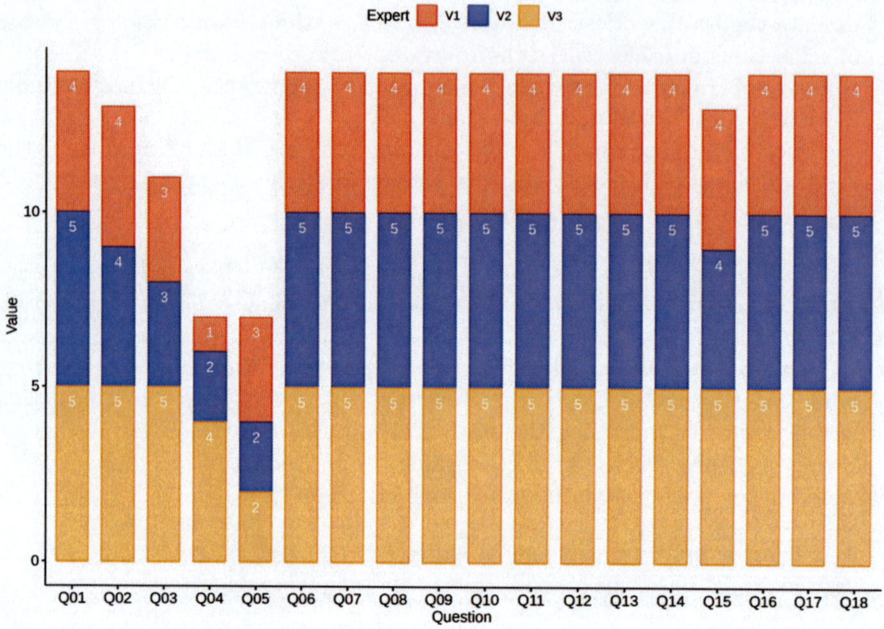

Fig. 2. Results of the expert consensus analysis for the 18 questions related to the main aspects of functionality, appearance, and relevance of the robot's use in therapy sessions.

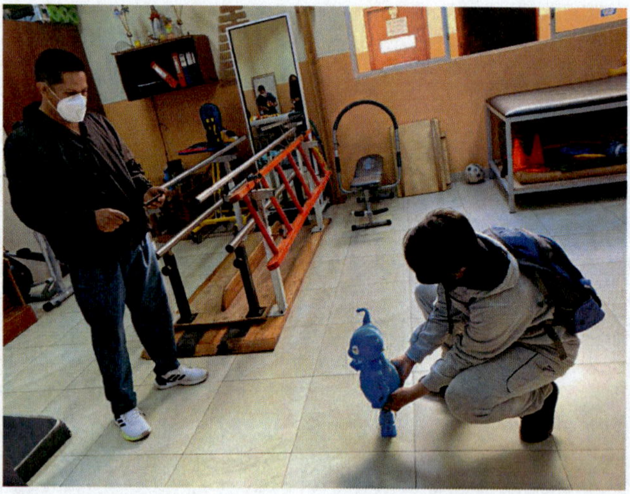

Fig. 3. Experts from the physiotherapy area interact with the robotic assistant during the validation process.

5 Discussion

Lucas-Bot is a robotic assistant that serves as a support in the development of relaxation activities before therapies with children with motor disabilities of different degrees of disability in a more dynamic way, encouraging them to perform specific movements that involve the functioning of the muscles of the arms and legs.

It is important to mention that the Lucas robotic assistant has a mobile device that allows the expert to interact with the robot, in which said device shows the different routines established for relaxation activities that can be executed randomly. The structure of the robot, as well as the movements it makes, capture the attention of children and encourage them to imitate its movements, causing a greater level of responsibility and attention during therapy, which is why experts mention that it is a very well structured resource to work in the physiotherapy area and that can be applied to any type of population due to the level of emotion it transmits, behaving like a partner throughout the session.

On the other hand, in this article, it was possible to verify that a group of 3 experts consider that the functionalities provided by the robotic assistant are adequate to carry out relaxation activities before therapy.

6 Conclusion

Lucas-Bot is a robotic assistant that focuses on the execution of different routines aimed at supporting the physiotherapy process with relaxation exercises prior to children with disabilities. The movement of its joints is controlled by a mobile device where experts can choose the activities to be developed at different levels of complexity.

In related research, DAREBOB is a robot that focuses solely on upper limb rehabilitation, while Lucas-Bot focuses on both upper and lower limbs, providing us with a wide variety of movements or rehabilitation routines. While DAREBOB only focuses on the upper part, it has 6 degrees of freedom, which allows for better intervention in that area by having the ability to perform a wide variety of movements. However, Lucas-Bot can combine various exercises and perform different routines, which is an advantage in its favor. Another similar research is the ZORA robot, which stands out for the level of interaction between user and robot, which seeks to encourage participation by motivating children to perform the activities. Similarly, the physical form of Lucas-Bot seeks to interest children in it, being a complementary companion during therapies, assisting throughout the process.

The use of the Lucas-Bot assistant was well received by experts and children in the accompaniment of therapies or prior to their development. However, in the validation process, some limitations were found such as the complexity of performing an exercise by children with a higher degree of disability, the level of attention, and the regulation in the execution times of an activity since some were able to achieve the goal faster than others. Therefore, it is considered in the future to add functions to the application that allow modifying parameters according to the child's profile to perform the different routines.

References

1. Area-Moreira, M., Bonilla, P.J.S., Mesa, A.L.S.: La transformación digital de los centros escolares. obstáculos y resistencias. Digit. Educ. Rev. (37), 15–31 (2020)
2. Baratta, A., Cimino, A., Gnoni, M.G., Longo, F.: Human robot collaboration in industry 4.0: a literature review. Procedia Comput. Sci. **217**, 1887–1895 (2023). https://doi.org/10.1016/J.PROCS.2022.12.389
3. Buitrago, J.A., Bolaños, A.M., Bravo, E.C.: A motor learning therapeutic intervention for a child with cerebral palsy through a social assistive robot. Disab. Rehab. Assist. Technol. **15**, 357–362 (2020). https://doi.org/10.1080/17483107.2019.1578999
4. Cifuentes, C.A., Pinto, M.J., Céspedes, N., Múnera, M.: Social robots in therapy and care. Curr. Robot. Rep. **1**(3), 59–74 (2020). https://doi.org/10.1007/S43154-020-00009-2, https://link.springer.com/article/10.1007/s43154-020-00009-2
5. (CONADIS), C.: Estadísticas de Discapacidades (2023). https://www.consejodiscapacidades.gob.ec/estadisticas-de-discapacidad/. Accessed 20 Feb 2023
6. Heredia, J.P.S., Sailema, W.C., Ku, X.Q., Rodriguez, M.T., Duffaut, A.C., Jacome, L.: Tecnología asistiva para la comunicación y movilidad de personas con discapacidad motriz. Informática y Sistemas Revista de Tecnologías de la Informática y las Comunicaciones **3**, 1–14 (2019). https://doi.org/10.33936/ISRTIC.V3I2.1958
7. van den Heuvel, R.J., Lexis, M.A., de Witte, L.P.: Zora robot based interventions to achieve therapeutic and educational goals in children with severe physical disabilities. Int. J. Soc. Robot. **12**, 493–504 (2020). https://doi.org/10.1007/S12369-019-00578-Z/TABLES/5, https://link.springer.com/article/10.1007/s12369-019-00578-z
8. Landis, J.R., Koch, G.G.: The measurement of observer agreement for categorical data. Biometrics, 159–174 (1977)
9. Li, G., Cai, S., Xie, L.: Cooperative control of a dual-arm rehabilitation robot for upper limb physiotherapy and training. In: IEEE/ASME International Conference on Advanced Intelligent Mechatronics, AIM 2019-July, pp. 802–807 (2019). https://doi.org/10.1109/AIM.2019.8868784
10. Soleimani, M.A., Zohoor, H., Yakhdani, A.R.F., Heravi, M., Mohammadi, E.: Designing, prototyping, and controlling a portable rehabilitation robot for the shoulder physiotherapy and training. In: ICRoM 2019–7th International Conference on Robotics and Mechatronics, pp. 281–284 (2019). https://doi.org/10.1109/ICROM48714.2019.9071844
11. Zahariev, R., Valchkova, N., Wagatsuma, H.: Service robots for special education of children with disabilities: robotized systems for social applications. In: ACM International Conference Proceeding Series, pp. 300–306 (2020). https://doi.org/10.1145/3407982.3408023, https://dl.acm.org/doi/10.1145/3407982.3408023

Design and Implementation of an Exploration Algorithm for Autonomous Robots

Anderson José Sánchez Padilla and A. Hernando González(✉)

Universidad Autónoma de Bucaramanga, Bucaramanga, Colombia
{asanchez478,hgonzalez7}@unab.edu.co

Abstract. This paper presents the design and implementation of an exploration algorithm for differentially driven autonomous robots. The algorithm is validated in a controlled environment by positioning the robot to perform a scanning task and determining the location of metal objects in a two-dimensional plane. This prototype is the first phase of a research project aimed at designing collaborative robots for landmine detection. The paper presents the kinematic model of a differential robot, the control system for trajectory planning and obstacle avoidance.

Keywords: Differential kinematics · PID Control · Robot control · trajectory planning · obstacle avoidance

1 Introduction

For the navigation of mobile robots in a space without colliding and towards the same destination point, numerical algorithms must be developed for planning routes along which the robots must move, avoiding obstacles in which various methods are found, such as the so-called swarms of active Brownian particles [1], potential fields that models obstacles as electrical charges, which generate repulsion fields to stay away from obstacles and find clear paths and attraction fields that lead the agent towards the goal [2]. Due to the geometry of the obstacles, local minima could occur when generating the trajectories that would take the robot along erroneous paths, in [3] presents a local minima avoidance strategy by increasing the propulsion energy of the robots when a certain threshold is reached. Local minimum and decreases when it leaves that point. For the correct tracking of the routes, the position of the robot must be known at all times, for which GPS sensors can be used, however, this limits the robot to work in open spaces, it is also common to find connection problems with the satellite presenting important delays in the process, one way to avoid this is to develop a way to know the orientation of the mobile robots through the encoders located on the wheels applying the technique known as odometry [4], the xy position of the robot is obtained and the orientation, storing data in a trajectory vector and the realization of the graph corresponding to the realized trajectory.

To reach these trajectories, some way of calculating the speeds at which each tire of the robot must move is necessary, taking into account that it must follow the route

C. Montenegro et al. (Eds.): ICMTT 2023, LNNS 774, pp. 103–112, 2024.
https://doi.org/10.1007/978-3-031-43733-5_10

as well as carry the correct orientation, this is known as level control means or local planning; one method is to summarize the trajectories as summations of circular arcs and it is analyzed in a time interval between two speed calculations of the route planning algorithm, in each calculation cycle candidate speeds are defined that are appropriate [5]. For each tire, to guarantee the speed of the tires, the low-level control is designed, which can be a PID control, a polar coordinate control or an LQR that aims to reach the speed given by the local planner. Guarantees control of the robot [6]. Graph-based algorithms start from the decomposition into cells of the space according to the type of graph to be used, one way to do this analysis is to assign a value to the starting cell, commonly 0, and from this calculate the value of the closest cells according to the distance between them and the target cell, assigning to the cells occupied by obstacles with a value of infinite cost [7]. These cells can be seen represented in nodes that join the cells with less cost (Floyd-Warshall algorithm), or in matrices that reach the target cell using the closest algorithm criterion (A* algorithm).

2 Differential Mobile Robot

The mobile robot developed is of the differential type and has two motors that govern the wheels (right and left) independently, controlled by an ATmega2560 microcontroller. The operation of the mobile robot is carried out in a controlled environment and on a flat surface seeking to eliminate slippage on the wheels. The speed measurement is carried out with incremental optical encoders coupled to the shaft of each motor, the encoder delivers two square signals 90° out of phase with each other, commonly, by analyzing the activation states of each signal and the combinations between them, it is possible to obtain the speed increase or decrease value and the direction of rotation of the wheel with high precision. The motors are controlled by means of an L298N driver capable of supplying up to 2 A of current to each motor, it communicates wirelessly through bluetooth at 2.5 GHz, for interaction with the environment it is equipped with ultra sound sensors that measure distances, all this powered by a LiPo battery of 11.1 V and 2.2 Ah.

2.1 PID Control

To ensure trajectory tracking, it is necessary for the robot to move its wheels at certain speeds that allow it to draw straight and curved paths and turn. Therefore, a PID controller must be implemented to guarantee the speed of the wheels [8]. To tune the controller, it is necessary to have the actuator transfer function that relates the delivered PWM and the developed speed ω. The function approximated to a first-order system will be used, represented in Eq. (1).

$$G(s) = \frac{1.6116e^{-0.0837s}}{1 + 0.0453s} \tag{1}$$

The controller is tuned using the Lambda method, because it presents a fast and without overshoots and the delay is eliminated using the Pade approximation. With the tuning of the controller, the PID constants of the controller are obtained: $K_c = 0.31$, $T_d = 0.09$, $T_i = 0.02$, obtaining the controller that relates the speed in [rad/s] and the PWM. The discretization of the controller is performed with a sampling period of $t_m = 0.1s$, to implement it in a microcontroller, obtaining the following discrete equation for the controller.

$$\begin{aligned} C(s) &= 0.31\left(1 + \frac{1}{0.02\,s} + 0.09\,s\right)e(s) \\ C(k) &= 1.611\left(e[k] + \frac{1}{0.02}e[k-2] + 0.09e[k-1]\right) \end{aligned} \tag{2}$$

3 Differential Mobile Robot Trajectory Following Control

Starting from the kinematic model of the differential mobile robot presented in Eq. (3), it is analyzed that $q(\dot{t})$ is the vector which stores the velocities to be calculated, therefore an expression must be found to determine the velocity of the actuators (Eq. 4). Equations (3) and (4) can be determined from Fig. 1 considering the reference frames and variables described therein.

$$\underbrace{\begin{bmatrix} \dot{p}_x \\ \dot{p}_y \\ \end{bmatrix}}_{h(t)} = \underbrace{\begin{bmatrix} \cos(\phi) & 0 \\ \sin(\phi) & 0 \\ & \end{bmatrix}}_{J} \underbrace{\begin{bmatrix} \mu \\ \omega \end{bmatrix}}_{q(t)} \tag{3}$$

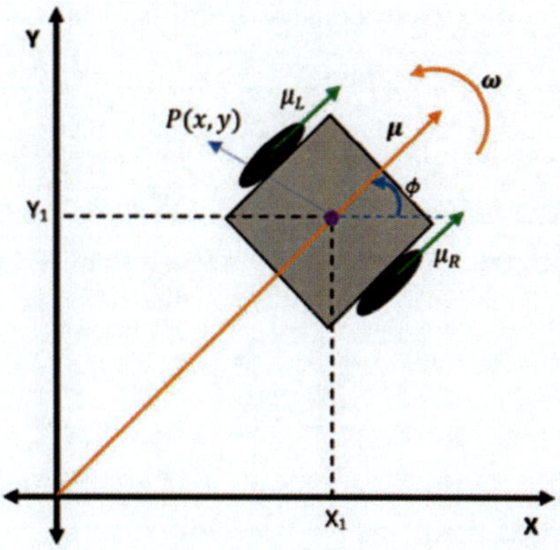

Fig. 1. Reference frames of a differential locomotion robot

Through the matrix clearance, the equation that determines the linear and angular velocities of the differential mobile robot is found.

$$q(\dot{t}) = J(q(t))^{-1}h(\dot{t}) \tag{4}$$

The trajectory is represented by the variable h in Eq. (4) for which it is necessary to find the derivative of the path to follow $\dot{h(t)}$ the algorithm is designed by means of a solution scheme based on the errors between the path desired and the actual path.

$$h_{\dot{e}} = h_{\dot{d}} - \dot{h} \tag{5}$$

$$\dot{q(t)} = J(q(t))^{-1}(\dot{h_d} - \dot{h_e}) \tag{6}$$

The goal of trajectory tracking control is for the robot to autonomously perform movements that are previously planned to follow a target trajectory. For this, a constant K is found that acts on the path error and stabilizes the robot on a work point. Using the Lyapunov stability criterion, the value of matrix K is found by identifying the candidate function (7) and its derivative with respect to time (8).

$$v(h_{ex}) = \frac{h_e^T h_{ex}^2}{2} \tag{7}$$

$$\dot{v} = h_e^T \dot{h_e} \tag{8}$$

Applying the Lyapunov criterion to Eq. (6) the model described in Eq. (9) is found:

$$\dot{q(t)} = J(q(t))^{-1}\left(\dot{h_d} + Kh_e\right) \tag{9}$$

$\dot{h_d}$ is the desired speed to traverse the path, $\dot{p_d}$ indicates the desired speed to reach a point within the path, in this way there will be a desired speed for each point as the robot approaches it, then $\dot{h_d} \neq \dot{p_d}$ (when considering this, a new term appears, the desired speed error δ_e).

$$\delta_e = h_{\dot{d}} - p_{\dot{d}} \tag{10}$$

The stability range of the K matrix for the path following model is determined, it is found that the range of values that the K matrix can take to ensure the stability of the system is given by

$$K > \frac{|\delta_e|}{|h_e|} \tag{11}$$

The constant K must be positive definite and greater than the ratio between the velocity error at the desired point and the trajectory error.

3.1 Computational Simulations of the Differential Mobile Robot Trajectory Tracking Control Algorithm

For the validation of the algorithms, a virtual environment was programmed in Python which allows validating the operation of the controllers described, in this virtual environment the robot is in a 3D space, the path to be followed is traced in blue and the path in red followed (Fig. 2).

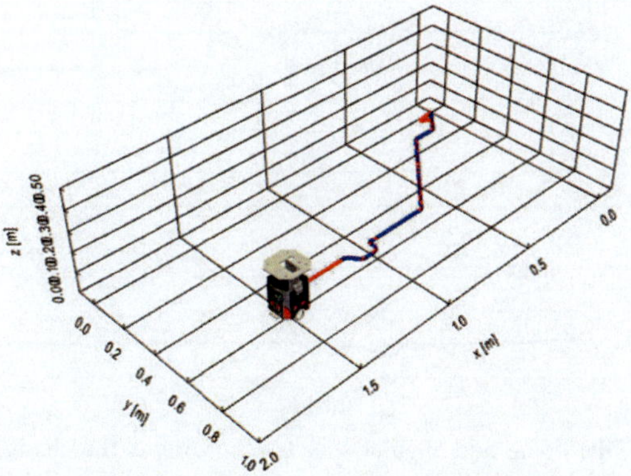

Fig. 2. Simulation in 3D environment of trajectory tracking

Simulation tests were carried out, varying the value of the *K* matrix for a given path, until establishing the value that provides the best response, eliminating the error in the stable state and generating speeds without oscillations. Figure 3 shows the trajectories generated for different constants of the K matrix. The trajectory applied as setpoint is shown in green, the trajectory followed by the algorithm with K = 8 in blue, the trajectory followed with K = 3 in red, and the trajectory followed with K = 12 in black.

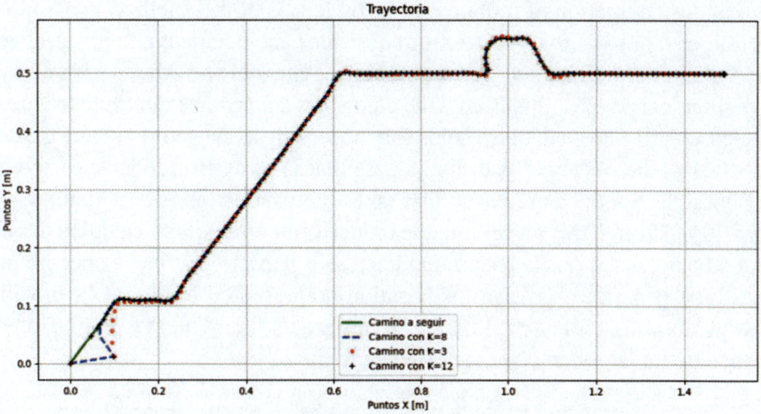

Fig. 3. Simulation in 3D environment of trajectory tracking

Figure 4 shows the linear and angular velocities developed, with the value K = 12 and K = 8 there are less oscillations in the calculated velocities. By increasing the gain K, the oscillations in the speeds decrease and the robot takes more time to complete the trajectory, the establishment time increases with the gain value, for K = 3 the time it takes the robot to complete the trajectory is 12 s, for the values K = 12 and K = 8 the settling times are approximately the same value. Since the response of the speeds generated with K = 12 had smaller oscillations in addition to stabilizing at zero, this value was taken as a parameter for the K matrix.

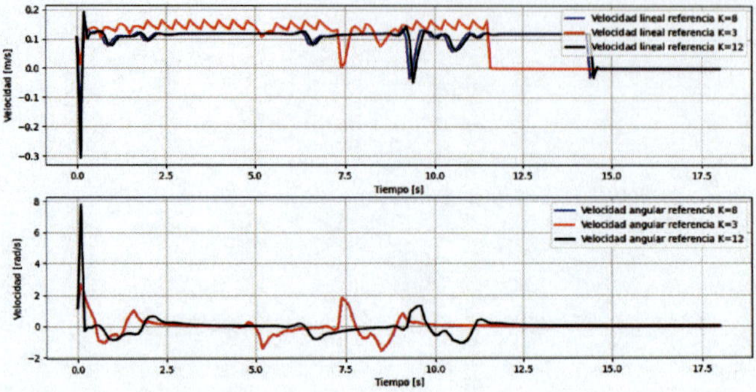

Fig. 4. Variation of the linear and angular speeds according to the change of the matrix K.

4 Obstacle Avoidance

For the execution of the algorithms that generate the paths free of obstacles to the goal, the map on which the robot moves must be known, this includes the location of the obstacles for later modeling using the planning algorithm.

4.1 Algorithm Based on Potential Fields

Potential field is a method of generation and planning of trajectories commonly used for autonomous navigation of mobile robots, by means of this method imaginary fields of attraction or repulsion that radiate from obstacles are calculated, being stronger near them, and weaker or null far away of the obstacles, the goal and other agents also radiate attractive imaginary fields, this is how the algorithm determines the route to follow free of obstacles towards the goal, these fields vary according to the geometry and the distance to the obstacles, the objective is to find a path that is away from obstacles and has the smallest possible Euclidean distance that takes the robot from a start point to another goal point [9]. Through this algorithm, the induced force values are calculated each time the robot advances, the obstacles are modeled as a particle that has a specific position and size, these generate repulsive fields and are represented by Eq. (12), in which the repulsive potential is represented by U_o, η represents the repulsive gain, d represents the distance to the target and q is the distance to the object.

$$U_o = \begin{cases} \frac{1}{2}\eta\left(\frac{1}{d} - \frac{1}{q}\right)^2, & d \leq q \\ 0 & d > q \end{cases} \tag{12}$$

The goal, which is the objective point to which the robot must move, generates an attractive potential field, this drives the robot to move towards said field, the attractive field is represented by Eq. (13).

$$U_g = \frac{1}{2}\gamma d(q, q_{goal}) \tag{13}$$

The attractive potential is represented by U_g, the parameter γ represents the attractive profit, d represents the distance to the goal as a function of q_{goal} the distance to the goal and q the potential value of the box in which the robot is located. Finally, the force induced in each step of the robot is calculated by adding the repulsive forces and the attractive forces. This applies to the rest of the scenario that is not occupied by the goal or some obstacle.

$$U_f = U_g + U_o \tag{14}$$

Figure 5 shows the change in the potential field according to the change in the parameters η and γ. The parameters established for the algorithm were $\eta = 2.5$, $\gamma = 5$, and d which is calculated using the Euclidean equation.

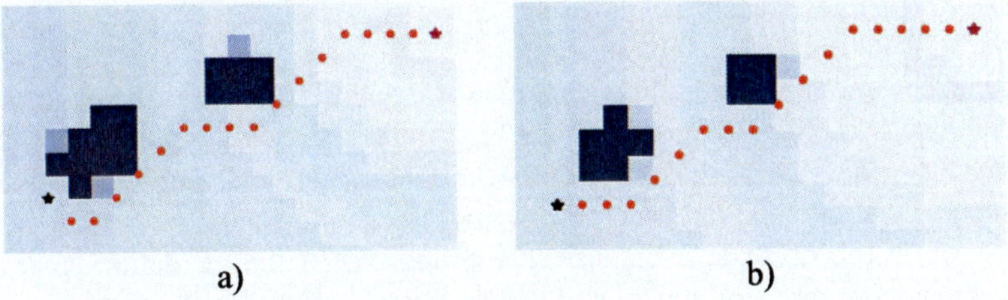

Fig. 5. Potential fields. a) $\eta = 1.1$ and $\gamma = 3$, b) $\eta = 2.5$ and $\gamma = 5$.

4.2 Graph-Based Algorithm: A*

The A* algorithm is a search algorithm that is part of visibility graphs, A* will find a solution if it exists, through an evaluation function

$$f(n) = g(n) + h(n) \tag{15}$$

where h(n) represents the heuristic value of the node to be evaluated from the current n to the end and g(n) is the actual cost of the path traveled to reach that node, h(n) is calculated using the Eq. (16)

$$h(n) = \sqrt{(x_2 - x_1)^2 + (y_2 - y_1)^2} \tag{16}$$

The A* algorithm navigates through the environment matrix checking which cells are free of obstacles until reaching the target cell, then calculating the lowest cost cells that lead to the target cell [10]. Figure 6 shows the behavior of the algorithm with different parameters of cost and length of the cells. It is observed that the shorter the grid length, the algorithm presents zigzag sections near the target point that are not viable for a differential robot, since in these sections the trajectory tracking error is increased.

5 Results

For the comparison of the trajectory generation algorithms, the same obstacles and goal will be assigned to find the route to the target point using both algorithms, determining the route with the shortest distance and that is feasible for the robot's kinematic configuration.

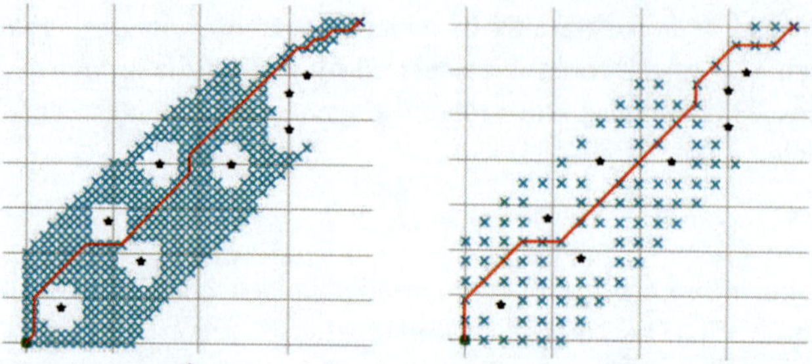

a) b)

Fig. 6. Algorithm A*. a) Grid length 5 cm, b) Grid length 10 cm

Taking into consideration the comparisons made for both algorithms in the same scenario, it can be concluded that the most efficient algorithm is that of potential fields, since it generates fewer deviations in the trajectory (curls) having mostly straight sections and in most cases the potential fields algorithm finds the shortest path.

To perform the obstacle avoidance, the conditions of the validation environment must be known (target point, location of the obstacles) in this way the route that reaches the objective without colliding with the objects is calculated by means of the trajectory generation algorithm through the validation environment, this algorithm generates a vector that stores the X and Y positions of the mentioned route, this route is applied as a set-point to the trajectory tracking algorithm which calculates the speeds so that the robot can move along the route, generating two vectors that store the values of angular and linear speed, these vectors are applied as set-point for the control of the actuators that calculate the PWM necessary for the wheels to rotate at the specified speeds and thus be able to move along the path. Specified path.

5.1 Exploration

For the exploration, the robot starts on the stage towards changing objectives and during the movement it reads the sensors, to identify the location of obstacles and present a 2D Cartesian graph that identifies the position of the obstacles or metallic objects found, in the execution of this task. The robot has no prior knowledge of the location of the obstacles on the stage. Figure 7 shows the scenario and 2D map with the location of the objects detected in the scenario (yellow color). To determine the location of metallic objects or obstacles, the robot has proximity sensors and metal detectors, when receiving a signal from one of these two sensors, the robot records the position in which it is, followed saves the X and Y coordinates of the object or metal in a vector to later present a 2D map with their location. The precision with which it determines the location of the object is $\cong 5$ [mm], due to the precision with which the validation environment was built.

For validation, the robot is located in the scenario, a target point is established and the algorithm for generating paths of potential fields is executed, this generates a vector that stores the points of the route, when moving it is taken continuously reading the sensors and if an object is found the robot stops, records the position of the object and again

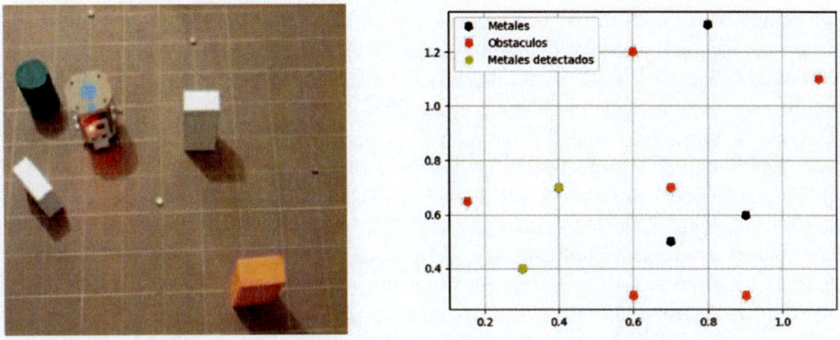

a) b)

Fig. 7. Mobile robot. a) Real scenario, b) 2D map with the location of the objects

runs the potential fields algorithm to find another route to the target point. The route vector is applied as a set-point to the trajectory tracking algorithm which generates the necessary speeds for the robot to follow the route, these speeds are entered as a set-point for the control of the actuators which calculate the PWM necessary for the wheels to follow the determined speeds to travel through the route. Next, the 2D map made from the exploration is presented, in which the metallic objects and obstacles are identified. Figure 7.b represents in black the metals distributed on the stage, in red the obstacles and in yellow the metals that the robot detected during the exploration.

6 Conclusions

The speed control of the robot wheels must be precise since this will prevent unwanted angular speeds from occurring due to a small deviation in the speed of any of the wheels, this leads to having a high-resolution speed measurement sensor and/or precision. For exploration, the path generation algorithm is executed several times, interrupting the robot's trajectory, and taking time to perform the algorithm's calculations, this for each object found in the environment, however, for obstacle avoidance, the path is previously known. Location of the objects for which the algorithm is executed once at the beginning and the robot moves along the path found by executing the trajectory tracking and actuator control algorithms. For the generation of trajectories in differential traction systems, the physical characteristics (geometry and construction) must be considered, since they are linked to the type of trajectories that the robot can follow; an initial and a final configuration cannot be joined by any path, the geometric and kinematic constraints of the system impose conditions that only some paths will meet.

References

1. Badesa, F.J., et al.: Metodos de control basados en campos potenciales y de fuerza para robotica de rehabilitacion. Actas de las XXXVI Jornadas de Autom. Bilbao, 332–339 (2015)
2. Tibaduiza, D.A.: Planeamiento de Trayectorias de un robot móvil. Trabajo de investigación de maestría. Directores: RM Ángel, J. Barrero. Universidad Industrial de Santander (2006)

3. Cuchango, H.E.E., et al.: Algoritmo para la planeación de trayectorias de robots móviles empleando enjambres de partículas brownianas. Vis. Electrónica **5**(1), 4–14 (2011)
4. Morales Muñoz, E.: Localización y mapeo simultáneo por robots móviles con ruedas basados en mapas de ocupación y matrices dispersas (2018)
5. Fox, D., Burgard, W., Thrun, S.: Controlling synchro-drive robots with the dynamic window approach to collision avoidance. In: Proceedings of IEEE/RSJ International Conference on Intelligent Robots and Systems. IROS 1996. IEEE, pp. 1280–1287 (1996)
6. Cornejo, J., et al.: Trajectory tracking control of a differential wheeled mobile robot: a polar coordinates control and LQR comparison. In: 2018 IEEE XXV International Conference on Electronics, Electrical Engineering and Computing (INTERCON), pp. 1–4. IEEE (2018)
7. Kumpel, J., Sparbert, J., Hofer, E.-P.: Numerical path optimization for path planning with cell decomposition methods. In: 2001 European Control Conference (ECC), pp. 1822–1827. IEEE (2001)
8. Meng, J., et al.: Two-wheeled robot platform based on PID control. In: 2018 5th International Conference on Information Science and Control Engineering (ICISCE), pp. 1011–1014. IEEE (2018)
9. VascaK, J.: Navigation of mobile robots using potential fields and computational intelligence means. Acta Polytech. Hung. **4**(1), 63–74 (2007)
10. Janet, J.A., Luo, R.C., Kay, M.G.: The essential visibility graph: An approach to global motion planning for autonomous mobile robots. In: Proceedings of 1995 IEEE International Conference on Robotics and Automation, pp. 1958–1963. IEEE (1995)

Strategic Management Innovation

Strategic Planning to Improve the Production Systems of an MSME: Case Study of a Toy Store

Jorge Buele[1](✉), José Antonio Lozada-Cepeda[2], and Belén Ruales[1]

[1] SISAu Research Group, Facultad de Ingeniería, Industria y Producción, Universidad Indoamérica, Ambato 180103, Ecuador
{jorgebuele,belenruales}@indoamerica.edu.ec

[2] Escuela de Posgrados, Pontificia Universidad Católica del Ecuador, Quito 170525, Ecuador
joselozada@worldvision.com

Abstract. Competitiveness in the sales and services sector is becoming more and more intense. Therefore, the implementation of methodologies and strategies is necessary for the achievement of objectives, business growth, and the improvement of production systems. This article presents the performance of a strategic plan (SP) to identify the shortcomings of a company belonging to the MSME group. A toy store developing in Ecuadorian business is presented as a case study. The initial analysis uses tools such as the external PESTEL analysis, the Servqual model, and an analysis of current competitiveness. A new business model (Business Model Canvas) is defined to implement the plan, and strategic initiatives and projects are established. Since this study is projected for five years, experimental results are obtained for the first year, which shows a reduction in the gaps between customer expectations and the service offered. There is also evidence of minor quality mismatches in the Servqual model and a better competitive position than others. These objectives are related to increasing the company's performance and combining quality, cost, and logistics on time.

Keywords: MSME · Production Systems · Strategic Planning · Toy Store

1 Introduction

Today, businesses are classified according to their number of employees and annual revenues [1, 2]. The most common type of business in Latin America are micro, small, and medium-sized enterprises (MSMEs) [3]. In fact, in countries such as Ecuador, MSMEs represent 99.55% of the total number of companies. According to data from the National Institute of Statistics and Census (INEC), 90.78% of microenterprises, 7.22% of small enterprises, and 1.55% of medium-sized enterprises in Ecuador. Of this percentage, 9 of 10 are dedicated to marketing or providing services. The growth in the number of companies makes it difficult for them to compete, regardless of their activity. Suppose an organization wants to maintain good competitiveness in the long term. In that case, it must use formal analysis and decision-making procedures framed within the framework of a strategic planning process. This led to the development of two levels of competitiveness: internal and external [4]. The former deals with the organization's ability to

C. Montenegro et al. (Eds.): ICMTT 2023, LNNS 774, pp. 115–124, 2024.
https://doi.org/10.1007/978-3-031-43733-5_11

achieve maximum performance from available resources and transformation processes. While external competitiveness is more oriented to the fulfillment of the organization's objectives in the context of the market or the sector to which it belongs. Therefore, in this last dimension of competitiveness, exogenous variables of a commercial nature must be considered. Once the company has reached a level of external competitiveness, it must be prepared to maintain it in the future, generating innovation and seeking new market opportunities [5].

Competitiveness could be expressed in several dimensions in a business context. However, for most researchers, the two main ones are quality and marketing. Based on this statement, in [6], in this research, several tools and models will be developed to obtain a first diagnosis of the company to measure the experiment's impact. The quality function deployment (QFD) is a matrix constructed to show how customer requirements relate directly to the ways and methods companies can use to meet those requirements. Combining the QFD with a Kano analysis model is ideal for understanding the relationship between the product or service and the degree of customer satisfaction. Another tool is the Servqual model, a business research technique that measures service quality by understanding customers' expectations and how they appreciate the service. A Mystery Shopper study is also developed in this research as an unbiased information-gathering method for evaluating service quality. In addition, empathy will make it possible to establish the customer segments that a company wants to reach by identifying and characterizing them in variables that define their profile.

This topic has been the subject of considerable research all over the world. Its application can be found in all types of enterprises, so [7] shows its contribution to the medical field. The article presents the challenges faced by healthcare institutions when considering its implementation. In [8] evaluates the relationship between strategic planning and competitiveness in small companies in Bogota. The data showed a positive influence on the competitiveness of MSMEs. In turn, [9] analyzes the implementation of strategic planning in SMEs in the textile sector in Barranquilla. These experimental results show a moderate influence on the planning and evaluation of this methodology and the possibility of improvement during its implementation. Even the participation of social networks in strategic planning can be evidenced, as shown in [10]. The findings show increased sales, customer retention, and marketing improvement. The literature review shows that the use of this methodology influences companies. Therefore, this research seeks to know its impact on a particular company in the Ecuadorian toy sector.

This document has five sections, including the introduction in Sect. 1.and the methodology planning in Sect. 2. Then, Sects. 3 and 4 present the implementation of the methodology and the preliminary results obtained accordingly. Finally, conclusions are provided in Sect. 5.

2 Materials and Methods

2.1 Initial Company Diagnosis

The company to be studied in this research is an MSME whose business is selling and distributing toys in Ambato, Ecuador. In recent years, to increase their competitiveness reduced the profit percentage of their products and increased the variety. Their main marketing strategy for the last 25 years has been customer referrals. This company has four differentiated types of customers: children, toy collectors, people looking for a gift, and resellers to microenterprises. To understand in more detail, the customer segment, the target the organization maintains, and its profile. The criteria for calculating criticality are presented as an annex in the following link: https://github.com/JBueleEC/Strategic_Planning.

2.2 External PESTEL Analysis

The external environment is affected by multiple factors, one of which is that importing products from the Asian continent is costly for small Ecuadorian companies. Therefore, international economic and commercial decisions will greatly affect the company. In the economic field in recent years, there has been a bilateral trade cooperation agreement between Ecuador and China signed on December 12, 2018, which facilitates the import of goods. In addition, since 2017, there has been a reduced tax rate of 5% that augurs better import costs.

In the social sphere, prominent global toy marketers are reducing their sales due to the growing preference of children for cell phones and computer games. In addition, a new trend of virtual reality and augmented reality applications has displaced traditional toys. The impact of these phenomena can be seen in countries with large toy industries, such as Spain or the United States, which have reduced their income from toy sales by 45% in the last five years. On the other hand, Ecuador is facing a severe economic recession that, far from being close to the end, finds the country with few sources of employment and little economy. Policies such as reducing public spending, privatizations, accepting the impositions of the International Monetary Fund (IMF), and increasing political instability reduces the purchase of non-essential products such as toys. Added to this is the economic crisis generated by the COVID-19 pandemic and its repercussions on Latin American economies.

2.3 Actual Competitiveness

Customer requirements are analyzed to calculate competitiveness, described in Table 1. These attributes are also compared with a Servqual model to understand them more deeply. The opinions of adult customers were obtained through surveys, and some focus groups were established. In addition, the mystery shopper technique was developed to analyze the offerings of competing brands.

2.4 Servqual Model Data Analysis and Interpretation

This organization serves an average of 150 clients per month without considering the last three months of the calendar year because of high demand. To calculate the sample a confidence level of 95% and a maximum permissible error of 5% were chosen for 109 participants. The questionnaire comprises 22 questions grouped into the five most representative dimensions for obtaining statistical data. With this, a numerical value is calculated to identify the client's expectations in contrast to their perceptions of the company. This is presented in Table 2, and the gap between what was specified and what was obtained is also obtained. The problems are mainly in the treatment and attention to customers. It is also observed that the speed of service, the modernity of its facilities, its appearance and the attractiveness of its products leave a great dissatisfaction. These areas will be improved in the plan.

Table 1. Customer requirements list.

N°	Customer requirements
A	Convenient prices
B	Quality products
C	Internet sales
D	Good service
E	Friendly and courteous treatment
F	Analyze the product they sell
G	Variety of products
H	Assistance during the purchase
I	Cleanliness of the store
J	Speed
K	Fulfillment of deliveries
L	Large quantity of products
M	After-sales problem solving
N	A special treatment to loyal customers
O	Friendly atmosphere
P	Games for children
Q	Recognized brands
R	Social and environmental management

Table 2. Gaps between customer expectations and the service offered.

Item	Dimensions	Service expected	Service perceived	Gap
1	Tangibles	5.6	4.7	−0.9
2	Reliability	5.2	3.8	−1.4
3	Sensitivity	5.2	3.3	−1.9
4	Assurance	5.0	3.5	−1.5
5	Empathy	4.9	2.5	−2.4

3 Implementation of the Strategic Plan

3.1 Business Model Canvas

Based on the information above, strategic planning meetings were held with all the people who make up the company. The model and strategy described in Fig. 1 were agreed upon. As part of developing this methodology, the organization's mission, vision, and values were established; these have yet to be shown at the organization's request.

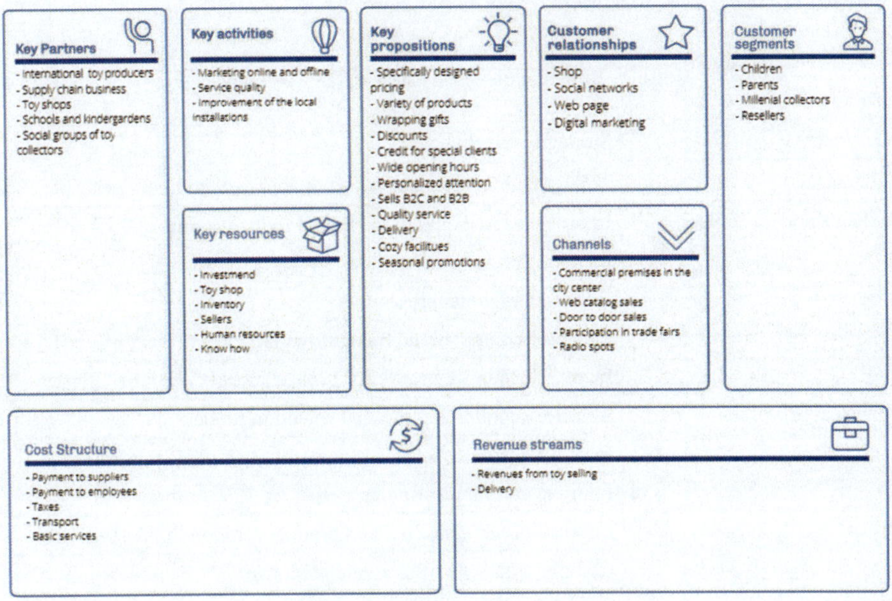

Fig. 1. Business Model Canvas.

3.2 Strategic Projects and Initiatives

Based on the established technical guidelines and the data provided by the Servqual model, strategic projects and initiatives are set to improve the service offered to its customers. The initiatives and projects listed to be developed throughout the implementation of the strategic plan are described below.

- Training for sales personnel.
- Increased investment in e-commerce.
- Establishment of standard processes and procedures for customer service.
- Implement 5S philosophy principles within the store.
- Training of staff in teamwork.
- Elaboration and development of a branding plan.
- Restructuring of the marketing area

- Increased customer knowledge.
- Training staff in the correct use of information and communication.
- Restructuring the distribution of the commercial premises
- Creation of a welcoming place
- Improvement of the inventory and warehousing system
- Implementation of processes for the selection of suppliers and products.

To understand the importance and impact that each strategic initiative will have on the scope of the strategic objectives, Table 3 shows the alignment of the strategic initiatives with the strategic objectives established by the institution.

Table 3. Strategic objectives and strategic initiatives.

Perspective	Strategic objectives
Financial	28% growth in revenues based on business competitiveness
Clients and customers	Improve service quality index and reduce claims to 0.1%
	Increase service quality index by 28%
	Increased regional presence
	Increase regional brand recognition by 25%
	Increase customer loyalty by 17% or more
Internal processes	Improve supplier and product selection process
	Improve customer service process by 2 points (Servqual)
	Improve communication processes by 12% or more
	Improve order, visibility, and product rotation in the store
	Comply with 90% of strategic planning
	Develop new promotional channels focused on e-commerce
Learning and growth	Build high-performing, empathetic and sales-focused teams
	Manage integrated and timely business information
	Improve the work environment
	Improve infrastructure and environment

3.3 Training of Sales Staff

Implementing the strategic plan requires obtaining financial and time resources more significant than the time established to develop this research. The implementation of the first strategic business project was developed because the sales training initiative is considered a priority project for the company's management after having conducted the analysis using the QFD (house of quality).

To prepare the employees in the sales area, interviews were conducted with customers about the significant shortcomings observed in sales and customer service. Based on the analysis of the interviews and the strategic areas for improvement, a commercial expert defined and developed a training agenda. The topics included the company's history, sales team, customer needs, product knowledge, competitive analysis, and motivation. Sales planning and organization, sales techniques and tools, and customer service with quality and warmth were also analyzed. Subsequently, the training program began, with daily feedback for those attending the program on the topics reviewed during the day.

3.4 Logistics and Supply Chain Management

Logistics and supply chain management are crucial aspects of improving the efficiency of this business. Strategic initiatives can be established to achieve an efficient combination of quality, cost, and timely delivery that focus on improving supply chain management, logistics, and distribution. Inventory tracking tools are proposed to control the products available in the warehouse better and optimize supplier orders. Delivery routes were also optimized to reduce costs and improve delivery to customers. In addition, it is proposed to strengthen relationships with suppliers, establishing long-term agreements to ensure the availability of stock and diversity of toys. A sustainable resource and waste management practices can also be adopted to reduce environmental impact and improve the company's image.

3.5 Improving Production Systems

Improving production systems is another critical aspect of increasing the efficiency and quality of the company's production processes. Strategic initiatives focused on reducing import and delivery times were developed. This allows for improved quality management and optimization of the use of resources. To achieve these objectives, Lean Manufacturing and continuous improvement can be promoted to identify and eliminate waste in production processes and improve product quality. Implementing an online sales platform also allows the automation of processes to increase efficiency and reduce sales times.

Finally, periodic audits should be conducted to evaluate performance and improvement in production systems and their impact on customer satisfaction and business profitability. In this way, opportunities for improvement can be identified, and a constant evolution of the company's production processes can be maintained.

4 Preliminary Results

4.1 Servqual Model

To determine the changes produced, new statistical data collection was carried out through surveys conducted after the implementation of the priority projects of the strategic plan. The perception obtained will then be analyzed using the Servqual model and the competitive position analysis offered by the house of quality. Table 4 shows the values obtained and the comparison with the previous ones. In addition, Fig. 2 shows a comparison by category with the initial results.

Table 4. Gaps between customer expectations and the service offered.

	Dimensions	Expected service	Previous value	Previous gap	Current value	Current gap
1	Tangibles	5,6	4,7	–0,9	5,1	–0,6
2	Reliability	5,2	3,8	–1,4	4,5	–0,7
3	Sensitivity	5,2	3,3	–1,9	4,7	–0,5
4	Assurance	5,0	3,5	–1,5	4,2	–0,8
5	Empathy	4,9	2,5	–2,4	2,6	–2,3

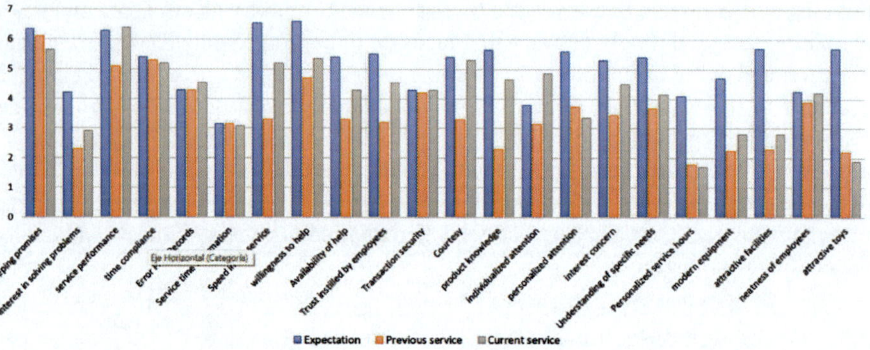

Fig. 2. Quality mismatches in the Servqual model after implementation.

4.2 Competitive Position

After implementing the first prioritized process, the gaps have been reduced. This is because the level of service quality perceived by customers has increased, mainly in aspects related to customer services, such as courtesy, individualized attention, and willingness to help the customer, among others. These issues were addressed during the training workshop developed for the company's sales personnel. A new competitive position analysis based on customer perception is also carried out. Figure 3 shows the competitive assessment according to the customer before and after the initial implementation of the strategic plan. Another essential aspect was developing a web page and sales through social networks that improve interaction with the user. This methodologically demonstrates the strong relationship between service quality and competitiveness and verifies the effectiveness of the strategic plan developed.

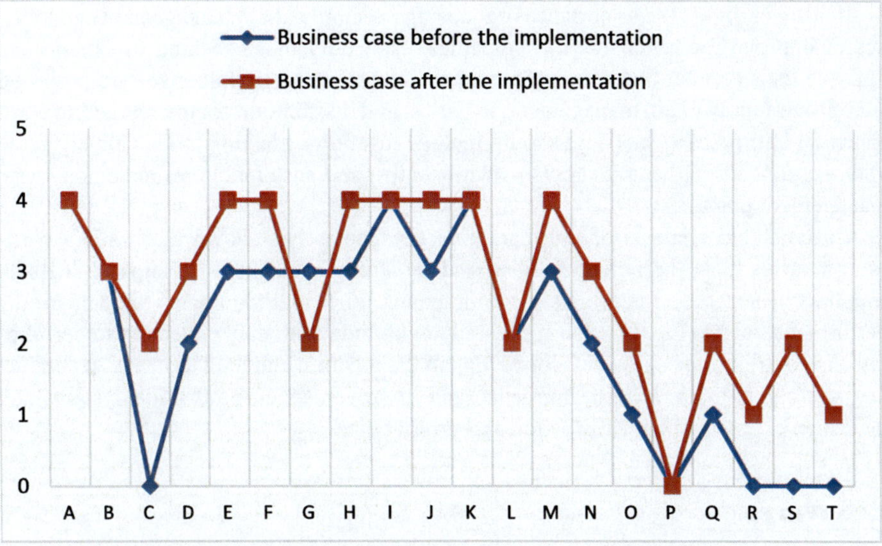

Fig. 3. Comparison of competitive position before and after implementation.

5 Conclusions

The research highlights the relevance of designing and implementing methodologies that strengthen companies in Latin America and around the world, especially SMEs. It has been shown that the gap between expectations and perception of customer service quality is reduced by developing a strategic plan and incorporating it properly. To do this, it is essential to have a well-designed plan and adequate allocation of resources, as well as a committed and trained team to implement the plan. It is also essential to constantly monitor the results and make the necessary adjustments to achieve the established objectives. Although no standardized methodologies are defined to measure competitiveness, using the Servqual and QFD models demonstrated the feasibility of evaluating this parameter. However, it is essential to remember that the Kano model analyzes service attributes since customer service expectations vary over time. This means that features that were previously attractive quality (competitive advantages) tend to become basic requirements expected by the customer.

The service quality assessment showed that improvements are needed in empathy, customer treatment, and the perception of security provided to the customer during service. This assessment will direct the company's strategic projects to the most affected areas. These projects include increasing customer awareness, marketing, and improving sales strategies. Based on Servqual and QFD market research, this plan will generate a higher value proposition for the customer, translated into the strategy map, strategic projects and initiatives, and the metrics aligned to them. The measurement of the business impact of the partial implementation of this proposal showed a positive increase in the quality of service perceived by customers and, therefore, an increase in business competitiveness.

It also highlights the importance of focusing on supply chain management and logistics to improve the efficiency of companies, especially those seeking to expand and improve their competitiveness in the market. Several strategic initiatives are proposed to improve supply chain management, logistics, and distribution for toy marketing companies in Latin America. These should include inventory tracking tools, delivery route optimization, strengthening supplier relationships, and sustainable resource and waste management practices.

Current field analysis of toy marketing companies has shown that Latin American countries have the potential to expand by strengthening partnerships with Asian suppliers. Therefore, it is proposed to implement other methodologies to compare the performance of SMEs. It is also suggested to continue with data collection to corroborate the contributions of the developed plan in the medium and long term and to conduct periodic audits to evaluate performance and improvement in production systems and their impact on customer satisfaction and profitability.

References

1. Franklin, T.-O., Tubon-Nunez, E.E., Carrillo, S., Buele, J., Franklin, S.-L.: Quality management system based on the ISO 9001:2015: study case of a coachwork company. In: Iberian Conference on Information Systems and Technologies, CISTI (2019). https://doi.org/10.23919/CISTI.2019.8760816
2. Rodríguez Gavilanes, D.: Incidencia del riesgo de crédito en los niveles de liquidez de las mutualistas de ahorro y crédito del Ecuador. Publicare. **2**, 1–9 (2022). https://doi.org/10.56931/pb.2022.21_6
3. Zottele, E.: Latin American MSMEs and the belt and road initiative. In: A New Blue Ocean, pp. 57–76. Palgrave Macmillan, Singapore (2021). https://doi.org/10.1007/978-981-15-7687-4_4
4. Niaki, M.K., Nonino, F.: Impact of additive manufacturing on business competitiveness: a multiple case study. J. Manuf. Technol. Manag. **28**, 56–74 (2017). https://doi.org/10.1108/JMTM-01-2016-0001
5. Aryal, G., Mann, J., Loveridge, S., Joshi, S.: Exploring innovation creation across rural and urban firms: analysis of the National survey of business competitiveness. J. Entrep. Public Policy. **7**, 357–376 (2018). https://doi.org/10.1108/JEPP-D-18-00026
6. Bobadilla, V., del P.C., Zapata, M.I.B., Brito, A.E.P.: La mercadotecnia, factor clave para la competitividad organizacional. Gestión Soc. **6**, 15–30 (2013)
7. Robles Salguero, R.E., Serrano Mantilla, H.B., Serrano Mantilla, G.L., Gaibor Vera, F.M., Armijo Borja, G.M., Fernández Lorenzo, A.: Retos de la planificación estratégica en instituciones de salud. Rev. Cuba. Investig. Biomed. **36**, 1–6 (2017)
8. Mora-Riapira, E.H., Vera-Colina, M.A., Melgarejo-Molina, Z.A.: Strategic planning and competitiveness levels of SMEs in the business sector in Bogotá. Estud. Gerenciales. **31**, 79–87 (2015). https://doi.org/10.1016/j.estger.2014.08.001
9. Garcia Guiliany, J.E., Duran, S.E., Cardeño Pórtela, E., Prieto Pulido, R., Garcia Cali, E., Paz Marcano, A.: Proceso de planificación estratégica: Etapas ejecutadas en pequeñas y medianas empresas para optimizar la competitividad. Espacios. **38** (2017)
10. Cheing, A., Hong, E.H., Kuek, T.Y., Chai, B.H.B., Cham, T.H.: Social media effectiveness indicators of microenterprise strategic planning. Asian J. Bus. Res. **10**, 150–166 (2020). https://doi.org/10.14707/ajbr.200079

International Scientific Production on Smart Cities and Innovation in Bibliometric Description of Data Indexed in Scopus

Rômulo Oliveira Barros[1], Monique de Menezes Urra[2], Fabiana Pinto de Almeida Bizarria[3(✉)], and Flávia Lorenne Sampaio Barbosa[4]

[1] Instituto Federal do Piauí – IFPI, Teresina, Brazil
[2] Universidade Federal do Piauí – UFPI, Teresina, Brazil
[3] Pontifícia Universidade Católica de Minas Gerais and Collaborating Professor in the Professional Master's Program in Public Management at the Federal University of Piauí – PPGP, Teresina, Brazil
fabiana.almeida.flf@gmail.com
[4] Universidade Federal do Piauí – UFPI, Teresina, Brazil
flsbarbosa@ufpi.edu.br

Abstract. Smart cities have aroused interest in the media and in various fields of knowledge, due to the search for solutions to various urban and social problems, suggesting analyses related to innovation, especially in the context of public policies. Thus, a bibliometric research was carried out on the relationship between innovation and cities in the world, considering the Scopus database, directed to scientific articles published in English between 2012 and 2022. As a result, 394 scientific articles were found, published by 977 different authors, in 252 journals, as well as the contributions of the 10 most cited articles in the period were described, in addition to three others selected by the relevance of the authors. It is concluded that the study of innovation in cities is on the rise, especially from 2018, with emphasis on interdisciplinary and transdisciplinary studies with a focus on sustainability.

Keywords: public management · innovation · cities · smart cities · bibliometric review

1 Introduction

The movement of society for better public services and for greater participation in the creation of public policies, drives cities to be increasingly focused on the welfare of people, on the citizen's quality of life (Menezes; Amorim & Fernandes 2022), especially within the context of decentralization of public policies in Brazil (Cavalcante 2011).

In this sense, innovation suggests the involvement with the search for solutions to urban problems that cannot be solved by traditional approaches or through increased public spending (Torfing 2019). These are increasingly complex and structuring challenges, while there are no previously defined solutions, which are called wicked problems (Cavalcante & Cunha 2017).

C. Montenegro et al. (Eds.): ICMTT 2023, LNNS 774, pp. 125–136, 2024.
https://doi.org/10.1007/978-3-031-43733-5_12

The concept of wicked problems emerged in the 1960s, associated with the conceptions of smart cities, when scientific management actions produced unexpected effects (Briggs 2007, Goodspeed 2015). They are problems that are difficult to define clearly and, therefore, different stakeholders have varying versions of the problem; they occur in unique contexts and have many interdependencies and multiple determinants; they are socially complex, requiring the participation of several public institutions, at different levels of government, which require joint analysis (Briggs 2007, Goodspeed 2015).

Smart cities have emerged as a research theme, both in scientific literature and in the media in general, and present debates favorable to approaches that aim at the solution of social problems experienced by cities. The definition and, the adjective, "smart", gives the dynamics of contributing to the expansion of the competitiveness of the local community through innovation, while improving the quality of life of its citizens through public services capable of meeting the needs of the population, as well as a cleaner environment (Appio, Lima & Paroutis 2019).

In this sense, this article performs a bibliometric review of the international scientific literature on the relationship between innovation and Smart Cities, seeking to answer the following questions: which innovative initiatives in the scope of cities are being developed and studied in the world? Which themes, authors and journals are most relevant in research on innovation in cities? What is the trend in the amount of scientific production on the subject over the years? Which countries and cities have contributed the most to the production of scientific articles on the subject? Which are the collaborating countries from the authors of these researches?

2 Methodological Procedures

To present a description of the international scientific production based on scientific articles that address the subject innovation aimed at solving problems related to cities around the world, we defined the Scopus database (Elsevier 2022b), using the proxy of the Coordination for the Improvement of Higher Education Personnel (CAPES), accessed through the CAPES Periodical Portal.

The database was consulted on April 28, 2022, using the following search term: innovation AND (city OR cities). The results obtained were then filtered, limiting the search to articles published in journals, in final version (excluding articles in press), between 2012 and 2022, in English, and in which the search terms were present in the title. By consolidating the initial terms and filters, the full search term was as follows: TITLE (innovation AND ("city" OR "cities")) AND PUBYEAR > 2011 AND PUBYEAR < 2023 AND (LIMIT-TO (SRCTYPE, "j")) AND (LIMIT-TO (PUBSTAGE, "final")) AND (LIMIT-TO (OCTYPE, "ar")) AND (LIMIT-TO (LANGUAGE, "English")).

After the database search, the complete records were exported in BibTeX and comma separated values (CSV) format. These data were loaded into the Bibliometrix software (Aria & Cuccurullo 2017), from where CSV files generated by the analyses programmed in that software were obtained. These CSV files were analyzed using the R language (The R Foundation 2022), with the R package collection tidy verse (Wickham et al. 2019), and the RStudio development environment (Rstudio Team 2022).

Next, contributions from the 10 most cited articles in the period were described, as well as three others selected for their relevance, seeking to present an integrative analysis of the emerging themes described by the authors.

3 Results

Figure 1 shows the trend in international scientific production on the subject investigated since 2012, indexed in the Scopus database. There has been a significant increase in the annual average number of publications since 2018, such that in the first four months of 2022 the number of publications is already higher than the annual average between 2012 and 2017. Over the last 10 years, 394 scientific articles have been published, in 252 different journals and written by 977 different authors in total. Table 1 presents the evolution of the number of publications in the 10 journals with the greatest contribution related to the themes of innovation and cities. These journals add up to 82 articles published since 2012, which represents 20.8% of the total publications in the last 10 years.

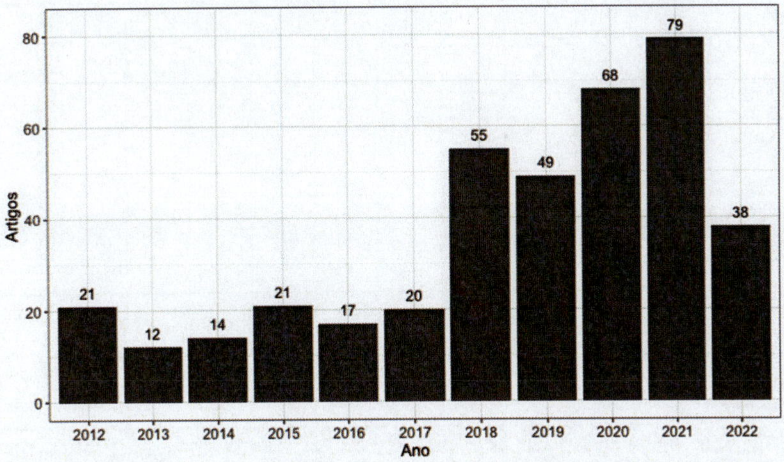

Fig. 1. Annual trend of number of publications in the area of innovation and cities. Source: research data

The most referenced journal is Sustainability (ISSN: 2071–1050), focused on interdisciplinary studies on sustainability, open access, published by Multidisciplinary Digital Publishing Institute (MDPI 2022). This journal has an impact factor of 3.251 and a Scopus citation score (CiteScore) of 3.9 and a h-index of 85. The second journal in the ranking of this research is the Journal of Cleaner Production (ISSN: 0959–6526), published by Elsevier, characterized by its transdisciplinary in addressing issues involving environmental sciences, with impact factor 9.297, CiteScore of 13.1 and h-index 200 (Elsevier 2022a).

The number of publications in the Sustainability journal are more constant over the years than in any other of the 10 selected journals. Between 2018 and the beginning of 2022, in no year did the journal stop publishing articles. On the other hand, in the Journal of Cleaner Production there were no publications in the years 2019 and 2020, pandemic years of Covid-19. This last journal requires a fee for publishing articles, which is around three thousand dollars (Elsevier 2022a).

Table 1. The 10 journals with the most publications in the area of innovation and cities since 2012.

Journal	Year of publication											
	2012	2013	2014	2015	2016	2017	2018	2019	2020	2021	2022	Total
Sustainability					1		1	5	4	8	3	22
Journal of Cleaner Production		1					3			6	2	12
Cities	1				2	2	1	1	1		1	9
Int. J. of Envion. Research And Public Health								1	2	2	2	7
Journal of Coastal Research								3	4			7
Regional Studies	1			1		1				1	2	6
City, Culture and Society				2			2	1				5
Techn. Forecasting and Social Change								3		1	1	5
Urban Studies		2				1		1	1			5
European Planning Studies		1							1	2		4

Source: research data.

Tables 2 and 3 illustrate the list of the 20 authors with the highest volume of publications between 2012 and 2022, and also the researchers with the most citations in the same period, adding up all articles published.

Table 2. The 20 most cited authors and the 20 authors with the most publications.

The 20 authors with the most citations				The 20 authors with the most published articles			
Author	**Citations**	**Articles**	**Countries**	**Author**	**Citations**	**Articles**	**Countries**
Zygiaris S.	336	1	Grécia	Wang X.	103	9	Estados Unidos
Caragliu A.	173	3	Itália	Fan F. *	171	7	China
Fan F. *	171	7	China	Li J.	137	7	China
Adler P.	156	2	Estados Unidos	Rodríguez-Pose A.	115	7	Reino Unido
Baccarne B.	156	2	Bélgica	Chen Z.	3	5	China
Florida R.	156	2	Canadá	Lee N. *	145	5	Reino Unido
Schuurman D.	156	2	Bélgica	Wang J.	12	5	China
De Marez L.	154	1	Bélgica	Wang S.	74	5	China
Mechant P.	154	1	Bélgica	Zhang J.	92	5	China
Mellander C.	154	1	Suécia	Chen Y.	19	4	China
Del Bo C.F.	151	2	Itália	Li Y.	15	4	China
Shearmur R.	146	1	Canadá	Zhang Y.	33	4	Estados Unidos
Appio F.P.	145	1	França	Bramwell A.	20	3	Estados Unidos
Lee N. *	145	5	Reino Unido	Caragliu A.	173	3	Itália
Lima M.	145	1	França	Hawken S. *	127	3	Austrália
Paroutis S.	145	1	Reino Unido	Ke H.	32	3	China
Li J.	137	7	China	Liu X.	78	3	China
Hawken S. *	127	3	Austrália	Ma Y.	14	3	China
Taylor Buck N.	120	1	Reino Unido	Qi Y.	32	3	Hong Kong
While A.	120	1	Reino Unido	Song Q.	32	3	Hong Kong

Source: research data
Legend: * - authors who are among the 20 most cited and 20 most productive, but whose articles are not among the 10 most cited in the last 10 years

Regarding the authors, we observe a predominance of Chinese researchers among those with the most published articles. However, European authors dominate the number of citations. Based on this information, it is inferred that the most predominant innovative actions in this study involve the Chinese territory.

In Table 3, the 10 most cited articles within the studied period are listed. It can be observed that, with few exceptions, the authors present in the lists of the most cited and most productive articles are also present in the authorship of the 10 most cited articles. Table 4 presents authors well represented in the survey.

The authors Fan, F. (China), Lee N. (United Kingdom) and Hawken, S. (Australia) are present in both the list of the 20 most productive and the 20 most cited (Table 2 and Table 3) and, however, their articles are not among the 10 most cited in the last 10 years. This reference is considered for the selection of texts for reading and presentation of contributions to the academic field.

Table 3. Top 10 most cited articles between 2012 and 2022.

Authors	Title	Year	Citations
Zygiaris S	Smart City Reference Model: Assisting Planners to Conceptualize the Building of Smart City Innovation Ecosystems (Zygiaris 2013)	2013	336
Florida R., Adler P., Mellander C	The city as innovation machine (Florida, Adler & Mellander 2017)	2017	154
Schuurman D., Baccarne B., De Marez L., Mechant P	Smart ideas for smart cities: Investigating crowdsourcing for generating and selecting ideas for ICT innovation in a city context (Schuurman et al. 2012)	2012	154
Shearmur R	Are cities the font of innovation? A critical review of the literature on cities and innovation (Shearmur 2012)	2012	146
Appio F.P., Lima M., Paroutis S	Understanding Smart Cities: Innovation ecosystems, technological advancements, and societal challenges (Appio Lima & Paroutis 2019)	2019	145
Caragliu A., Del Bo C.F	Smart innovative cities: The impact of Smart City policies on urban innovation (Caragliu & Del Bo 2019)	2019	134
Taylor Buck N., While A	Competitive urbanism and the limits to smart city innovation: The UK Future Cities initiative (Taylor Buck & While 2015)	2015	120
Abella A., Ortiz-de-Urbina-Criado M., De-Pablos-Heredero C	A model for the analysis of data-driven innovation and value generation in smart cities' ecosystems (Abella, Ortiz-De-Urbina-Criado & De-Pablos-Heredero 2017)	2017	118
Ning L., Wang F., Li J	Urban innovation, regional externalities of foreign direct investment and industrial agglomeration: Evidence from Chinese cities (Ning, Wang & Li 2016)	2016	90
Cohen B., Almirall E., Chesbrough H	The city as a lab: Open innovation meets the collaborative economy (Cohen, Almirall & Chesbrough 2016)	2016	85

Source: research data.

Table 4. The most cited articles by Fan, F., Lee N. and Hawken, S.

Authors	Title	Year	Citations
Fan F., Lian H., Liu X., Wang X	Can environmental regulation promote urban green innovation efficiency? An empirical study based on Chinese cities (Fan et al. 2021)	2021	60
Lee N	Migrant and ethnic diversity, cities and innovation: Firm effects or city effects? (Lee 2015)	2015	54
Praharaj S., Han J.H., Hawken S	Urban innovation through policy integration: Critical perspectives from 100 smart cities mission in India (Praharaj, Han & Hawken 2018)	2018	66

Source: research data.

4 Top 10' Contributions

In the 2013 article Smart City Reference Model: Assisting Planners to Conceptualize the Building of Smart City Innovation Ecosystems, Zygiaris proposes a reference model that can be used by cities as the base document for creating their own smart city plans. The author argues that his model, being a collection of various notions of smart cities, can be used to assess the innovative capacity of cities that use it, identifying complementarities and inconsistencies in their own plans (Zygiaris 2013).

Zygiaris' (2013) model is composed of six layers: layer 1 is the green city, which involves commitments to use technologies to reduce greenhouse gas emissions; layer 2, interconnection, which involves the broadband infrastructure that interconnects the city, in the figure of its equipment and its inhabitants; layer 3, instrumentation, related to the infrastructure to constantly monitor the metrics of the city, such as sensors; layer 4, open integration, where the city shares the data it produces; layer 5 is the application layer, which integrates the previous layers; and layer 6, which is the innovation layer, related to the creation of business opportunities related to the implementation of the other layers.

The focal point of the reference model proposed by Zygiaris (2013) is the contribution of cities to a sustainable, green and smart growth of the planet. In addition, the author suggests that the model helps provide a common understanding among stakeholders about the priorities of investments to be made.

In the same sense of creating models to aid Zygiaris' smart city planning, authors Abella, Ortiz-De-Urbina-Criado and De-Pablos-Heredero (2017) suggest, in the article A model for the analysis of data-driven innovation and value generation in smart cities' ecosystems, a three-phase model to demonstrate how data generated by smart cities can create value for citizens and society. The suggested phases are: analysis of the degree of reuse of open data, analysis of the innovation realized from the data, and finally, analysis of the social and economic impact of the data-driven products and services developed by the smart city ecosystem. The authors argue that the model suggests using smart city data to create innovative services that improve the quality of life of the inhabitants by meeting their real demands (Abella, Ortiz-De-Urbina-Criado & De-Pablos-Heredero 2017).

In the article The city as innovation machine, published in 2017, Richard Florida, Patrick Adler, and Charlotta Mellander initially review the literature on innovation, giving special emphasis to the thinkers Marx and Schumpeter. Then the authors address the influence of the territorial aspect in fostering innovation. According to the authors, innovation, as well as entrepreneurship, exists in territorial agglomerations. And, in this sense, location is an important factor in determining the innovation potential of a region (Florida, Adler & Mellander 2017).

Based on this argument, Florida, Adler & Mellander (2017) finish their article by proposing that the region and the city are the center of the innovation process. According to the authors, innovation and entrepreneurship are more essentially spatial processes rather than firm- or individual-based. For Florida et al., place has replaced industrial cooperation as the key economic and social unit of the modern knowledge economy.

Schuurman et al. (2012) conducted a case study of the city of Ghent, Belgium, where a participatory process (crowdsourcing) for selecting projects to use information technology for innovation in that city.

The essence of Smart ideas for smart cities: Investigating crowdsourcing for generating and selecting ideas for ICT innovation in a city context is the account of the Belgian city's experience in the participatory involvement of society in choosing projects that concern subjects at the same time so close (cities) and so "distant" to many, technology. The authors argue that crowdsourcing proved to be a useful and effective tool in the context of innovation in that city, even though the process did not result in extremely innovative ideas. And, for this reason, that the authors also consider that this tool should be combined with others, such as urban living labs (Schuurman et al. 2012).

The article Are cities the font of innovation? A critical review of the literature on cities and innovation, published in 2012 by the Canadian Shearmur, presents a counterpoint to the previously mentioned articles. For the author, it is a mistake to attribute all the innovation that is observed in cities to cities alone (Shearmur 2012).

Shearmur's (2012) argument is that population movements take and bring innovative ideas from diverse places in a country or region to a geographically limited space, the cities. As a consequence, the ideas and knowledge that are shared in the city are not necessarily generated in that location, nor do they necessarily generate consequence in that specific place, but they spread along with the migration process. In this way, Sheamur (2012) argues that the most pressing issue involving geographical aspects and innovation revolves around the consequences of the distribution of innovation across territories and communities (Shearmur, 2012). This phenomenon is what other authors call geographic spillover of innovation (Ning, Wang & Li 2016).

Understanding Smart Cities: Innovation ecosystems, technological advancements, and societal challenges is an introduction to a special edition of scientific articles focused on studies on smart cities. The authors, Appio, Lima, and Paroutis (2019) analyze the topic based on previous studies, make an overview of the main topics discussed in the articles that would be part of the special issue, and, at the end, offer a research agenda that points to an urgent need for the development of a science of smart cities. The agenda includes addressing the conceptualization of cities as complex adaptive systems (Appio, Lima & Paroutis 2019).

In the article Smart innovative cities: The impact of Smart City policies on urban innovation authors Caragliu and Del Bo (2019) seek to analyze whether smart city initiatives promote innovation in the urban environment. Through quantitative study, the authors sought association between the number of patents and intensity of smart city policies in 309 European cities (Caragliu & Del Bo 2019).

In the research, the authors concluded that those cities that invest above the European average in smart city policies tend to innovate more, especially in high technology. The authors consider that the research findings provide understanding about the mechanisms through which smart city policies favor the economic development of cities. Moreover, according to the authors, these policies increase the "stock" of knowledge of places and improve city performance in terms of economic growth, quality of life, sustainability, and population well-being (Caragliu & Del Bo 2019).

The article Competitive urbanism and the limits to smart city innovation: The UK Future Cities initiative is a case study of the "TSB Future Cities Demonstrator Competition" in the UK. In it the authors highlight the challenges for governments to transform the spatial imaginary of a technologically enhanced smart city into meaningful urban projects (Taylor Buck & While 2015).

In their studies Taylor Buck and While (2015) noted difficulties in coordination between national government innovation policies and local government priorities. Case studies like this one bring alert for future initiatives around the world, especially in places with decentralization of political competences between federative levels, such as Brazil (Góes & Mello 2018).

The same phenomenon was observed by Praharaj, Han, and Hawken (2018) in their study of 100 smart cities cases in India. In Urban innovation through policy integration: Critical perspectives from 100 smart cities mission in India, the authors reported that different levels of government produced regulations that sometimes overlapped, hindering the development of smart city policies. According to Praharaj, Han & Hawken (2018), the large amount of policies, plans, and regulations from the various states and local agencies are not integrated and sometimes overlap under divergent views about policies, hindering the development of cities.

Also on the subject of public policymaking, Cohen, Almirall, and Chesbrough (2016) explore a number of tensions that are emerging as innovators and entrepreneurs seek to partner with local governments and citizens in an effort to improve quality of life and promote local economic growth.

According to the authors, many solutions to these tensions reside in the local bureaucracies responsible for service delivery and public asset management. In this sense, for them, the focus of the local bureaucracy should be on organizing and planning the structures that lead to the "orchestration" of the innovation ecosystem in smart cities, rather than trying to deliver services directly (Cohen, Almirall & Chesbrough 2016).

Ning, Wang & Li (2016) produced an empirical study on the role of foreign direct investment on innovation in cities in China, an emerging economy. Through statistical models, the authors concluded that foreign direct investment plays an important role as an external source of knowledge to increase local innovation capacity. However, the spatial externalities of this type of investment are restricted to the city where this investment occurred, not favoring its spillover. In addition, according to the authors, industrial clusters moderate the externalities of foreign direct investment in urban innovation, since they absorb this knowledge within cities and can favor spillover in nearby regions (Ning, Wang & Li 2016).

Authors Fan, Lian, Liu, and Wang (2021) used spatial measurement mathematical models to investigate the mechanism by which environmental regulations affect the efficiency of sustainability-related innovation at the regional level in 235 cities in China. In the end, the authors make three recommendations: that China should continue to adhere to innovation-based development, especially on environmental issues; the central government should formulate intense environmental regulations to generate stimulus for innovation in pollution-reducing technologies while enhancing international competitiveness and technological development; control the scale, quality, and proportion of foreign capital; and improve the training of industry workers.

The role of diversity in innovation in the urban environment was addressed by Lee (2015). The author, focusing on cities in the UK, analyzes the impact of cultural diversity in firms on their innovation capabilities. At the same time, they investigate whether the diversity of present in cities causes innovation in firms, or whether the opposite occurs.

The findings of the study point to the fact that more migrant or migrant-owned firms tend to introduce new products and processes in the city. It is therefore suggested that diversity within firms benefits innovation more than the diverse environment of the cities themselves. However, according to the study, although the effect of the amount of migrants in cities is not greater, it still exists. Cities with more migrants tend to be more innovative than less diverse cities (Lee 2015).

5 Conclusion

It is concluded that the study of innovation in cities is growing in terms of publications in journals, especially from 2018. The most relevant journals are those focused on interdisciplinary and transdisciplinary studies, with a focus on sustainability.

The top three journals are Sustainability, the Journal of Cleaner Production and Cities. Of the 20 most cited authors, only six are not European. The three most cited authors are Zygirais S. (Greece), Caragliu, A (Italy) and Fan, F. (China). In terms of productivity, on the other hand, only nine authors are not from China. The most productive authors are Wang X. (United States), Fan F. (China) and Li J. (China).

Theoretical models aimed at helping the planning of smart city policies were addressed by two of the most relevant authors. For these authors, models such as these serve for decision-making on investments and for comparing cities in terms of evolution in the implementation of smart city policies.

Another recurring theme was the need for integration between policies and planning at different governmental levels, to reconcile diverse views on innovation to meet local demands. Furthermore, an important theme debated in the literature during the period concerns the spreading of innovation actions to regions beyond the geographical limits of a single city.

References

Abella, A., Ortiz-De-Urbina-Criado, M., De-Pablos-Heredero, C.: A model for the analysis of data-driven innovation and value generation in smart cities' ecosystems. Cities **64**, 47–53 (2017). https://doi.org/10.1016/j.cities.2017.01.011

Appio, F.P., Lima, M., Paroutis, S.: Understanding smart cities: innovation ecosystems, technological advancements, and societal challenges. Technol. Forecast. Soc. Chang. **142**, 1–14 (2019). https://doi.org/10.1016/j.techfore.2018.12.018

Aria, M., Cuccurullo, C.: Bibliometrix : an R-tool for comprehensive science mapping analysis. J. Informetrics, **11**(4), 959–975 (2017). https://doi.org/10.1016/j.joi.2017.08.007

Briggs, L.: Tackling Wicked Problems: a Public Policy Perspective. Australian Public Service Commission, Canberra (2007)

Caragliu, A., Del Bo, C.F.: Smart innovative cities: the impact of smart city policies on urban innovation. Technol. Forecast. Soc. Chang. **142**, 373–383 (2019). https://doi.org/10.1016/j.techfore.2018.07.022

Cavalcante, P., Cunha, B.Q.: É preciso inovar no governo, mas por quê? In: Cavalcante, P., et al. (org.). Inovação no setor público: teoria, tendências e casos no Brasil. Brasília: Ipea, 15–32. (2017). Disponível em: http://repositorio.ipea.gov.br/handle/11058/8785. Acesso em: 28 abril 2022

Cohen, B., Almirall, E., Chesbrough, H.: The city as a lab: open innovation meets the collaborative economy. Calif. Manage. Rev. **59**(1), 5–13 (2016). https://doi.org/10.1177/0008125616683951

Elsevier.: Journal of Cleaner Production. Journal of Cleaner Production. (2022a) Disponível em: https://www.journals.elsevier.com/journal-of-cleaner-production. Acesso em: 28 abr. 2022

Elsevier.: Scopus. Start exploring. (2022b). Disponível em: https://www-scopus.ez117.periodicos.capes.gov.br/search/form.uri?zone=TopNavBar&origin=searchadvanced&display=basic#basic. Acesso em: 28 abr. 2022

Fan, F., Lian, H., Liu, X., Wang, X.: Can environmental regulation promote urban green innovation efficiency? an empirical study based on Chinese cities. J. Cleaner Product. **287** (2021). https://doi.org/10.1016/j.jclepro.2020.125060

Florida, R., Adler, P., Mellander, C.: The city as innovation machine. Reg. Stud. **51**(1), 86–96 (2017). https://doi.org/10.1080/00343404.2016.1255324

Goodspeed, R.: Smart cities: moving beyond urban cybernetics to tackle wicked problems. Cambridge J. Regions Econ. Soc. **8**(1), 79–92 (2015). https://doi.org/10.1093/cjres/rsu013

Lee, N.: Migrant and ethnic diversity, cities and innovation: firm effects or city effects? J. Econ. Geogr. **15**(4), 769–796 (2015). https://doi.org/10.1093/jeg/lbu032

MDPI.: Sustainability. Sustainability (2022). Disponível em: https://www.mdpi.com/journal/sustainability. Acesso em: 28 abr. 2022

Menezes, M., Amorim, E.E.R., Fernandes, K.: Inovação nas cidades: o papel dos laboratórios urbanos vivos. In: Encontro Internacional Participação, Democracia e Políticas Públicas, 5., 2022, Natal. Anais eletrônicos [...] Natal: UFRN, 2022 (2022). Disponível em: https://www.pdpp2022.sinteseeventos.com.br

Ning, L., Wang, F., Li, J.: Urban innovation, regional externalities of foreign direct investment and industrial agglomeration: evidence from Chinese cities. Res. Policy **45**(4), 830–843 (2016). https://doi.org/10.1016/j.respol.2016.01.014

Praharaj, S., Han, J.H., Hawken, S.: Urban innovation through policy integration: critical perspectives from 100 smart cities mission in India. City Cult. Soc. **12**, 35–43 (2018). https://doi.org/10.1016/j.ccs.2017.06.004

Rstudio Team.: RStudio: Integrated Development Environment for R. 2022. (2022). Disponível em: http://www.rstudio.com

Schuurman, D., Baccarne, B., De Marez, L., Mechant, P.: Smart ideas for smart cities: investigating crowdsourcing for generating and selecting ideas for ICT innovation in a city context. J. Theor. Appl. Electron. Commer. Res. **7**(3), 49–62 (2012). https://doi.org/10.4067/S0718-18762012000300006

Shearmur, R.: Are cities the font of innovation? a critical review of the literature on cities and innovation. Cities **29**(2), S9–S18 (2012). https://doi.org/10.1016/j.cities.2012.06.008

Taylor Buck, N., While, A.: Competitive urbanism and the limits to smart city innovation: the UK future cities initiative. Urban Stud. **54**(2), 501–519 (2015). https://doi.org/10.1177/0042098015597162

The R Foundation. R: The R Project for Statistical Computing. 2022 (2022). Disponível em: https://www.r-project.org/. Acesso em: 14 fev

Torfing, J.: Collaborative innovation in the public sector: the argument. Pub. Manag. Rev. **21**(1), 1–11 (2019). https://doi.org/10.1080/14719037.2018.1430248

Wickham, H., et al.: Welcome to the tidyverse. J. Open Source Softw. **4**(43), 1686 (2019). https://doi.org/10.21105/joss.01686

Zygiaris, S.: Smart city reference model: assisting planners to conceptualize the building of smart city innovation ecosystems. J. Knowl. Econ. **4**(2), 217–231 (2013). https://doi.org/10.1007/s13132-012-0089-4

Experience Marketing Management for Commercial Success in a Shopping Mall in Peru

Elsa Denisse Paredes-Rivadeneyra[1], Franklin Cordova-Buiza[2,3](✉), and Claudia Amparo Torres-Linares[1]

[1] Research and Innovation Department, Universidad Privada del Norte, Lima, Peru
claudia.torres@upn.edu.pe

[2] Faculty of Business Sciences, Universidad Continental, Huancayo, Peru
fcordova@continental.edu.pe

[3] Business Faculty, Universidad Privada del Norte, Lima, Peru
franklin.cordova@upn.edu.pe

Abstract. This article aims to establish the relationship between experience marketing and the commercial success of Megaplaza Independencia, a mall located in Lima, the capital of Peru. The research has a non-experimental design, of correlational scope. For data collection, the survey was applied, the instrument was the questionnaire with the Likert scale. Likewise, the documentary analysis technique was used for the review of the financial statements of the Shopping Center, both validated by Cronbach's Alpha. The results show that experience marketing strategies such as having a variety of available brands, organizing events and sales promotions, allow obtaining commercial success, which is reflected in increased sales revenue, greater affluence and increased leasable area. This research concluded that there is a weak positive relationship between experience marketing and the commercial success of Megaplaza Independencia.

Keywords: Experiential marketing · shopping center · consumer · retail

1 Introduction

According to the International Monetary Fund, it considers that outside of the pandemic, in the last 25 years the world economy grew an average of 3.6% per year. Peru with a growth of 4.7% has been considered as a reference of successful economic management for the region, generating a scenario of growth and low inflation [1].

This better positioning of the Peruvian economy has allowed the construction and consolidation of many shopping centers nationwide, going from eight to sixty in the last ten years, with 20 operators, highlighting among them: Cencosud, Inmuebles Panamericana, Administradora-Jockey Plaza, Centenario Retail S.A.C, Inversiones Castelar, Open Plaza, Mall Aventura Plaza [2–4].

Despite the initial studies in the scientific literature by books and articles, theoretical studies are scarce, hence the present research seeks to improve the existing knowledge about Experience Marketing strategies in shopping centers; [5] mentions that Megaplaza

C. Montenegro et al. (Eds.): ICMTT 2023, LNNS 774, pp. 137–147, 2024.
https://doi.org/10.1007/978-3-031-43733-5_13

Independencia, located in Lima the capital of Peru, is considered the most referenced shopping center and with greater affluence in Peru; but it could grow to a real dimension if it considers the application of Experience Marketing strategies.

The objective of the research is to establish the relationship between experience marketing and the commercial success of Megaplaza Independencia in the period 2014–2018. The dimensions of Experience Marketing are considered as the variety of available brands, events and sales promotion, and the dimensions of Commercial Success as sales revenue, affluence and leasable area.

2 Literature Review

Previous research highlights the importance of planning and managing the customer experience by studying the influence of products on their senses [6–8].

[9] presents a study in the city of Lima for ESAN University, where he argues that the critical factors to consider for the success of a shopping center are access roads, tenant mix, location, centralized management, design and infrastructure, availability of anchor stores, consumer credit, economic stability and growth, population, complementary and additional services, and good relationship with stakeholders.

[10] proposes five different strategies based on experiential areas such as perceiving, feeling, thinking, acting, relating; these are the areas that make us perceive such experiences through stimuli and sensations. [11] agrees with this theory and highlights that consumers have gone from being isolated to being connected, i.e. from being uninformed to being informed and from being passive to being active. [12] states that experiential marketing is the set of sensations, feelings, cognitions and behavioral response evoked by stimuli related to a brand, and which are part of the design and identity of a brand.

[10] for his part states that experiences in addition to having a nexus with the company, the brand and the customer's life, align personal facts and the purchase situation with a broader social environment, which complements [13] saying that entrepreneurs offer the same services, the advantage lies in the way they lead their customer to delight in an unforgettable experience.

[14] and [15] explained that the generation of experiences is of utmost importance in the relationship with the audiences of an organization, presents the concept of audience development as a continuous, planned and managed process that seeks to build long-term relationships between an organization and its existing or potential audiences. This process brings together marketing, education and programming activities that seek to act proactively to increase the engagement of existing audiences, attract new ones and offer increasingly enriching experiences.

A shopping mall is a facility generally managed by a single operator, has been planned and built for the purpose of retail and has its own parking lot [16, 17]. Three main and basic physical configurations of shopping malls can be identified, which are: Malls (large shopping centers), open malls and hybrid malls (possessing characteristics of both) [18–20]. Studies by [21] analyze the effect of the consumer's emotional shopping experience on the perception of benefits and corporate reputation of department stores in Mexico and Chile; their results show that the functional and symbolic benefit perceived by consumers significantly influences the reputation of department stores.

3 Methods

The research had a non-experimental design, and a descriptive correlational analysis will be made. The unit of analysis was the Megaplaza Independencia Shopping Mall, with a population of 402 stores; a statistically representative sample of 196 stores of the Megaplaza Independencia Shopping Mall was established. The technique to collect data on both variables (experience marketing and commercial success) was the survey, which was applied to the selected sample in 2018, with the purpose of establishing the relationship between experience marketing and commercial success of Mega plaza Lima Norte, the instrument was the questionnaire, using the psychometric Likert scale (categories: strongly agree, somewhat agree, neither agree nor disagree, somewhat disagree, strongly disagree, strongly disagree).

Likewise, the documentary analysis technique was used, in a longitudinal cut research, and the instrument used was the documentary review, which presented information from the financial statements (EEFF) of Megaplaza Independencia Shopping Center from the years 2014 to 2018. The dichotomous questions were evaluated by the nominal scale, yes and no, referring to a series of conditions of the financial statements. Both instruments were validated by Cronbach's Alpha.

4 Results

The results of the survey applied to 196 tenants of the Megaplaza Independencia Shopping Mall are presented and analyzed below, with the intention of establishing the relationship between experience marketing and the commercial success of the aforementioned mall in 2018, to then be analyzed in conjunction with the results obtained from the documentary review (of the Financial Statements) in the period 2014–2018. The results are shown from the tabulated data and the application of the SPSS statistic.

To obtain the reliability of the instrument, the analysis of the data was performed from the criteria of form and content, through Cronbach's alpha coefficient [22] suggest the following recommendations for evaluating Cronbach's alpha coefficients:

Next, the results obtained:

Table 1. General criterion for interpreting Cronbach's Alpha coefficient.

Alpha coefficient	Description
Alpha coefficient > .9	Excellent
Alpha coefficient > .8	Well
Alpha coefficient > .7	Acceptable
Alpha coefficient > .6	Questionable
Alpha coefficient > .5	Poor
Alpha coefficient < .5	Unacceptable

Fuente: George & Mallery (2003).

1.-Experience marketing strategies (Table 1).

Table 2. Summary of case processing used in the SPSS program for the instrument.

		No	%
Cases	Valid	196	98,0
	Excluded*	4	2,0
	Total	200	100,0

Notes: * List deletion is based on all procedure variables.

In Table 2, the summary of the process of the cases for this variable is presented, as shown, it is valid in 98% and only 2% excluded. Which can be interpreted as a good result.

Table 3. Reliability statistics.

Alfa de Cronbach	Number of elements
,797	6

The questionnaire was validated according to Cronbach's Alpha to know the reliability of the Variable Experience Marketing (Table 3). This result shows that Cronbach's alpha coefficient is eighty percent (80%), according to [22] the coefficient is good. 2.-Commercial success

Table 4. Summary of case processing used in the SPSS program for the instrument.

		No	%
Cases	Valid	196	98,0
	Excluded*	4	2,0
	Total	200	100,0

Notes: * List deletion is based on all procedure variables.

Table 4 shows the summary of the process of the cases for this variable, where it is observed that it is valid in 98% and only 2% excluded. Which can be interpreted as a good result.

Regarding the validity according to Cronbach's Alpha to know the reliability of the variable commercial success (Table 5) (Table 6).

This result shows that Cronbach's alpha coefficient is at eighty percent (80%), according to [22] the coefficient is good. The validity of the instrument was also verified through the judgment of experts, giving its approval with the following results:

In summary, the validity of the survey-type instrument allowed greater reliability for the collection of information and to determine the correlation between both variables.

Table 5. Reliability statistics.

Alfa de Cronbach	Number of elements
,804	6
Applicable	Applicable

Table 6. Results of validation by expert judgment.

Indicators	Expert 01	Expert 02	Expert 03
1. Clarity	Excellent	Very good	Excellent
2. Objectivity	Excellent	Excellent	Excellent
3. News	Excellent	Excellent	Very good
4. Organization	Excellent	Excellent	Excellent
5. Sufficiency	Very good	Excellent	Very good
6. Intentionality	Very good	Excellent	Excellent
7. Consistency	Excellent	Excellent	Excellent
8. Methodology	Excellent	Excellent	Excellent
9. Relevance	Excellent	Excellent	Excellent
Final decision	Applicable	Applicable	Applicable

Following this, the results obtained from the correlation according to Spearman's Rho are presented, with the purpose of determining if there is a linear relationship between the two variables at the ordinal level, in other words, to know if the relationship is statistically significant.

To contrast both variables and their dimensions, the following rule is taken into account:

H_0 = There is a significant relationship between experience marketing and the commercial success of Megaplaza Independencia in the period 2014–2018.

H_1 = There is no significant relationship between experience marketing and the commercial success of Megaplaza Independencia in the period 2014–2018.

A significance level $\alpha = 1\%$ was used.

If the p-value $> = 0.01$, we accept the null hypothesis (H0). If the p-value < 0.01 we reject the null hypothesis (H0) and accept the alternative hypothesis (H1). Consequently, for the correlation between both variables experience marketing and commercial success (Table 7) applying Spearman's Rho, to validate the degree of association of the two variables. A p-value ≥ 0.01 was obtained (Table 8).

According to the proposed rule, as shown in Table 7, the correlation coefficient 0.279 was obtained, being greater than 0.01 ($rS = 0.279 > 0.01$), we accept the null hypothesis (H0) and reject the alternative hypothesis (H1); therefore, then it is affirmed that there is a relationship between experience marketing and commercial success of

Table 7. Correlation of experience marketing and commercial success.

			Marketing	Commercial
Rho de Spearman	Experience Marketing	Correlation coefficient	1,000	,279*
		Sig. (bilateral)	.	,000
		No.	196	196
	Commercial Success	Correlation coefficient	,279*	1,000
		Sig. (bilateral)	,000	.
		No.	196	196

Notes: * The correlation is significant at level 0.01 (bilateral)

Table 8. Correlation variety of available brands and commercial success.

			March	Commercial
Rho de Spearman	Variety of brands available	Correlation coefficient	1,000	,293*
		Sig. (bilateral)	.	,000
		No.	196	196
	Commercial Success	Correlation coefficient	,293*	1,000
		Sig. (bilateral)	,000	.
		No.	196	196

Notes: * The correlation is significant at level 0.01 (bilateral)

Megaplaza Independencia in the period 2014–2018. As pointed out by [23] in the interpretation of Spearman's Correlation Coefficient, a statistically significant linear relationship, between weak positive and mean and directly proportional, was found between both variables.

The correlation between the variety of available brands and commercial success (Table 9). A value rS = 0.293 was obtained, which is interpreted according to [23] as a positive correlation between weak and medium (rS = 0.25–0.50).

Regarding the correlation between the effectiveness of events and commercial success (Table 9). It turned out to be a value of 0.234, which is interpreted according to [23] as a weak positive correlation.

On the other hand, the correlation between the effectiveness of sales promotion and commercial success (Table 10), with a value of 0.173, according to [23] is a positive correlation between very weak and weak (rS = 0.10–0.25).

As for the results of the documentary review of the Financial Statements in the period 2014–2018, they were obtained through the construction of a "documentary review

Table 9. Correlation of event effectiveness and commercial success.

			Efficiency	Commercial
Rho de Spearman	Effectiveness of events	Correlation coefficient	1,000	,234*
		Sig. (bilateral)	.	,001
		N	196	196
	Commercial Success	Correlation coefficient	,234*	1,000
		Sig. (bilateral)	,001	.
		No.	196	196

Notes: * The correlation is significant at level 0.01 (bilateral)

Table 10. Correlation of effectiveness of sales promotion and commercial success.

			Sales	Commercial
Rho de Spearman	Effectiveness of Sales Promotion	Correlation coefficient	1,000	,173*
		Sig. (bilateral)	.	,015
		N	196	196
	Commercial Success	Correlation coefficient	,173*	1,000
		Sig. (bilateral)	,015	.
		No.	196	196

Notes: * The correlation is significant at level 0.05 (bilateral)

form" that facilitated the organization of the information to perform its analysis. The balance and interpretation were key in the research, it should be noted that the financial management of the CC. Megaplaza Independencia, provided the main guideline to define the dimensions of analysis to be used, since this agency applies the horizontal, vertical and financial ratios analysis.

The most significant results were: the mall has a wide variety of stores and brands available to customers (80%), the mall has a wide variety of stores and brands available to customers (60%), the events held in the mall are of great value (80%), the mall has a very well programmed and structured calendar (80%), the mall's promotions are effective and contribute to the business strategy (80%), the programming of promotions is effective and helps to increase sales (60%), the mall's management contributes to increase sales (60%), the mall has increased sales this year compared to previous years (80%), the mall's management generates more customers (100%); the mall has increased the number of customers this year compared to previous years (80%), the management of the mall is related to the evolution of the leasable area (40%), the mall has had an evolution of the leasable area compared to previous years (100%).

The analysis of the financial statements allows highlighting the importance of the implementation of strategies for the success of the commercial management and therefore of the business management of the company; this is observed with the implementation of strategies such as, national and international music concerts, live presentations for television, culinary presentations, dances, events for children and adults, fashion shows, photo shoots, contests, parades with characters, vehicle exhibitions, exhibition of collections, book fairs, Christmas fairs, school fairs, among others, in order to provide a greater number of customers. On the other hand, it is observed that the mall has had an evolution of its leasable area compared to previous years, this is demonstrated by the expansion of its second and third level, investments in expansion works, the implementation of new operations, including transnational firms, the creation of modules in the aisles and courtyards, the constant renovations of facades, floors, lighting, which seek to increase the consumption of its visitors, and that buyers are better served.

5 Discussion

From the results of this study, we obtained the acceptance of the general hypothesis. This research concluded that there is a weak positive relationship between experience marketing and the commercial success of Megaplaza Independencia. in the period 2014–2018, indicating that the correlation is significant at the 0.01 level (bilateral).

These results are in line with [9] who point out that the commercial success of shopping centers, especially in Latin America, has a strong relationship with the marketing implemented by the company, being experience marketing an innovative strategy that changes the paradigm of traditional marketing, and that focuses on recognizing the current needs and behaviors of consumers in order to meet the expectations and requirements of today's society.

Companies and brands that use experience marketing strategies will be the ones to succeed, since consumers feel overwhelmed, tired of conventional environments, and what they are looking for are innovative formats that allow them to perform daily tasks, but in an efficient and enjoyable way. These new strategies lead to a closer, more personal and lasting relationship between the brand and the consumer. [9] point out that the critical success factors in shopping malls in Lima and in Peru have to do with the tenant mix, accessibility, and promotion.

The results also agree with [24] who indicates that experience marketing seeks to present products in an attractive way, coupled with a novel and impactful environment, and media for advertising; these factors can generate commercial success if they are applied effectively; this is confirmed by [25] who points out that the strategies applied in experience marketing have evolved and have been adjusted to technological progress, consumer tastes, fashion, the spirit of the brand, as well as the environment, product design and social environment. [26] says that experience marketing makes use of technological tools for promotions and events.

The results are in agreement with what [27] points out, for this author experiential marketing is a business strategy, it is an opportunity for growth, and should make use of market segmentation, the resources of the population, the needs of potential consumers, and anchor stores and events to make a commercial space attractive. The above is related to what is expressed by [28] who points out that the success of a company is not only to manage the resources it possesses, but it is also necessary to make known the products it sells and the role that the company represents in society as an economic entity.

In this sense, it is important to recognize that shopping centers are no longer just an industry and have become referents for society worldwide, in Peru their growth has been significant, as evidenced by [29] noting that gross sales, number of visitors and leasable area have increased. The association also points out that the heart of every successful business lies in marketing, and relies heavily on a good marketing strategy, which takes into consideration advertising, tastes, promotions, feelings, sensations, cognition, public relations and sales.

[30] highlights the importance of relating accounting and marketing, therefore, with regard to the documentary observation of the Financial Statements, in relation to the established dimensions, it was observed that the commercial management of the MegaPlaza Independencia shopping center has been improving, in the sense that it has been maintaining a wide variety of stores and brands available to customers, It has been carrying out events of great value, with very well structured and programmed calendars, as well as offering increasingly effective promotions, which contribute to the increase in sales, which year after year and thanks to the economic situation of the country, allows the increase in turnover and generates a greater influx of customers to this commercial space, which allows the evolution of the leasable area. This agrees with the figures mentioned by [29] and supports the work done by [9] on the critical success factors in shopping centers in Metropolitan Lima and Callao, these authors indicated that Megaplaza Independencia orients its strategies to the expansion of its infrastructure, having in its expansion plans to cover about 25. 000m2, focused on achieving and maintaining the highest number of visitors, and with this the existence of higher probability of sales, and therefore higher income.

6 Conclusion

This research concluded that there is a relationship, at a weak positive level, between experience marketing and the commercial success of Megaplaza Independencia in the period 2014–2018. Therefore, the application of experience marketing strategies in shopping centers is recommended, with emphasis on implementing a variety of available brands, organizing events and sales promotions, in order to achieve commercial success, which is reflected in increased sales revenue, greater affluence and increased leasable area.

References

1. International Monetary Fund, Fiscal Policy in Latin America: Lessons and legacies of the global financial crisis (2015)
2. Lust, J.: The rise of a capitalist subsistence economy in Peru. Third World Q. **40**(4), 780–795 (2019)
3. Ramírez-Rondán, N.R.: Balance sheet and currency mismatch: evidence for Peruvian firms. Empirical Econ. **57**(2), 449–473 (2019)

4. Cordova-Buiza, F., Calderon-Sanchez, J., Chuzon-Canicela, L.: Tourist satisfaction level: evaluation in a peruvian adventure tour operator. IBIMA Bus. Rev. (2022)
5. Arellano, R.: https://www.pqs.pe/economia/malls-que-prefieren-peruanos-arellano-marketing. Accessed 14 Mar 2017
6. Klein, R., La ciudad y el turismo. experiencias desde la gestión del street art. Sociologia Rev. da Faculdade de Letras da Universidade do Porto, 54–71 (2018)
7. Zha, D., Foroudi, P., Jin, Z., Melewar, T.C.: Making sense of sensory brand experience: constructing an integrative framework for future research. Int. J. Manag. Rev. **24**(1), 130–167 (2022)
8. Cordova-Buiza, F., Gabriel-Campos, E., Castaño-Prieto, L., García-García, L.: The gastronomic experience: motivation and satisfaction of the gastronomic tourist—the case of Puno city (Peru). Sustainability (Switzerland) **13**(16) (2021)
9. Regalado, O., Fuentes, C., Aguirre, G., García, N., Miu, R., Vallejo, R.: Factores críticos de éxito en los centros comerciales de Lima Metropolitana y el Callao, (1° ed). Lima: Universidad ESAN (2009)
10. Schmitt, B.: Experiential Marketing. Ediciones, Barcelona (2006)
11. Lenderman, M.: Experiential Marketing: the Brand Revolution. Spain, ESIC, Madrid (2008)
12. Aguilera, M.: La gestión de cuentas en la agencia de comunicaciones de marketing. Madrid: ESIC (2016)
13. Kotler, P.: Marketing para Latinoamérica. (11a ed.). México-DF: Pearson Educación (2007)
14. Cuenca-Amigo, M., Makua, A.: Audience development: a cross-national comparison. Acad. Rev. Latinoamericana de Administración **30**(2), 156–172 (2017)
15. García, M.D.L.Á.O., Expósito, M.M., Verdugo, M.C.: La expansión de las redes sociales. Un reto para la gestión de marketing. Contabilidad y Negocios **10**(20), 59–69 (2015)
16. Trevinal, A.M., Stenger, T.: Toward a conceptualization of the online shopping experience. J. Retail. Consum. Serv. **21**(3), 314–326 (2014)
17. Datta, V.: A conceptual study on experiential marketing: importance, strategic issues and its impact. Int. J. Res. Granthaalayah **5**(7), 26–30 (2017)
18. Rodríguez, C.: Razones para la emoción en el consumo de moda. Vivat Academia **19**(137), 57-76A (2016)
19. Maklan, S., Klaus, P.: Customer experience: are we measuring the right things? Int. J. Mark. Res. **53**(6), 771–772 (2011)
20. Katrandjiev, H., Velinov, I.: Investigation of online visual merchandising (OVM) elements of apparel web stores. Econ. Altern. **1**, 84–98 (2014)
21. Marinao-Artigas, E., Valenzuela-Fernández, L., Barajas-Portas, K.: Comprender la evaluación afectiva en el comercio minorista: perspectiva de los consumidores. Acad. Rev. Latinoamericana de Administración **32**(4), 541–565 (2019)
22. George, D., Mallery, P.: SPSS for Windows Step By step: A Simple Guide and Reference, 11.0 update (4a. ed.). Boston: Allyn & Bacon (2003)
23. Hernández, R., Fernández, C., Baptista, P.: Metodología de la Investigación (4a. ed.), México D.F: Mc Graw Hill Interamericana (2006)
24. Martínez, E.: Diseño de estrategias mercadológicas para lograr las ventajas competitivas del Centro Comercial Zamora. (Tesis de pregrado), Universidad Autónoma Tomás Frías, Oruro, Bolivia (2007)
25. Lancis, N.: Marketing de experiencias en el centro comercial Mejor vivirlo a que te lo cuenten, (Tesis de pregrado). Universidad Zaragoza, Zaragoza, España (2014)
26. Lara, M.: Marketing Experiencial: Una nueva tendencia del marketing. (Tesis de maestría). Universidad de Oviedo, Oviedo, España (2013)
27. Collazos, C.: Propuesta de mejora de la estrategia comercial del centro comercial Open Plaza Chiclayo, (Tesis pregrado). Universidad Católica Santo Toribio de Mogrovejo, Chiclayo, Perú (2015)

28. Moon, M.A., Khalid, M.J., Awan, H.M., Attiq, S., Rasool, H., Kiran, M.: Consumer's perceptions of website's utilitarian and hedonic attributes and online purchase intentions: a cognitive–affective attitude approach. Spanish J. Mark.-ESIC **21**(2), 73–88 (2017)
29. Asociación de Centros Comerciales y de Entretenimiento del Perú [ACCEP]. https://www.accep.org.pe/. Accessed 14 Feb 2015
30. Murphy, D.S.: Contabilidad y marketing: buscando un denominador común. Contabilidad y negocios **7**(13), 5–12 (2012)

Influence of Corporate Social Responsibility on Industry's Green Product and Process Innovation: Sustainable Development as a Mediator

Parvaneh Saeidi[1], Sayedeh Parastoo Saeidi[2], Sayyedeh Parisa Saeidi[3](✉), Seyyed Alireza Saaeidi[4], and Mercedes Galarraga Carvajal[5]

[1] Research Center in Business, Society and Technology, ESTec, Universidad Tecnologica Indoamerica, Quito, Ecuador
parvanehsaeidi@uti.edu.ec

[2] Facultad de Emprendimiento, Negocios y Economía, Universidad Espíritu Santo, Guayaquil, Ecuador
parastoosaeidi@uees.edu.ec

[3] Faculty of Management, Universiti Teknologi Malaysia, Johor Bahru, Malaysia
saeidi55@gmail.com

[4] Faculty of Management, Kharazmi University, Tehran, Iran

[5] Facultad de Administración y Negocios, Universidad Tecnológica Indoamérica, Quito, Ecuador
mercedesgalarraga@uti.edu.ec

Abstract. While there are wide-ranging studies on corporate social responsibility (CSR), innovation, and sustainable development (SD), few studies have tried to make a connection between those three organizational factors in the industry section. Therefore, this study tried to investigate how does CSR with its four main components contribute to industry's green product and process innovation through sustainable development. This study has a comprehensive view to compare the SD to the existing studies by considering all three aspects of SD, namely, economic, social and environmentally sustainable development. In fact, it is going to be answered that: whether economic, social and environmentally sustainable developments play as mediators in between CSR and green innovation (product and process) in industry? Data were collected during June-October 2020 from 217 manufacturing industries from Iran where CSR has not been appropriately dealt within theoretical as well as practical terms. AMOS structural equation modeling (AMOS-SEM) method was applied to analyze data. Following the Baron and Kenny approach in testing hypotheses, it was found that CSR-process innovation is a fully mediated relationship by environmentally sustainable development, and partially mediated by economic and social sustainable development. Whereas CSR-product innovation is fully mediated by social sustainable development and partially mediated by economic and environmentally sustainable development. In the second part of the data analysis, customers' expectations, and financial issues respectively, were known as the highest and lowest motivational factors in Iranian manufacture industry, for implementing CSR. The results help Iranian managers to understand their current status, capabilities, strengths and weaknesses in sustainable development. It also recommends that since CSR dimensions had a positive

C. Montenegro et al. (Eds.): ICMTT 2023, LNNS 774, pp. 148–160, 2024.
https://doi.org/10.1007/978-3-031-43733-5_14

effect on all sustainable development aspects, they should be integrated in the organizational strategies for having better sustainable development, and higher green product and process innovation in the industry.

Keywords: Corporate social responsibility · Innovation · sustainable development

1 Introduction

Research over the past twenty years has discovered notable insight regarding the outcome that CSR, innovation, sustainability and sustainable development have on each other, as well as performance of firms. It is now well established from a variety of studies that being responsible for the environment can boost profits ethically and economically. The consumer drive towards using less plastic and avoiding environmental damage means that companies are under pressure to make changes. Companies that don't embrace the green movement are likely to lose customers who, now, expect eco-friendly products. Data from several studies suggest that implementing CSR in firms improves creativity in firms to offer new production processes and products (Gallego-Álvarez, Manuel Prado-Lorenzo, & García-Sánchez 2011; Tsai & Gururajan 2005). Subsequently, according to Resource-based view (RBV) theory, with CSR improving relationships with stakeholders enhancing the firm's reputation among internal and external stakeholders, enables sustainable competitiveness for firm (Brammer & Pavelin 2006; Vilanova, Lozano, & Arenas 2009). Consequently, it can be claimed that sustainable competitiveness as a valuable intangible asset supports firms to achieve sustainable developments in different areas of business, science and technology (Balkyte & Tvaronavičiene 2010).

A substantial sum of research has been conducted on the outcome that CSR has on innovation and sustainable development, but few writers have been able to draw on any systematic research into the mutual relationship between those three variables in a single study. To that regard, Martín, and Rubio (2017) and Ratajczak & Szutowski (2016) have specified that the correlation among CSR, SD and innovation is yet to be conclusive, and one main question has remained which is the role of SD as a mediator in between CSR and innovation. Consequently, this research will aim to close the gap of knowledge from the existing literature regarding the correlation among CSR and firm performance. Subsequently, the new question put forward in this research is: 'is sustainable development a mediator in the relationship between CSR and innovation?

Among the current available literature on CSR and firm performance, a common factor is that all literature has been done on countries from developed economies, for example, Europe and the United States of America. For this reason, it seems appropriate to use Iran for our research sample as Iran is from a developing economy and therefore it will provide further insight into the topic of CSR and firm performance (Chapardar & Khanlari 2011; Nejati & Ghasemi 2012; Saeidi, Sofian, Saeidi, Saeidi, & Saaeidi 2015). Furthermore, for a country such as Iran, CSR and firm performance has never been appropriately acknowledged, both in the practical business environment and for academics theoretically. By providing data from a developing economy this research will expand the scope of existing research into a worldwide context. Considering the above discussion and following Baron and Kenny (1986) approach in testing the mediating effect, the seventeen hypotheses are developed in this study.

2 Literature and Hypotheses Development

2.1 Corporate Social Responsibility (CSR) and Innovation

CSR is a positive strategy which when implemented can preserve the social, cultural, and economic parts of an organisation (Raimi 2017). Whilst the importance of CSR is increasingly significant among the corporate sector, it is only in the last ten years that it has been regarded as a notable corporate strategy (Hou 2019). In particular, when manufacturing firms engaged in CSR strategies, they noticed increased market share, reduced emissions, waste and pollution, and energy saving, according to Awan et al. (2017). Subsequently, various stakeholders and ecologists encourage manufacturing businesses to engage in green management and CSR strategies (Yu & Huo 2019). Consequently, many scholars started to observe the correlation among CSR and firm performance, directly (Awaysheh, Heron, Perry, & Wilson 2020; Cho, Chung, & Young 2019; Nyeadi Joseph, Ibrahim, & Sare Yakubu 2018; Sekhon Amritjot & Kathuria Lalit 2019; Torugsa, O'Donohue, & Hecker 2012) or indirectly by considering customers' loyalty, employees' satisfaction, customer satisfaction, firm reputation and competitive advantage as mediators in that relationship. Compare this to the existing wide range of studies on CSR, very little studies have considered innovation as a possible outcome of CSR implementation. The correlation among CSR and innovation has received attention among academics in the past ten years by using innovation as a key factor when understanding the link between a company's economic performance and CSR (Rexhepi, Kurtishi, & Bexheti 2013; Visser 2010). Nidumolu et al. (2009) and later Wu, Liu, Chin, and Zhu (2018) acknowledged CSR as a fundamental driver of innovation, so that CSR may encourage corporate competition by inspiring innovation. In (2018), Briones Peñalver et al. found that innovation is affected positively by CSR and this effect is mediated by cooperation. In a recent study by Zhou, Wang, and Zhao (2020). From this study, it was proven that there was a positive correlation between CSR and service innovation and product innovation among both service and manufacturing firms, which were positively attributed by employee involvement and supplier collaboration. Reaffirmed by Ratajczak and Szutowski in (2016), where an assumption was made on the influence of CSR on innovation performance, and respectively the innovations positive effect on the firms CSR. It has also been argued by other authors that CSR and innovation can be deemed as a foundation of business success (Rexhepi et al. 2013; Saeidi et al. 2015).

Although Gallego-Alvarez et al. (2011) conveyed an adverse outcome of CSR on innovation the positive impact of environmental mandatory regulations on innovation was broadly proved (Briones Peñalver et al. 2018; Guerrero-Villegas, Sierra-García, & Palacios-Florencio 2018; Luo & Du 2015; Rennings & Rammer 2011; Wu et al. 2018; Zhou et al. 2020). However, the correlation between CSR being encouraged in all parts of a business (not only environmental and obligatory), and innovation has not found a thorough conclusion (Lockett, Moon, & Visser 2006); so that after analyzing the data from 2009–2014 Spanish Community Innovation Survey, García-Piqueres and García-Ramos (2020) concluded that, although the optimistic outcome of CSR on innovation is well researched, they confirmed there are discrepancies dependent on the innovation type, and the CSR dimension Although academics are now accepting the directly related correlation among CSR and innovation, availability of empirical research is limited, and when it is, it is only in reference to the effect of CSR on innovation (Wagner 2010). Thus, the relationship continues to lack a thorough understanding due to the lack of research available on the relationship both factors have on one another (Ratajczak & Szutowski 2016; Rennings & Rammer 2011; Wagner 2010). Furthermore, Marin et al. (2017) concluded that although CSR is being regarded with more importance in relation to greater competitiveness and value, internationally among firms, CSR and innovation have not been thoroughly evaluated in terms of a circular relationship among the two factors.

Subsequently, this study aims to close the gap on the understanding of CSR and innovation by considering all four dimensions of CSR namely economic, legal, ethical and discretionary (Maignan & Ferrell 2000) and dividing innovation to two main groups; process innovation and product innovation. In doing so, two hypotheses will be examined in this study as below:

H1: Product innovation is influencing positively by CSR.

H2: Process innovation is influencing positively by CSR.

2.2 CSR, Sustainable Development (SD) and Innovation

Corporate social responsibility (CSR) is a positive strategy that is loosely defined as an organization's or brand's commitment to maximize long-term economic, societal, and environmental well-being through business practices, policies, and resources (Du, Bhattacharya, & Sen 2011). CSR has evolved over the last few decades from something considered to be philanthropic, to common practice in compulsory and obligatory actions (Carroll 2008). Moreover, becoming a successful business model that provides comparative advantage in many aspects (Battaglia, Testa, Bianchi, Iraldo, & Frey 2014; El-Garaihy, Mobarak, & Albahussain 2014; Porter & Kramer 2006). This is supported by the legitimacy-based view of the firm, which proposes that CSR activities empower organizations to develop and sustain legitimacy among society, and that it helps the firm to capitalize on its corporate image as well as other benefits such as higher competitiveness (Du et al. 2011; Walker, Zhang, & Ni 2019) and consequently, promoting corporate contributions to sustainable development (SD) (de Camargo, Mendonça, de Oliveira, Jabbour, & de Sousa Jabbour 2019; Lopes de Sousa Jabbour, Vazquez-Brust, Chiappetta

Jabbour, & Andriani Ribeiro 2020). The literature shows that the research fields of CSR, SD have been increasing for more than 30 years respectively (Fatma & Rahman 2015; Jamali, Jain, Samara, & Zoghbi 2020; Malik 2015), however the inclusion of both topics are comparatively far less and far more recent and still there is some debates on the relationship between the two (Ye, Kueh, Hou, Liu, & Yu 2020).

Nonetheless, the majority of scholars believe that CSR engagement and integration in an organization's strategic practices is a significant attribution to attaining sustainability and are supporting this claim that CSR strengthens companies' competitiveness and consequently SD (Ye et al. 2020). Furthermore, Miska, Szőcs, and Schiffinger (2018) call CSR as a main driver for sustainable development. Moreover, the European Commission stated that a significant role in contributing to sustainable development while at the same time enhancing Europe's innovativeness and competitiveness would be CSR (Commission 2005). Subsequently, one of the vital incentives for organizations to encourage CSR is the opportunity of refining and strengthening their innovation and competitive position (Du et al. 2011; Flammer 2018).

Despite numerous scholars acknowledging the relationship between CSR, SD, and innovation, the relationship has only been analyzed in one direction – from CSR to innovation (Wagner 2010), CSR to SD (Simionescu 2015) or SD to innovation (Shahzad, Qu, Javed, Zafar, & Rehman 2020). In this regard, Marin et al. (2017) and Ratajczak and Szutowski (2016) have specified that the correlation among CSR, SD and innovation remains to be inconclusive and one main question remains which is the role of SD as a mediator in between CSR and innovation.

In light of the above discussion, in addition to having a complete view to CSR by reflecting four dimensions of it (ethical, economic, legal, discretionary), all three components of SD namely environmental, economic and social sustainable development also have been considered to have a comprehensive view to SD. Accordingly, following the approach of Baron and Kenny in examining mediating effect, the following hypotheses will be examined in this study.

H3: Environmentally sustainable development is influencing positively by CSR.

H4: Economic sustainable development is influencing positively by CSR.

H5: Social sustainable development is influencing positively by CSR.

H6: Product innovation is influencing positively by Environmentally sustainable.

H7: Product innovation is influencing positively by Economic sustainable.

H8: Product innovation is influencing positively by Social sustainable.

H9: Process innovation is influencing positively by Environmentally sustainable.

H10: Process innovation is influencing positively by Economic sustainable.

H11: Process innovation is influencing positively by Social sustainable.

H12: CSR and product innovation relationship is mediated by environmentally sustainable development.

H13: CSR and product innovation relationship is mediated by economic sustainable development.

H14: CSR and product innovation relationship is mediated by social sustainable development.

H15: CSR and process innovation relationship is mediated by environmentally sustainable development.

H16: CSR and process innovation relationship is mediated by economic sustainable development.

H17: CSR and process innovation relationship is mediated by social sustainable development.

3 Methodology

3.1 Sample and Population

The firms deemed appropriate and therefore involved for this study are only those firms in the manufacturing sector located in big cities of Iran. The study used a structured questionnaire to collect data. 673 primary and secondary questionnaires were sent to owners and managers and during two months 274 were received back and finally, 217 usable responses (32%) were analyzed.

3.2 Measurement of Variables

Corporate social responsibility (CSR): Four linked factors are considered to be the most appropriate for CSR, as documented by Carroll (1999), which have been later developed by Maignan and Ferrell (2000). They discovered that the most sufficient model encompassed all four CSR dimensions, these being; ethical, economic, discretionary, and legal. Therefore, Maignan and Ferrell's CSR measurement was chosen to measure firm's judgement of every CSR aspect aforementioned. A 29 item, 5-point Likert scale was used to ensure all four CSR aspects were captured in the study, where '1 = strongly disagree' and '5 = strongly agree'.

Sustainable Development (SD): Three linked factors are considered to be the most appropriate for sustainable development, as documented by Bansal (2005). The author encompassed all three SD dimensions, namely; envieonmental, economic, and social. Therefore, Bansal SD measurement was chosen to measure firm's level of SD. A 22 item, 5-point Likert scale was used to ensure all three SD aspects were captured in the study, where '1 = strongly disagree' and '5 = strongly agree'.

Innovation: To test the effect of CSR on innovation, we use the two out of five different innovation types identified in the study of Wang and Ahmed (2004): product and process innovation. It has allocated four questions each for product and process innovation. Measuring innovation level of participant firms based on the managers' perception in eight items with five-point Likert scales, where '1 = strongly disagree' and '5 = strongly agree'.

3.3 Confirmatory Factor Analysis

In order to create the results, Confirmatory Factor Analysis (CFA) and structural models in AMOS were executed to examine the findings. CFA was executed to check standardized factor loading, validity, and reliability of the variables and concepts. The results showed that all the items are presented, relating to their specific constructs with factor loading, after dropping two items from CSR and three items from SD, due to low factor loading. A suitable framework recommended by Hair, Black, Babin, Anderson, and Tatham (2006) was found with GFI = 0.94, AGFI = 0.85 and NFI = 0.93 RMR = 0.014 and RMSEA = 0.058. In addition, convergent validity was checked, and the outcomes highlighted that all the constructs have sufficient convergent validity (AVE above 0.50). The investigation also found satisfactory value of discriminant validity for all the factors which are above 0.70. As the last step, reliability of the factors also was checked and all the factors provided acceptable CR above 0.70 (Hair et al. 2006). With having a proper CFA, hypotheses were tested through a structural model.

3.4 Structural Model (Mediation Test)

Baron and Kenny (1986) procedure is employed in order to test the hypothesis. Four steps were discussed by them to establish mediation. First, it is necessary for a noteworthy connection between the independent and dependent variable to be clarified. Second, a noteworthy connection between the independent variable and the hypothesized mediating variable must be made. Third, there must be a significant correlation made between the mediating variable and the dependent variable. Fourth, in order to ascertain that the connection of the mediating variable has full effect, there should be no noteworthy mediation found between the dependent and dependent variables. Therefore, with reference to these four steps, the outcome obtained proves there to be a beneficial and noteworthy direct correlation among independent, mediator and dependent variables with the basic fits of CFI = 0.898; GFI = 0.915; RMR = 0.021.

After testing indirect consequences of CSR on innovation, it was revealed that, no noteworthy correlation among CSR on product innovation can be found through social SD (β 0.104 $p > 0.05$). It is presented that the correlation between CSR and product innovation is completely mediated by social SD, while it is partially mediated by environmentally SD and economic SD. Accordingly, H12-H14 are supported. However, no noteworthy correlation among CSR on process innovation can be found through environmental SD (β 0.237 $p > 0.05$). It is presented that the correlation between CSR and process innovation is completely mediated by environmental SD, while it is partially mediated by social SD and economic SD. Accordingly, H15-H17 are supported (Table 1).

Table 1. Hypotheses testing (Direct and Indirect)

H		Direct effect	P	Indirect effect	P	Result
H1	CSR → Product innovation	0.294	***			supported
H2	CSR → Process innovation	0.306	***			supported
H3	CSR → Environmentally SD	0.491	***			supported
H4	CSR → Economic SD	0.204	***			supported
H5	CSR → Social SD	0.433	***			supported
H6	Environmentally SD → Product innovation	0.343	***			supported
H7	Economic SD → Product innovation	0.106	***			supported
H8	Social SD → Product innovation	0.331	***			supported
H9	Environmentally SD → Process innovation	0.296	***			supported
H10	Economic SD → Process innovation	0.670	***			supported
H11	Social SD → Process innovation	0.308	***			supported
H12	CSR → Environmentally SD → Product innovation			0.109	*	Partial mediated
H13	CSR → Economic SD → Product innovation			0.176	*	Partial mediated
H14	CSR → Social SD → Product innovation			0.203	0.104	Full mediated
H15	CSR → Environmentally SD → Process innovation			0.197	0.237	Full mediated
H16	CSR → Economic SD → Process innovation			0.202	*	Partial mediated
H17	CSR → Social SD → Process innovation			0.140	*	Partial mediated

4 Discussion and Conclusion

CSR can be recognized as an instrument for firms to willingly apply social and environmental apprehensions into their activities and their collaboration with stakeholders (Rajmanthri 2005). In this regard, some scholars claim that CSR, as a business or marketing strategy, acts as a bridge to connect the arena of business to innovation and sustainable development (Du et al. 2011; Flammer 2018; Ltd., November 2018; Miska et al. 2018). However, much uncertainty still exists around the relationship between CSR, SD and innovation (Marin et al. 2017; Ratajczak & Szutowski 2016). This study therefore sets out to fill the existing gap by assessing the role of all sustainable developments' dimensions (environmentally, economic and social) as mediators between CSR and innovation, as well as examining direct effects among CSR, SD and innovation. By following the Baron and Kenny approach in testing the mediating effect of SD in between CSR and innovation, statistical results have been divided to four parts.

To conclude from the first part of the study's research model (H1&H2), it was found that higher levels of both product and process innovation can be identified when there is a higher level of CSR. Therefore, this study further supports and emphasizes the positive and crucial link between CSR and innovation that has been previously highlighted by other related academics (Briones Peñalver et al. 2018; Guerrero-Villegas et al. 2018; Luo & Du 2015; Rennings & Rammer 2011; Wu et al. 2018; Zhou et al., 2020). This finding is contrary to a previous study done by Gallego-Álvarez et al. (2011) who stated a negative effect of CSR on innovation. Resource-based view (RBV) states that competitive advantage caused by CSR ultimately boosts companies' willingness to transform (Tsai & Gururajan 2005). Thus, it can be argued that CSR as an intangible asset assists firms to attain higher innovation directly and indirectly (Matthews 2003).

Additionally, it can be confirmed from the second part of the statistical result of this study (H3–H5) that CSR is a business or marketing strategy that acts as a bridge to connect the arena of business to sustainable development (Ltd., November 2018; Miska et al. 2018). According to RBV, since CSR can improve firm–stakeholder relationships and enhance the firm's reputation among internal and external stakeholders (Brammer & Pavelin 2006), it is able to bring sustainable competitiveness for firms (Vilanova et al. 2009). Consequently, it can be claimed that sustainable competitiveness as a valuable intangible asset supports firms to achieve sustainable developments in different areas of business, science and technology (Balkyte & Tvaronavičiene 2010). The gained statistical result of this study supports evidence from previous observations, with a high level of confidence, with showing positive effect of CSR on all aspects of sustainable development among involved manufacturing firms.

To conclude from the third part of the study's research model (H7–H11), it was found that higher levels of sustainable development encourage firms to be more active in both product and process innovation. These results reflect those of Knut et al. (2011) who also found that sustainability encourages innovation because sustainably motivated business inspires the optimal utilization of resources, efficient production with less energy consumption, innovation of improved products, as well as, processes, and better relationships with suppliers in terms of the development of new efficient products. All these activities inspire new ways of doing things, innovation, and furthermore thus keeping in line with sustainable practices,

Finally, it can be declared from the outcome of this study that SD can be identified as having a mediator role between CSR and innovation (H12–H17). Furthermore, it can be gathered from the statistical results that among all three dimensions of SD, only environmental SD plays as a full mediator between CSR and process innovation, while this relationship is partially mediated by economic and social SD. This confirms the results obtained from prior research where environmental care is a motivation for being creative in manufacturing processes (Knut et al. 2011). But the result is not the same for CSR and product innovation, so this relationship is fully mediated by social SD and partially mediated by environmental and economic SD. This finding supports previous research into this brain area which link CSR, meeting society's needs and product innovation (Matthews 2003).

With all available evidence taken into consideration, from this study and previous literature, it can be argued that CSR should be considered as an effective factor for firms to achieve SD and being more creative by producing new processes and products. The potential returns from CSR extend far beyond that of economic benefit, but also effect corporate reputation, customer satisfaction and innovation making it a highly valuable resource for firms (Knut et al. 2011; Matthews 2003; Saeidi et al. 2015). In reality, firms that hold high regard for CSR obtain higher levels of competitive advantage within the market (Saeidi et al. 2015). Subsequently, according to RBV theory, organizational competitiveness leads to sustainable superior performance (Guo 2007). In summary, CSR, in high levels, has the ability to reinforce the encouraging outcome of being environmentally friendly on innovation, by improving dimensions of SD.

Moreover, practical and relevant recommendations can be drawn upon from this paper. It has been further demonstrated in this study that strategies need to be enforced that go beyond market competition of general products and services. Furthermore, current business practices should incorporate strategies taking into account factors such as ethical values and environmental impact across all levels of the organization, to reach high levels of sustainability and ultimately success of a firm in being creative in offering new manufacturing processes and products. When incorporating ethical, economic, legal, and discretionary responsibilities, CSR can be empowered to create strong economic, environmental, and social stability within companies simultaneously (Carroll 1979; Funaru & Iosip 2010). Therefore, firms can increase their innovation performance by growing SD levels as a mediator. It is acknowledged that the data from this study solely takes into account Iranian firms and therefore creates a notable limitation to the research. Preferably, a comparison would be made between a variety of firms based in different economic and cultural regions globally.

References

Awan, U., Kraslawski, A., Huiskonen, J.: Understanding the relationship between stakeholder pressure and sustainability performance in manufacturing firms in Pakistan. Procedia Manufac. **11**, 768–777 (2017). https://doi.org/10.1016/j.promfg.2017.07.178

Awaysheh, A., Heron, R.A., Perry, T., Wilson, J.I.: On the relation between corporate social responsibility and financial performance. Strateg. Manag. J. **41**(6), 965–987 (2020)

Balkyte, A., Tvaronavičiene, M.: Perception of competitiveness in the context of sustainable development: facets of "sustainable competitiveness." J. Bus. Econ. Manag. **11**(2), 341–365 (2010)

Bansal, P.: Evolving sustainably: a longitudinal study of corporate sustainable development. Strateg. Manag. J. **26**(3), 197–218 (2005)

Baron, R.M., Kenny, D.A.: The moderator-mediator variable distinction in social psychological research: conceptual, strategic and statistical considerations. J. Pers. Soc. Psychol. **51**(6), 1173–1182 (1986)

Battaglia, M., Testa, F., Bianchi, L., Iraldo, F., Frey, M.: Corporate social responsibility and competitiveness within SMEs of the fashion industry: evidence from Italy and France. Sustainability **6**(2), 872–893 (2014)

Brammer, S.J., Pavelin, S.: Corporate reputation and social performance: the importance of fit. J. Manage. Stud. **43**(3), 435–455 (2006)

Briones Peñalver, A.J., Bernal Conesa, J.A., de Nieves Nieto, C.: Analysis of corporate social responsibility in Spanish agribusiness and its influence on innovation and performance. Corp. Soc. Responsib. Environ. Manag. **25**(2), 182–193 (2018)

Carroll, A.B.: A three-dimensional conceptual model of corporate performance. The Academy of Management Review, vol. 4, pp. 4497–530 (1979)

Carroll, A.B.: Corporate social responsibility: evolution of a definitional construct. Bus. Soc. **38**(3), 268–295 (1999)

Carroll, A.B.: A history of corporate social responsibility: concepts and practices. In: Crane, A., Moon, D.M.J.M., Siegel, D.S. (ed.). In r. The Oxford handbook of corporate social (Ed.), pp. 19–46. Oxford University Press, New York (2008)

Chapardar, H., Khanlari, R.: Iranian Corporations and Corporate Social Responsibility: An Overview to Adoption of CSR Themes. SAGE Open (2011). http://sgo.sagepub.com/content/early/2011/11/18/2158244011430988

Cho, S.J., Chung, C.Y., Young, J.: Study on the relationship between CSR and financial performance. Sustainability **11**(2), 343 (2019)

Commission, E.: Working together for growth and jobs. A new start for the Lisbon strategy. Retrieved from Brussels. COM 24, Belgium (2005)

de Camargo, J.A., Mendonça, P.S.M., de Oliveira, J.H.C., Jabbour, C.J.C., de Sousa Jabbour, A.B.L.: Giving voice to the silent: a framework for understanding stakeholders' participation in socially-oriented initiatives, community-based actions and humanitarian operations projects. Ann. Oper. Res. **283**(1), 143–158 (2019)

Du, S., Bhattacharya, C.B., Sen, S.: Corporate social responsibility and competitive advantage: overcoming the trust barrier. Manage. Sci. **57**(9), 1528–1545 (2011)

El-Garaihy, W.H., Mobarak, A.-K.M., Albahussain, S.A.: Measuring the impact of corporate social responsibility practices on competitive advantage: a mediation role of reputation and customer satisfaction. Int. J. Bus. Manage. **9**(5), 109 (2014)

Fatma, M., Rahman, Z.: Consumer perspective on CSR literature review and future research agenda. Manag. Res. Rev. **38**(2), 195–216 (2015). https://doi.org/10.1108/MRR-09-2013-0223

Flammer, C.: Competing for government procurement contracts: the role of corporate social responsibility. Strateg. Manag. J. **39**(5), 1299–1324 (2018)

Funaru, M., Iosip, A.: corporate social responsibility. environmental applications in Romania. REVISTA ECONOMICĂ **1**(52), 127 (2010)

Gallego-Álvarez, I., Manuel Prado-Lorenzo, J., García-Sánchez, I.M.: Corporate social responsibility and innovation: a resource-based theory. Manag. Decis. **49**(10), 1709–1727 (2011). https://doi.org/10.1108/00251741111183843

García-Piqueres, G., García-Ramos, R.: Is the corporate social responsibility–innovation link homogeneous?: Looking for sustainable innovation in the Spanish context. Corp. Soc. Responsib. Environ. Manag. **27**(2), 803–814 (2020)

Guerrero-Villegas, J., Sierra-García, L., Palacios-Florencio, B.: The role of sustainable development and innovation on firm performance. Corp. Soc. Responsib. Environ. Manag. **25**(6), 1350–1362 (2018)

Guo, C.: Is Sustainable competitive advantage an achievable holy grail: the relevance gap between academia and business. J. Bus. Manage. **13**(2) (2007)

Hair, J., Black, B., Babin, B., Anderson, R., Tatham, R.: Multivariate Data Analysis, 6th editio edn. Prentice-Hall, Upper Saddle River, NJ (2006)

Hou, T.C.: The relationship between corporate social responsibility and sustainable financial performance: firm-level evidence from Taiwan. Responsib. Environ. Manag. **26**, 19–28 (2019)

Jamali, D., Jain, T., Samara, G., Zoghbi, E.: How institutions affect CSR practices in the Middle East and North Africa: a critical review. J. World Bus. **55**(5), 101127 (2020). https://doi.org/10.1016/j.jwb.2020.101127

Knut, H., et al.: The second annual sustainability and innovation survey, MITSloan management review, USA (2011)

Lockett, A., Moon, J., Visser, W.: Corporate social responsibility in management research: focus, nature, salience and sources of influence. J. Manage. Stud. **43**(1), 115–136 (2006)

de Sousa, L., Jabbour, A.B., Vazquez-Brust, D., Chiappetta Jabbour, C.J., Andriani Ribeiro, D.: The interplay between stakeholders, resources and capabilities in climate change strategy: converting barriers into cooperation. Bus. Strateg. Environ. **29**(3), 1362–1386 (2020)

Ltd., A.A.: Customer Perceptions of Corporate Social Responsibility (CSR) (2018). https://ukdiss.com/examples/customer-perception-corporate-social-responsibility.php?vref=1:

Luo, X., Du, S.: Exploring the relationship between corporate social responsibility and firm innovation. Mark. Lett. **26**(4), 703–714 (2015)

Maignan, I., Ferrell, O.C.: Measuring corporate citizenship in two countries: the case of the United States and France. J. Bus. Ethics **23**, 283–297 (2000)

Malik, M.: Value-enhancing capabilities of CSR: a brief review of contemporary literature. J. Bus. Ethics **127**(2), 419–438 (2015)

Marin, L., Martín, P.J., Rubio, A.: Doing good and different! The mediation effect of innovation and investment on the influence of CSR on competitiveness. Corp. Soc. Responsib. Environ. Manag. **24**(2), 159–171 (2017). https://doi.org/10.1002/csr.1412

Matthews, J.: Knowledge Management and Innovation: How are they related? In: Proceedings of the Knowledge Management Challenge 2003: sharing the latest in thinking and practice, pp. 135–144 (2003)

Miska, C., Szőcs, I., Schiffinger, M.: Culture's effects on corporate sustainability practices: A multi-domain and multi-level view. J. World Bus. **53**(2), 263–279 (2018)

Nejati, M., Ghasemi, S.: Corporate social responsibility in Iran from the perspective of employees. Soc. Responsib. J. **8**(4), 578–588 (2012). https://doi.org/10.1108/17471111211272552

Nidumolu, R., Prahalad, C.K., Rangaswami, M.R.: Why sustainability is now the key driver of innovation. Harv. Bus. Rev. **87**, 56–64 (2009)

Nyeadi Joseph, D., Ibrahim, M., Sare Yakubu, A.: Corporate social responsibility and financial performance nexus: empirical evidence from South African listed firms. J. Global Responsib. **9**(3), 301–328 (2018). https://doi.org/10.1108/JGR-01-2018-0004

Porter, M.E., Kramer, M.R.: The link between competitive advantage and corporate social responsibility. Harv. Bus. Rev. **84**(12), 78–92 (2006)

Raimi, L.:. Understanding Theories of Corporate Social Responsibility in the Ibero-American Hospitality Industry', Corporate Social Responsibility and Corporate Governance (Developments in Corporate Governance and Responsibility, Volume 11). In: Emerald Publishing Limited (2017)

Rajmanthri, S.: Effects of Corporate Social Responsibility (CSR) on organizational sustainability: a case study of the Sri Lankan Corporate Sector. Soc. Responsib. J. **1**(3/4), 149–152 (2005). https://doi.org/10.1108/eb045805

Ratajczak, P., Szutowski, D.: Exploring the relationship between CSR and innovation. Sustain. Account. Manage. Policy J. **7**(2), 295–318 (2016). https://doi.org/10.1108/SAMPJ-07-2015-0058

Rennings, K., Rammer, C.: The impact of regulation-driven environmental innovation on innovation success and firm performance. Ind. Innov. **18**(3), 255–283 (2011). https://doi.org/10.1080/13662716.2011.561027

Rexhepi, G., Kurtishi, S., Bexheti, G.: Corporate Social Responsibility (CSR) and innovation-the drivers of business growth? Procedia. Soc. Behav. Sci. **75**, 532–541 (2013). https://doi.org/10.1016/j.sbspro.2013.04.058

Saeidi, S.P., Sofian, S., Saeidi, P., Saeidi, S.P., Saaeidi, S.A.: How does corporate social responsibility contribute to firm financial performance? The mediating role of competitive advantage, reputation, and customer satisfaction. J. Bus. Res. **68**(2), 341–350 (2015). https://doi.org/10.1016/j.jbusres.2014.06.024

Sekhon Amritjot, K., Kathuria Lalit, M.: Analyzing the impact of corporate social responsibility on corporate financial performance: evidence from top Indian firms. Corporate Govern. Int. J. Bus. Soc. **20**(1), 143–157 (2019). https://doi.org/10.1108/CG-04-2019-0135

Shahzad, M., Qu, Y., Javed, S.A., Zafar, A.U., Rehman, S.U.: Relation of environment sustainability to CSR and green innovation: acase of Pakistani manufacturing industry. J. Clean. Prod. **253**, 119938 (2020). https://doi.org/10.1016/j.jclepro.2019.119938

Simionescu, L.N.: The relationship between Corporate Social Responsibility (CSR) and Sustainable Development (SD). Internal Audit. Risk Manage **10**(2) (2015)

Torugsa, N.A., O'Donohue, W., Hecker, R.: Capabilities, proactive CSR and financial performance in SMEs: empirical evidence from an Australian manufacturing industry sector. J. Bus. Ethics **109**(4), 483–500 (2012)

Tsai, H.-S., Gururajan, R.: Mobile business: an exploratory study to define a framework for the transformation process. Collaborative Decision Making In The Internet Era, 1–11 (2005)

Vilanova, M., Lozano, J.M., Arenas, D.: Exploring the nature of the relationship between CSR and competitiveness. J. Bus. Ethics **87**(1), 57–69 (2009). https://doi.org/10.1007/s10551-008-9812-2

Visser, W.: The age of responsibility CSR 2.0 and the new DNA of business. J. Bus. Syst. Govern. Ethics **5**(3) (2010)

Wagner, M.: Corporate social performance and innovation with high social benefits: a quantitative analysis. J. Bus. Ethics **94**(4), 581–594 (2010)

Walker, K., Zhang, Z., Ni, N.: The mirror effect: corporate social responsibility, corporate social irresponsibility and firm performance in coordinated market economies and liberal market economies. Br. J. Manag. **30**(1), 151–168 (2019)

Wang, C.L., Ahmed, P.K.: The development and validation of the organisational innovativeness construct using confirmatory factor analysis. Eur. J. Innov. Manag. **7**(4), 303–313 (2004). https://doi.org/10.1108/14601060410565056

Wu, W., Liu, Y., Chin, T., Zhu, W.: Will green CSR enhance innovation? A perspective of public visibility and firm transparency. Int. J. Environ. Res. Public Health **15**(2), 268 (2018)

Ye, N., Kueh, T.-B., Hou, L., Liu, Y., Yu, H.: A bibliometric analysis of corporate social responsibility in sustainable development. J. Clean. Prod. **272**, 122679 (2020). https://doi.org/10.1016/j.jclepro.2020.122679

Yu, Y., Huo, B.: The impact of environmental orientation on supplier green management and financial performance: the moderating role of relational capital. J. Clean. Prod. **211**, 628–639 (2019)

Zhou, H., Wang, Q., Zhao, X.: Corporate social responsibility and innovation: a comparative study. Ind. Manag. Data Syst. **120**(5), 863–882 (2020). https://doi.org/10.1108/IMDS-09-2019-0493

Sustainability Management

Improving Urban Food Markets as a Local Sustainable Meeting Point

Duarte Xara-Brasil[1](✉), Pedro Pardal[1], João Pedro Cordeiro[1], and Luísa Cagica Carvalho[1,2]

[1] Centro de Investigação em Ciências Empresariais (CICE), Instituto Politécnico de Setúbal, Setúbal, Portugal
{duarte.brasil,pedro.pardal,joao.cordeiro, luisa.c.carvalho}@esce.ips.pt
[2] CEFAGE, Évora University, Setúbal, Portugal

Abstract. Urban food markets are considered as catalyzers of regional development and central places for promoting sustainability through the short cycles between the farm and the plate. This paper aims to identify mechanisms for improving the performance of an urban food market, considering an extended internal and external analysis. This includes competitive pressure and opportunities and their specific constraints on maximizing its relations with the region's inhabitants and simultaneously promoting the local economy and sustainable behaviors and policies in the community. This is a case study carried out using a qualitative approach with different key players from one of the most important fish markets in the world – Mercado do Livramento in Setúbal (Portugal). The conclusions reinforce the need to preserve its authenticity, through the attraction, satisfaction, and improvement of loyalty rates of local customers, while also reaching a different public and increasing its regional, national, and international attractiveness. To a large extent, this action will involve investing in improving the value proposition of merchants and promoting exclusive experiences, focused on products and people, which is in line with the goals of sustainable development.

Keywords: Urban Food Markets · Sustainability · Marketing · Local Products · Consumer Behavior

1 Introduction

Municipal markets, urban markets, or urban food markets (UFMs) have been agents of economic development with great relevance throughout the world, due to their turnover and level of employment – in 2008 they represented about one million jobs and 40 billion euros of sales in the European Union. They are also relevant in terms of promoting local products, preserving local traditions and relationships, attractiveness of cities and regions and sustainability practices. Nevertheless, empirical studies applied to these contexts are still scarce and this theme is underexplored.

This research addresses the theme of UFMs in the context of promoting lifestyles and a more sustainable management of territories: it is intended to contribute to a better

C. Montenegro et al. (Eds.): ICMTT 2023, LNNS 774, pp. 163–172, 2024.
https://doi.org/10.1007/978-3-031-43733-5_15

knowledge of the role of UFMs as a local sustainable player improving its marketing orientation and practices, proposing the implementation of a structured strategic and operational marketing approach to a specific case study.

From a theoretical point of view, there is a lack of marketing research on UFMs, and particularly on the behavior and perceptions of local consumers – and other stakeholders – towards these spaces, and therefore this research works as a relevant contribution to a better understanding of these phenomena.

This research is based on a UFM - Livramento Market in Setubal – with important national and international relevance, in a region which is very rich in terms of local productions mainly in industries as agriculture, fishing and other productions, as wines, cheese, delicatessen, among others. This UFM is a very important meeting point for locals and tourists, and it's usually considered one of the most relevant fish markets in the world.

Our results highlight the importance of investing in the preservation of this market's identity and the valorization of its local stakeholders, as a mean to influence local sustainability, through the achievement of different Sustainable Development goals.

2 Theoretical Background

2.1 Urban Markets and Their Public

Retailing is a central economic activity in urban communities, with a great impact on the way cities are perceived by inhabitants and tourists [1]. Urban food markets are generally relevant actors in the urban landscape of many regions and cities and in the sustainability of territories, as they provide high quality local and traditional products while promoting leisure and cultural events [2–4]. These markets can assume different roles, facilitating consumer purchase processes with the congregation of various retailers in a single location; economies of scale and shorter marketing circuits; local regulation of prices and competition; and better customer services, such as extended schedules, among others [5].

The extent of competition in the retail industry requires managerial focus on understanding the market structure and trends, for developing sustainable, long-term strategies for survival [6].

2.2 Socially Responsible Consumption, Local, and Green Consumption

Sustainable development implies a growing commitment to environmentally cleaner technologies and production systems, as well as structured economic and social policies that lead to changes in the behavior of different agents, including consumers [7]. A growing number of consumers are becoming aware of the social, economic, and environmental consequences of their decisions and are modifying their preferences, attitudes, and behaviors [8].

The globalization of the economy has engendered greater difficulty in developing relevant marketing approaches [9]. There has been some change in the paradigm of agricultural and food production, around greater transparency, traceability, quality, less standardization, concentration and massification of conventional models [10, 11]. There is also some investment in local networks and short sales channels, with less weight of intermediaries and greater proximity to consumers often supported by local governments and national and international support programs [12, 13].

Local products are a central component of communities: their consumption is an increasingly relevant trend, associated with greater social justice and environmental sustainability [14]. There are intrinsic factors for choosing local products, such as taste, freshness, appearance, exclusivity/authenticity, and health benefits, but also reasons related to the willingness to support local communities [14].

3 Methodology

This research is focused on understanding the relevance of UFMs in the context of sustainable management of territories to contribute to a deeper knowledge of the role of UFMs as local sustainable players, improving their marketing orientation and practices. Our research is based at Livramento Municipal Market in Setúbal (LMS), one of the main municipal markets in Portugal and in this city. Inaugurated in 1876, it maintains relevance in the supply and preferences of the inhabitants and tourists who visit it.

According to the objectives of this study, we used focus groups as a qualitative and exploratory approach. This methodology has relevant advantages when we seek to capture complex meanings and their thematic dimensions, collecting qualitative data on a particular topic and getting more in-depth knowledge and new insights [15–17].

We organized four separate focus groups (FG), with LMS traders (merchants), consumers, specialists, and tourist agents. All focus group participants were selected according to their profile and experience. The sessions were held at Livramento Market (3) and online (1) in 2020. Nine market retailers participated in the traders focus group; four regular customers participated in the focus group with consumers; and the focus group with specialists included seven participants and took place online due to the nature of the actors and the travel difficulties related to the pandemic situation. In the tourist agents focus group, among the seven participants we selected restaurant managers, chefs and other businessmen, and tourism professionals with regular interaction with many tourists visiting the region.

A common session guide was drawn up for the different groups, based on the literature review and included topics, related to the current situation and future perspectives. These sessions were moderated by two researchers who welcomed the participants, conducted the meetings, posed questions, recorded the meetings, and took notes. In these meetings, researchers used an informal approach to maximize the participants' contributions [18, 19].

4 Findings and Discussion

The process of data analysis included the identification of the thematic dimensions – theoretically relevant or addressed concepts – the establishment of relationships between them and ascertaining the suitability of the reported situation with the existing models and theoretical framework through observed similarities and differences [20]. In this framework, the themes portrayed are (1) the market: social, cultural, and economic centrality, which unfolds into the sub-themes "social and cultural centrality", "supply diversity" and "shopping experience" and (2) the market as sustainability agent, which unfolds into the sub-themes "valuing production" and "promotion of sustainable behaviors and attitudes".

The issues addressed in these focus groups show an important alignment of the different stakeholders regarding the development of this municipal market in strategic and operational terms and are very much aligned with the sustainable development goals, which should be promoted by central and local governments, namely through investment in local products and local trade, which preserve local cultures, economic activities, and livelihoods.

4.1 Livramento Market: Social, Cultural, and Economic Centrality

The topic "social, cultural and economic centrality" includes the subtopics "social and cultural centrality", "diversity of offer" and "shopping experience".

Social and Cultural Authenticity

We observed the existence of a very relevant set of associations between the LMS and the topic of authenticity topic presented in the literature, which means that we are facing significant market opportunities. The feelings of the participants may be attested to by the expressions: "Livramento Market is a very strong brand" (FG Merchants), "This market is unique… it has nothing to do with others and cannot be misrepresented" (FG Consumers). Participants also related the feeling of their experiences and personal links, such as: "I accompanied my grandmother to the market. At that time, I sometimes felt uncomfortable because the merchants questioned me a lot and the visits were very long… but now I like it and feel a strong connection. I'm a regular customer" (FG Consumers). References to the antiquity of relationships, which span several generations of consumers, also highlight the social and cultural importance. This market is a living space which is relevant and remarkable for people; in addition to the feeling of belonging – to space and experiences – there is a feeling of pride ("our market") for its regional and national relevance (it is "unique"), as architectural heritage with commercial importance. Some participants have very strong ties with the spot and fear that the traditions may get lost in the future by architectural or conceptual interventions: "I hope that our market never loses its tradition… it's to remain as it is – you can bring all the events you want, but our market is our market" (FG Traders). The fear of a massive "touristification" is a frequent statement, due to its potential impact on the loss of traditions, diminishing the ties with local habitants and with the end consequence of lowering its potential as a tourist spot "When you travel, you look for the real places, the people" (FG Specialists).

Diversity and Excellence of the Offer (Local Products)

The Setúbal region is very rich in local products. In addition to having one of the main fishing ports in the country, it has agricultural, dairy and wine production, contributing to the relevance of this city as a significant gastronomic tourism destination.

Different stakeholders identify the excellence of the offer of the region, "we are lucky to have a region with an exceptional offer" (FG Traders), and mention that in the LMS they find this regional wealth of fresh products "when people want something good, they come to the market" (FG Merchants). They recognize that "there is an excellent offer in terms of seasonal products" (FG Consumers) and that "you can find products from the region that could not be found elsewhere" (FG Tourist Agents), as well as an effective opportunity to interact with the producers who "tell their story and with whom they can learn new ways to use them and make [things]".

Focus groups provided important insights that highlight this diversity and rich interpersonal interaction between inhabitants, "The stalls of the producers are the shop windows of the region, there are many products that can only be found here" (FG Consumers) and the relational component "There is proximity with the products and producers" (FG Specialists).

Shopping Experience
During focus group sessions, participants discussed topics associating the shopping experience with pleasure and experiences, in statements like "It is a pleasure to come here" and "The best experience in the market is our contact with producers" (FG Consumers). Local traders also observe these consumer reactions and perceptions "When they see the fish, they go crazy" (FG Traders). Also, for tourists, the Livramento Market is a central spot for experiencing the region and its flavors and traditions, where they can experience the whole region: "Cheese, wine, fish. They [tourists] are amazed… just follow the tile panel and tell the story" (FG Tourist agents).

4.2 The Market as an Agent of Sustainability

The topic "the market as an agent of sustainability" includes the sub-themes "valuating local production" and "promotion of sustainable behaviors and attitudes". This section includes the expectations of these stakeholders regarding the role that this UFM should play, not only as a promoter of the sustainable development of the region and its produce, but also as an attempt to influence attitudes towards sustainability and the purchasing and consumption behavior of the inhabitants.

Valuing' Local Productions
The globalization of food production and consumers' concerns have led to consumer demand for greater transparency and information on the provenance of food [21]. This may be reached through closer relations between producers and consumers in different retail formats, such as UFMs. Promoting local products may encourage sustainable agricultural practices, the conservation of traditional farming landscapes, and the reduction of the carbon footprint [22, 23]. In this context, the promotion of local food systems and short supply chains is a tool to increase the resilience of the local food system [24].

Various agricultural products from across this region and the potential of the LMS in stimulating them were highlighted. There is also the perception that there are very relevant growth opportunities, which respond to consumption behaviors and healthier and greener lifestyles. The focus groups outputs are in line with the conclusions of [25], who highlights the role of UFMs as relevant spots in terms of job fixation, the flow of local production, wealth benefits and attracting tourists.

Fresh and local products are highlighted as a strength in all focus groups. The story of the producers can be told, along with the products' distinctive characteristics. Many of these products have traditional, authentic production methods, and are particularly valued by many people.

The centrality of the market as a sustainable spot of the entire region was frequently identified in different focus groups. In the FG Specialists we had inputs such as "the market must be an anchor place in the region and that we must be aware of the actual trends"; "local products are a fashion trend". "Authenticity is an opportunity". Consumers highlighted the importance of product certification and the "need to increase the perceived quality of the products". Aligned with these priorities, local tourist agents highlighted the "importance of valuing and promoting local products such as sweet potatoes and tomatoes from Comporta, Alcacer rice, oysters, olive oil, muscatel, and biological products, but also vegetarian products" (FG Tourist agents).

Promotion of Sustainable Behaviours and Attitudes

Education is a powerful vehicle for promoting sustainable development. It is included in the United Nations Sustainable Development Goals (SDG4) and requires pedagogies and approaches to catalyze the necessary change [26]. Examples of education for sustainability can be found in the literature, targeting different objectives and educational levels, (e.g., [27]). On this specific topic, according to participants of our focus groups sessions, it was stated that "The market must assume a central educational role, for example with partnerships and activities with schools". It could also be observed that there are specific educational programs developed by UFMs, like the "Mercat de Barcelona".

Based on in the LMS experience, marketing campaigns are also a powerful tool for promoting sustainable behaviors, using different promotional practices as in-store activities focused on their product assortment, store atmosphere and traders [28], using storytelling and tastings as tools. The use of digital marketing has become a fundamental element in the dissemination and communication of sustainable actions, social strategies and actions that companies want to spread online [29].

In the focus group sessions, several participants gave feedback on this dimension, as "The market has to tell the story of this region, these products, these people" (FG Tourist Agents), which should lead to storytelling practices of sustainable behaviors and attitudes. They also evidenced the "Need for workshops and experimentation sites" (FG Traders) and that "People don't know the flavors. Flavor courses should be promoted" (FG Tourist agents). These "experiences" may be provided using digital and non-digital marketing platforms and could be developed based on the promotion of sustainability.

5 Conclusions and Final Remarks

Urban food markets are very relevant economic agents all over the world. They are central spaces for boosting the economy and communities, in terms of securing employment, selling local products, creating wealth, and providing tourist attractions with a very differentiated and comprehensive commercial offer. They contribute to the revitalization of the urban centers or neighborhoods where they are located, to improving the quality of life and to promoting healthier lifestyle habits.

There are UFMs of different shapes and sizes that have a place in the local community, generate jobs and well-being. They are unique commercial spots, characterized by the coexistence of many traditional small retailers – mainly fresh food retailers – in competition in a single location. As well as their historical relevance, markets have been losing part of their relevance due to demographic constraints, but also with the absence of a global and strategic vision or of an integrated management model, as well as a lack of innovation, training, and renewal of traders.

In recent years, there has been a renewal of some UFMs in different countries, aligned with their original features and adapting them to current consumption and comfort patterns and their (new) utility. Simultaneously, consumption of local products is an increasingly relevant trend and is generally associated with intrinsic quality factors, social justice and environmental sustainability and health benefits, and UFM are generally perceived as being specialized in those products.

The Livramento Market in Setúbal is one of the main symbols of the city and the region, where local inhabitants and tourists find the region's traditions and where you can experience the local culture. Its importance for the entire region is evident, especially to take advantage of its exceptional wealth of excellent products. At the same time, products and producers are highly valued by local consumers. It also highlights the promotion of links with other partners such as restaurants and hotels, and the market's role in the education of children and young people for citizenship.

Aligned with the literature and considering the outputs of the focus groups, it is clear that more attention should be focused on the merchandising and promotion of produce from the entire region, consolidating this location as a place of reference: exhibition of the region in terms of local and regional gastronomy and production including wines, olive oil and rice, but also with green, sustainable production.

The possible advantages of creating a greater emotional connection between consumers, the market and producers are also highlighted: a more pronounced orientation towards experiences and towards valuing the relationship (for example, through storytelling and proximity actions, such as visits, show cooking, etc.), will also include an extensive use of digital media. These marketing activities will benefit from the engagement and pride that merchants and consumers feel in being part of the community of (their) local market (Mercado do Livramento).

The transformations that may be carried out will contribute to enhancing its identity, not turning it into a tourist space, without a soul or life of its own. As mentioned, "our market is our market". A UFM so historically iconic as the Livramento Market, should be focused on the locals and, inherently, attract tourists.

The analysis highlights the benefits of developing a structured work plan for the local community, but also for tourists, who have specific needs related to a lack of knowledge of the language and products, length of stay in the city and limitations, for example in terms of the possibility of cooking food. The expansion of sustainability practices should be a priority, namely in its ecological footprint, the protection of endangered species, the forms of production, the recycling of waste and packaging materials and the use of renewable energies. It will also be important to study the issue of accessibility related to parking, schedules, and the possibility of online shopping and home delivery.

5.1 Limitations and Future Research

This case study was carried out in a specific food urban market with very relevant distinguishing characteristics. In fact, this specific Market is still a very important local and regional actor in terms of sales, performance of local products and economic opportunities for local retailers. The development of similar approaches in other markets, national and international, may also be very relevant, to allow a greater generalization of the results.

This study includes the perceptions of related stakeholders, such as traders, tourist agents and clients, but does not include contributions from other representative groups, such as non-clients of this market and tourists.

In future research, the intention is to carry out a quantitative study of the region's inhabitants about their behavior and perceptions of purchasing these products and in these places. Additionally, carrying out additional qualitative approaches, through in-depth interviews and observation, may bring new contributions.

The lack of marketing research on urban food markets, and particularly on the behavior and perceptions of local consumers towards these spaces, is an important opportunity to develop a different academic contribution to better understanding this retail format and to make a practical contribution to preserving and improving these sustainable economic actors that have an important impact on their regions.

Acknowledgment. This paper is financed by Instituto Politécnico de Setúbal. The authors would also like to thank Câmara Municipal de Setúbal for their support in data collecting activities. The fourth author is pleased to acknowledge: the financial support from Fundação para a Ciência e a Tecnologia under the Project UIDB/04007/2020.

References

1. Barata-Salgueiro, T., Guimarães, P.: Public policy for sustainability and retail resilience in Lisbon city center. Sustainability (Switzerland) **12**(22), 1–22 (2020). https://doi.org/10.3390/su12229433
2. Crespi-Vallbona, M., Domínguez Pérez, M., Miró, O.M.: Urban food markets and their sustainability: the compatibility of traditional and tourist uses. Curr. Issue Tour. **22**(14), 1723–1743 (2019). https://doi.org/10.1080/13683500.2017.1401983
3. Orfanakos, P.: Redevelopment and Reuse of the Urban Market "Modiano" in Thessaloniki. University of Thessaly, Greece (2018)
4. Svensson, G., Høgevold, N., Ferro, C., Varela, J.C.S., Padin, C., Wagner, B.: A triple bottom line dominant logic for business sustainability: framework and empirical findings. J. Bus. Bus. Market. **23**(2), 153–188 (2016). https://doi.org/10.1080/1051712X.2016.1169119
5. El Ripol, J.C.: Papel de los mercados municipales en la vertebración de las tramas urbanas. Distribución y Consumo **9**(44), 23–29 (1999)

6. Sinha, A.: Understanding supermarket competition using choice maps. Mark. Lett. **11**, 21–35 (2020). https://doi.org/10.1023/A:1008190508803
7. Peattie, K.: Green consumption: behavior and norms. Annu. Rev. Environ. Resour. **35**, 195–228 (2010). https://doi.org/10.1146/annurev-environ-032609-094328.(2010)
8. Gillani, A., Kutaula, S., Leonidou, L.C., Christodoulides, P.: The impact of proximity on consumer fair trade engagement and purchasing behavior: the moderating role of empathic concern and hypocrisy. J. Bus. Ethics 0123456789 (2019) https://doi.org/10.1007/s10551-019-04278-6
9. Nicholls, A., Huybrechts, B.: Sustaining inter-organizational relationships across institutional logics and power asymmetries: the case of fair trade. J. Bus. Ethics **135**(4), 699–714 (2016). https://doi.org/10.1007/s10551-014-2495-y
10. Sadler, R.C., Arku, G., Gilliland, J.A.: Local food networks as catalysts for food policy change to improve health and build the economy. Local Environ. **20**(9), 1103–1121 (2015). https://doi.org/10.1080/13549839.2014.894965
11. Goodman, D.: Rural Europe Redux? Reflections on alternative agro-food networks and paradigm change. Sociol. Rural. **44**(1), 3–16 (2004). https://doi.org/10.1111/j.1467-9523.2004.00258.x
12. Jablonski, B.B.R., Bauman, A., Thilmany, D.: Local food market orientation and labor intensity. Appl. Econ. Perspect. Policy **43**(3), 916–934 (2021). https://doi.org/10.1002/aepp.13059
13. Printezis, I., Grebitus, C.: Marketing channels for local food. Ecol. Econ. **152**(June), 161–171 (2018). https://doi.org/10.1016/j.ecolecon.2018.05.021
14. Skallerud, K., Wien, A.H.: Preference for local food as a matter of helping behaviour: insights from Norway. J. Rural. Stud. **67**, 79–88 (2018). https://doi.org/10.1016/j.jrurstud.2019.02.020
15. Breen, R.L.: A practical guide to focus-group research. J. Geogr. High. Educ. **30**(3), 463–475 (2006). https://doi.org/10.1080/03098260600927575
16. Morgan, D.L., Spanish, M.T.: Focus groups: a new tool for qualitative research. Qual. Sociol. **7**(3), 253–270 (1984) https://doi.org/10.1007/BF00987314
17. Wilkinson, S.: Focus group research. In Silverman (Ed.), Qualitative Research: Theory, Method, and Practice, pp. 177–199. Thousand Oaks, CA, Sage (2004)
18. Voinov, A., et al.: Modelling with stakeholders - Next generation. Environ. Model. Softw. **77**, 196–220 (2016). https://doi.org/10.1016/j.envsoft.2015.11.016
19. Palermo, V., Hernandez, Y.: Group discussions on how to implement a participatory process in climate adaptation planning: a case study in Malaysia. Ecol. Econ. **177**, 106791 (2019). https://doi.org/10.1016/j.ecolecon.2020.106791
20. Ryan, G.W., Bernard, H.R.: Techniques to identify themes. Field Methods **15**(1), 85–109 (2003). https://doi.org/10.1177/1525822X02239569
21. Feldmann, C., Hamm, U.: Consumers' perceptions and preferences for local food: a review. Food Qual. Prefer. **40**, 152–164 (2015)
22. Shindelar, R.: The ecological sustainability of local food systems. RCC Perspectives **1**, 19–24 (2015)
23. Madaleno, A., Eusébio, C., Varum, C.: Purchase of local food products during trips by international visitors. Int. J. Tour. Res. **20**(1), 115–125 (2018)
24. Stein, A.J., Santini, F.: The sustainability of "local" food: a review for policy-makers. Rev. Agric. Food Environ. Stud. **103**, 77–89 (2022). https://doi.org/10.1007/s41130-021-00148-w
25. Barreta, J.M.: Organização e gestão dos mercados municipais. Mudar e inovar para competir. GEPE - Gabinete de Estudos e Prospectiva Económica do Ministério da Economia (2002)
26. Agyeiwaah, E., Dayour, F., Otoo, F.E., Goh, B.: Understanding backpacker sustainable behavior using the tri-component attitude model. J. Sustain. Tour. **29**(7), 1193–1214 (2021)

27. Antón-Peset, A., Fernandez-Zamudio, M.A., Pina, T.: Promoting food waste reduction at primary schools A case study. Sustainability **13**(2), 600 (2021)
28. Sheth, J.N.: Future of brick and mortar retailing: how will it survive and thrive? J. Strateg. Mark. **29**(7), 598–607 (2021)
29. Saura, J.R., Palos-Sanchez, P., Rodríguez Herráez, B.: Digital marketing for sustainable growth: Business models and online campaigns using sustainable strategies. Sustainability **12**(3), 1003 (2020)

An Analysis of Airline GRI and SDG Reporting

Eljas Johansson(✉)

Gdańsk University of Technology, ul. Narutowicza 11/12, 80-233 Gdansk, Poland
eljas.johansson@pg.edu.pl

Abstract. This study aims to increase our understanding of the Global Reporting Initiative's (GRI) topic-specific disclosures and the sustainable development goals (SDGs) addressed in the global passenger airline industry's sustainability reporting (SR). Based on a quantitative content analysis of the industry's sustainability reports from the financial year 2019 (FY19), this study reveals that airlines focused more on reporting environmental issues, especially emissions, than economic or social dimensions, demonstrating this emission-intensive industry's responsiveness to stakeholders' information needs. However, a closer look at the reported impacts shows that many topic-specific disclosures and SDGs, which industry associations have not identified as relevant to the industry, were also mentioned across the reports. Moreover, the results indicated a broader use of SDGs in Asia-Pacific reports than in European. The results are expected to interest practitioners and academics in assessing and developing the industry's SR.

Keywords: Airline industry · GRI · SDGs

1 Introduction

Since its inception in 1999, the GRI has become the most widely accepted sustainability reporting (SR) framework for organizations to disclose information on their significant impacts in the three sustainability areas, i.e., economic, environmental, and social [1]. The framework has been updated numerous times, and recently, the GRI has also guided organizations to report their commitments to SDGs [2].

Nevertheless, companies may struggle to report on their sustainability performance and commitments to SDGs, partly because they have difficulties assessing what is relevant to their industry. The proof is that the GRI is now developing a sector program that provides certain industries with specific disclosure standards and their correspondence to SDGs [3]. Although some high-impact sectors, such as oil and gas and coal, have been provided with sector-specific standards, there is still a long list of sectors with high impact and a long SR tradition that this program does not cover. One industry that has lacked sector-specific guidance [4] is the airlines.

Accordingly, this paper aims to analyze the GRI disclosure and SDG coverage in the global airline industry, which may help relevant stakeholders assess and develop sector-specific SR. Considering airlines' significant impacts on all three sustainability areas, this paper focuses specifically on this industry. After all, it has a central role in

C. Montenegro et al. (Eds.): ICMTT 2023, LNNS 774, pp. 173–184, 2024.
https://doi.org/10.1007/978-3-031-43733-5_16

enabling global tourism and trade, consequently supporting millions of jobs worldwide; however, it also contributes over 2% to anthropogenic CO2 emissions [5].

Despite the industry's significant impacts, scholarly research on airline SR has remained very limited [4]. Although the recent literature contains global overviews and comparative assessments of the scope of GRI-specific topics covered in airlines' reports [6, 7], such analysis has not yet been performed at the exact disclosure level, and the SDGs in this connection have not been examined either. This study seeks to narrow this gap by addressing the following research questions: **RQ1:** Which GRI's topic-specific disclosures and SDGs are reported in airlines' sustainability reports, and to what extent? **RQ2:** Are there differences in the count of GRI disclosures and SDGs reported across regions?

2 Background of Sustainability Reporting, GRI and SDGs

Greater public demand for sustainability disclosure started emerging in the '90s when corporations also recognized SR initiatives bringing them a competitive advantage [8]. Since then, corporate SR has proliferated. According to KPMG's [9] recent survey, the proportion of large and mid-cap firms (known as N100) practising SR increased from 24% to 80% between 1999 and 2020.

In general, SR can be defined as reporting publicly on an organization's economic, environmental and social impacts [10], often released in stand-alone reports or incorporated into annual reports. Despite this consensus, businesses may struggle to assess and demonstrate their impacts in these areas, which is also reflected in the use of numerous SR frameworks [11], of which the GRI is the most prominent [9].

The GRI. The GRI is an independent international organization whose initial aim was to standardize, simplify and globalize SR that lacked comparability [1]. Its GRI guidelines, developed in 1999, quickly became the most widely accepted SR framework worldwide [1], with 67% of N100 reporters using it in 2020 [9]. In 2016, the GRI transitioned from guidelines to set the first global standards for SR. These standards [12] comprise Universal and Topic-specific standards. Universal Standards specify the reporting principles and disclosures related to the organization's context that all organizations claiming compliance with the GRI must follow. Organizations that claim compliance with the 2016 Standards can choose to report under the 'Core' or 'Comprehensive' option; the latter reflects a greater degree of the GRI application. The 2016 Standards also allow organizations to choose to report specific economic, environmental, and social impacts without looking to provide a full picture of their impacts using the 'GRI-referenced' claim. Topic-specific Standards, in turn, are organized into three series of disclosures: 200 (Economic topics), 300 (Environmental topics), and 400 (Social topics), from which the reporting organizations are expected to report on material topics, i.e., issues reflecting the organizations' most significant impacts. It is worth noting that the GRI has offered sector-specific supplements for many industries, including airport operators [4]. However, it has not offered sector-specific supplements or standards for airlines yet.

SDG Reporting. SDG reporting can be defined as publicly reporting how organizations address the SDGs [10]. The SDGs were introduced as part of the 2030 Agenda for

Sustainable Development in 2015 when the United Nations member states agreed to create a global agenda for sustainable development based on 169 targets grouped into 17 SDGs, tackling a range of issues relevant to sustainable development, such as ending world poverty and combating climate change [13]. Soon after the SDGs were introduced, the GRI and UN Global Compact published several guides [2] to help organizations to integrate SDGs into their non-financial reports, especially those following the GRI Standards. According to KPMG [9], SDG reporting among the N100 reporters increased from 39% to 69% between 2017 and 2020.

The GRI and SDG Reporting in the Airline Industry. In the absence of sector-specific guidance, the International Transport Association (IATA) noticed the airlines' need for specific guidance on GRI reporting. As a result, IATA [14] developed an Airline Sustainability Reporting Handbook (ASRH) based on industry input through a series of surveys, workshops and background research to determine the most important topics for airlines to report on, their associated disclosures and relevant SDGs. Specific guidance is given for 4 economic, 9 environmental and 8 social GRI disclosures. ASRH also references the Air Transport Association Group's (ATAG) [15] 'Flying in Formation' report, which outlines how the industry plays a role in 15 of the 17 SDGs. The report assigns a score of 0–3 for each SDG to reflect their direct relevance to the sector. The three-bar scores are assigned to the following SDGs: 5: Gender equality, 7: Affordable and clean energy, 8: Decent work and economic growth, 9: Industry innovation and infrastructure, 10: Reduced inequalities, 12: Responsible consumption and production, 13: Climate action. Which GRI disclosures and SDGs are addressed in the global airline industry's SR and to what extent this reporting corresponds to the abovementioned industry recommendations will be the interest of the remaining paper.

3 Methodology

The data used in this study were compiled from several sources. First, the initial sample comprising data about airlines' SR activity was obtained from Johansson [16]. This data contained a list of 61 global passenger airline companies, compiled from IATA's 2019 annual report and Skytrax airline ranking organization's lists, which were active in SR in FY19 before COVID-19 disrupted the industry. Based on this data, 37 companies were found to reference the GRI and 39 companies the SDGs in their SR.

Next, these companies' sustainability reports were reviewed by applying quantitative content analysis [17] to identify and measure the extent to which the airlines had reported GRI disclosures and SDGs. More specifically, each report was downloaded from the Internet and screened against a coding scheme corresponding to the topic-specific disclosures appearing in the GRI 200, GRI 300, and GRI 400 series and the 17 SDGs. It is worth noting that the GRI updates its standards continuously, and the coding scheme in this study was based on disclosures published in the GRI 2016 Standards and the updates made to them until 2019 [3, 18]. To address RQ1, all identified topic-specific disclosures and SDGs were recorded in an Excel spreadsheet, whose frequencies and proportions of sustainability reports mentioning them were measured.

To address RQ2, each airline was assigned a categorical code based on their domicile in IATA's regional classification (i.e. Africa & Middle East, Asia Pacific, China & North

Asia, Europe, the Americas) to test whether there existed differences in the count of reported GRI disclosures and SDGs across regions. The analysis was carried out using SPSS (v. 28). This study opted to test for statistically significant differences using the non-parametric Kruskal-Wallis test because unequal sample sizes across groups were considered to affect the robustness of the equal variance assumption [19].

4 Results

The Use of the GRI. Thirty-seven of the 61 reporting companies used the GRI. Twenty of them claimed compliance with the GRI Standards (1 at the Comprehensive level and 19 at the Core level), and 10 made a 'GRI-referenced' claim. One of these reports had not specified the exact topic-specific disclosures and could not be taken to further analysis. The remaining seven companies had only cited the GRI without using its disclosures. Accordingly, the sample scrutinized for GRI topic-specific disclosures was 29 reports. Twenty-four of these reports also addressed SDGs.

The basic descriptive statistics reveal that the number of topic-specific disclosures per report varied from 3 to 79, with mean and median values of 28.9 and 28. The standard deviation of 14.8 highlights a high dispersion of some disclosures within the 29 reports. Indeed, out of the 89 topic-specific GRI disclosures, all were used at least once, but 25 of them were mentioned in fewer than five reports. On the other end, the most frequently used disclosures were 305-1 (Direct [Scope 1] GHG emissions) with 25 mentions out of the 29 reports and 305-4 (GHG emission intensity) with 24 mentions. The complete breakdown of the reported disclosures is provided in Tables 1, 2 and 3, where disclosures in bold represent topics that IATA's ASRH identifies as material for airlines. The average coverage of all IATA-recommended GRI disclosures was 38%. Details on which disclosures in each GRI series were used and to what extent they corresponded to IATA-recommended disclosures are provided below.

The Use of the GRI Economic Disclosures. The number of economic disclosures per report ranged from 0 to 13, with mean and median of 4.9 and 4. As shown in Table 1, out of the 17 GRI 200 disclosures, 201-1 (Direct economic value generated and distributed) was the most used one, appearing in 79% of the reports, followed by 205-2 (Communication and training about anti-corruption policies and procedures) and 205-3 (Confirmed incidents of corruption and actions taken), with 62% and 55%, respectively. These three disclosures from the GRI 200 series were the only ones reported in more than half of the reports. The average coverage of the GRI 200 disclosures was 29%.

Regarding the IATA's four recommended disclosures in GRI 200 series, it is worth noting that only 205-3 was reported in over half of the reports. The average coverage of IATA-recommended GRI 200 disclosures was 30%

The Use of the GRI Environmental Disclosures. The number of environmental disclosures per report ranged from 1 to 32, with mean and median of 11.62 and 12. The use of these disclosures is summarized in Table 2. Out of the GRI 300 series' 32 disclosures, 11 were reported in more than half of the reports. Two of the most frequently reported environmental disclosures, 305-1 and 305-4 were also the most reported GRI disclosures, appearing in 86% and 83% of the reports, respectively. The average coverage of the GRI 300 disclosures was 36%.

While 5 out of 9 IATA's recommended environmental disclosures were reported in over half of the reports, others had very low frequencies. For example, 308-2 (Negative environmental impacts in the supply chain and actions) was reported only in 10% of the reports, and 303-4 (Water discharge) in 7%. The average coverage of IATA-recommended GRI 300 disclosures was 44%

The Use of the GRI Social Disclosures. The number of social disclosures per report ranged from 1 to 34, with mean and median of 12.38 and 11. Table 3 summarizes the use of the GRI 400 disclosures. Out of the 40 disclosures in this series, seven were reported in more than half of the reports. Overall, the average coverage of GRI 400 disclosures was 31%. The most frequently used disclosure was 405–1 (Diversity of governance bodies and employees Disclosure), with 69% of the reports mentioning it. Interestingly, IATA's ASRH does not include materiality guidance on this disclosure. While IATA-recommended 401-1 (New employee hires and employee turnover) and 403-2 (Hazard identification, risk assessment, and incident investigation) were each reported in 62% of the reports, some other IATA's recommended disclosures, like 403-10 (Work-related ill health), were reported only by 14%. The average coverage of IATA-recommended GRI 400 disclosures was 38%.

The Use of SDGs. 39 of the 61 reporting companies claimed to have connected their business activities to the SDGs in their sustainability reports. One of them did not specify which specific SDGs it had addressed and was excluded from further analysis. The frequency of each SDG and the proportion of the remaining 38 reports mentioning them are summarized in Table 4. The scores (0–3) within the brackets indicate aviation's significance across the SDGs, as identified by ATAG [15]. Those SDGs recognized as having full relevance to the industry are also highlighted in bold.

The SDGs addressed per report ranged from 2 to 17, with mean and median of 9.37 and 9. SDG13 (Climate action) was the most addressed, with 97% of the reports mentioning it. The second most frequently used SDG was SDG8 (Decent work and economic growth), with 95% of the reports mentioning it. SDG5 (Gender equality), SDG12 (Responsible consumption and production), and SDG3 (Good health and well-being) followed the other SDGs, constituting 89%, 84%, and 74%of the reports, respectively. It is worth noting that 6 out of the 7 SDGs recognized as directly relevant to the industry by ATAG were addressed in over half of the reports. Also, the only two Goals, SDG14 (Life below water) and SDG16 (Peace, Justice and Strong Institutions), which are not identified as having a direct aviation element, were among the three least addressed SDGs.

Table 1. Frequency of use of GRI 200 Economic Disclosures

		n	%
201-1	Direct economic value generated and distributed	23	79%
201-2	Financial implications and other risks and opportunities due to climate change	10	34%
201-3	Defined benefit plan obligations and other retirement plans	8	28%
201-4	Financial assistance received from government	5	17%
202-1	**Ratios of standard entry level wage by gender compared to local minimum wage**	**4**	**14%**
202-2	**Proportion of senior management hired from the local community**	**5**	**17%**
203-1	Infrastructure investments and services supported	10	34%
203-2	Significant indirect economic impacts	11	38%
204-1	**Proportion of spending on local suppliers**	**7**	**24%**
205-1	Operations assessed for risks related to corruption	10	34%
205-2	Communication and training about anti-corruption policies and procedures	18	62%
205-3	**Confirmed incidents of corruption and actions taken**	**16**	**55%**
206-1	Legal actions for anti-competitive behavior, anti-trust, and monopoly practices	11	38%
207-1	Approach to tax	1	3%
207-2	Tax governance, control, and risk management	1	3%
207-3	Stakeholder engagement and management of concerns related to tax	1	3%
207-4	Country-by-country reporting	1	3%

4.1 Comparison Between Regions in the Airline GRI and SDG Reporting

Kruskal-Wallis tests were performed to test whether significant differences existed in the extent of the reported GRI disclosure and SDGs across the regions classified by IATA. Due to a small number of cases (fewer than 5 observations), the group comprising Africa and the Middle East was excluded from the sample. Accordingly, the groups included in the tests were Asia Pacific, China and North Asia, the Americas, and Europe. Regarding the GRI disclosure, the tests revealed no significant difference in the total count of the GRI indicators, $\chi 2$ (3, n = 28) 3.635, p = 0.30, or in the use of GRI 200, $\chi 2$ (3, n = 28) 3.365, p = 0.34, GRI 300, $\chi 2$ (3, n = 28) 5.306, p = 0.15, or GRI 400 disclosures, $\chi 2$ (3, n = 28) 3.620, p = 0.31. Regarding the SDGs, the results revealed a statistically significant difference, $\chi 2$ (3, N = 35) 8.19, p = 0.04. A closer look at the data showed that the Asia Pacific region (n = 10) recorded a higher median score (Md = 12) in comparison to China & North Asia (n = 6, Md = 9.5), the Americas (n = 8, Md = 8.5) and Europe (n = 11, Md = 7). Dunn's pairwise tests were carried out to compare all pairs of groups, showing evidence (p = .043, adjusted using the Bonferroni correction) of a

Table 2. Frequency of use of GRI 300 Environmental Disclosures

		n	%
301-1	Materials used by weight or volume	9	31%
301-2	Recycled input materials used	5	17%
301-3	Reclaimed products and their packaging materials	3	10%
302-1	**Energy consumption within the organization**	**23**	**79%**
302-2	**Energy consumption outside of the organization**	**9**	**31%**
302-3	Energy intensity	18	62%
302-4	Reduction of energy consumption	20	69%
302-5	Reductions in energy requirements of products and services	15	52%
303-1	Interactions with water as a shared resource Disclosure	13	45%
303-2	Management of water discharge-related impacts	4	14%
303-3	Water withdrawal	7	24%
303-4	**Water discharge**	**2**	**7%**
303-5	Water consumption	4	14%
304-1	Operational sites owned, leased, managed in, or adjacent to, protected areas and areas of high biodiversity value outside protected areas	2	7%
304-2	Significant impacts of activities, products and services on biodiversity	5	17%
304-3	Habitats protected or restored	4	14%
304-4	IUCN Red List species and national conservation list species with habitats in areas affected by operations	2	7%
305-1	**Direct (Scope 1) GHG emissions**	**25**	**86%**
305-2	Energy indirect (Scope 2) GHG emissions	20	69%
305-3	Other indirect (Scope 3) GHG emissions	11	38%
305-4	**GHG emissions intensity**	**24**	**83%**
305-5	Reduction of GHG emissions	21	72%
305-6	Emissions of ozone-depleting substances (ODS)	6	21%
305-7	**Nitrogen oxides (NOx), sulfur oxides (SOx), and other significant air emissions**	**15**	**52%**
306-1	Water discharge by quality and destination	8	28%
306-2	**Waste by type and disposal method**	**22**	**76%**
306-3	**Significant spills**	**7**	**24%**
306-4	Transport of hazardous waste	4	14%
306-5	Water bodies affected by water discharges and/or runoff	4	14%
307-1	Noncompliance with environmental laws and regulations	16	55%

(*continued*)

Table 2. (*continued*)

		n	%
308-1	New suppliers that were screened using environmental criteria	6	21%
308-2	**Negative environmental impacts in the supply chain and actions**	**3**	**10%**

significant difference between the European and Asia Pacific airlines' SDG reporting. However, no evidence of a difference between the other pairs was detected.

5 Discussion and Conclusion

Using quantitative content analysis to review 61 airlines' FY19 sustainability reports, this study identified 29 companies reporting GRI topic-specific disclosures and 38 addressing SDGs in their reports. The analysis of these companies' reports revealed what topic-specific disclosures and SDGs were reported and to what extent.

The results showed that airlines focused more strongly on reporting environmental issues, especially GHG emissions, than economic or social dimensions [cf. 7]. This is expected, given that GRI Standards instruct reporters to focus on material topics. After all, airlines are big emitters and have been at the center of climate change debate in recent years. This issue also appears to resonate with the industry's SDG reporting, as the study found airlines most often stating their commitment to SDG13 (climate action).

Aside from the emissions, the overall topic-specific disclosure in reports showed limited alignment with the topics IATA [14] has identified as material for the industry. This is reflected in the fact that the average frequency of all IATA-recommended topic-specific disclosures in the reports was only 38%. Also, the array of disclosed information seemed highly dispersed across the reports, with numerous disclosures mentioned in low frequencies. On the one hand, this is expected as not all indicators are relevant to airlines. However, this raises the question of why some of these disclosures are reported. Although many reasons could exist, a reasonable explanation could be that some airlines cannot properly assess what issues are material with limited sector-specific guidance. Indeed, the past literature has suggested that the wide latitude provided by the GRI may lead to inconsistencies in the indicator disclosure [20]. Similarly, while most observed reports indicated companies' commitment to SDGs that are identified as fully relevant to the industry [15], many also claimed commitment to SDGs that have very little or no industry relevance.

Despite the above inconsistencies, the results indicate that the extent of GRI topic-specific disclosure is similar worldwide. This deviates from previous research [7], which has found differences between European and Asia-Pacific airlines. On the one hand, the sample size and the year of the reports could explain these deviating results. On the other hand, the results may also suggest a global institutionalization of airline SR, meaning that companies' sustainability reports start gradually looking similar in the same industry [21]. However, Asia-Pacific airlines were detected reporting more SDGs in their reports than their European counterparts. Although explanations for this could be sought from firm-specific [22] and institutional [10] factors, recent research [7] has indicated that

Table 3. Frequency of use of GRI 400 Social Disclosures

		n	%
401-1	**New employee hires and employee turnover**	**18**	**62%**
401-2	Benefits provided to full-time employees that are not provided to temporary or part-time employees	12	41%
401-3	Parental leave	9	31%
402-1	Minimum notice periods regarding operational changes	11	38%
403-1	Occupational health and safety management system	13	45%
403-2	**Hazard identification, risk assessment, and incident investigation**	**18**	**62%**
403-3	Occupational health services	12	41%
403-4	**Worker participation, consultation, and communication on occupational health and safety**	**11**	**38%**
403-5	Worker training on occupational health and safety	4	14%
403-6	Promotion of worker health	4	14%
403-7	Prevention and mitigation of occupational health and safety impacts directly linked by business relationships	4	14%
403-8	Workers covered by an occupational health and safety management system	5	17%
403-9	**Work-related injuries Disclosure**	**6**	**21%**
403-10	**Work-related ill health**	**4**	**14%**
404-1	Average hours of training per year per employee	18	62%
404-2	Programs for upgrading employee skills and transition assistance programs	16	55%
404-3	Percentage of employees receiving regular performance and career development reviews	11	38%
405-1	Diversity of governance bodies and employees Disclosure	20	69%
405-2	Ratio of basic salary and remuneration of women to men	10	34%
406-1	**Incidents of discrimination and corrective actions taken**	**7**	**24%**
407-1	Operations and suppliers in which the right to freedom of association and collective bargaining may be at risk	6	21%
408-1	Operations and suppliers at significant risk for incidents of child labor	6	21%
409-1	Operations and suppliers at significant risk for incidents of forced or compulsory labor	7	24%
410-1	Security personnel trained in human rights policies or procedures	2	7%
411-1	Incidents of violations involving rights of indigenous peoples	2	7%
412-1	Operations that have been subject to human rights reviews or impact assessments	4	14%

(*continued*)

Table 3. (*continued*)

		n	%
412-2	Employee training on human rights policies or procedures	6	21%
412-3	Significant investment agreements and contracts that include human rights clauses or that underwent human rights screening	3	10%
413-1	Operations with local community engagement, impact assessments, and development programs	10	34%
413-2	Operations with significant actual and potential negative impacts on local communities	3	10%
414-1	New suppliers that were screened using social criteria	8	28%
414-2	**Negative social impacts in the supply chain and actions taken**	**6**	**21%**
415-1	Political contributions	12	41%
416-1	Assessment of the health and safety impacts of product and service categories	15	52%
416-2	Incidents of noncompliance concerning the health and safety impacts of products and services	10	34%
417-1	Requirements for product and service information and labeling	4	14%
417-2	Incidents of noncompliance concerning product and service information and labeling	8	28%
417-3	Incidents of noncompliance concerning marketing communications	9	31%
418-1	**Substantiated complaints concerning breaches of customer privacy and losses of customer data**	**16**	**55%**
419-1	Noncompliance with laws and regulations in the social and economic area	9	9%

the Asia-Pacific region is generally quick in adopting new CSR practices, which may explain the difference in the early stages of SDG reporting.

This study provided one of the broadest snapshots of what GRI's topic-specific disclosures are included in airlines' sustainability reports. Assumably, this is also the first academic paper to provide an overview of the SDGs in such reports. All in all, these results contribute to our understanding of the current state of the industry's SR practices, which can help airlines and stakeholders benchmark the industry's SR and academia and NGOs in their research that can positively influence the development of sector-specific guidance. In this regard, further scholarly research could also explore relevant stakeholders' views on these reports, which was outside the scope of this study. It is also worth noting that while this paper revealed the GRI disclosures used, the actual disclosure may show noncompliance with the GRI Standards and limited substantiation of the SDGs. Thus, further research is needed to examine the quality of airline SR.

Table 4. Frequency of SDGs addressed in sustainability reports

SDGs		Relevance	n	%
1:	End poverty in all its forms everywhere	(1)	14	37%
2:	End hunger, achieve food security and improved nutrition and promote sustainable agriculture	(1)	7	18%
3:	Ensure healthy lives and promote well-being for all at all ages	(2)	28	74%
4:	Ensure inclusive and equitable quality education and promote lifelong learning opportunities for all	(2)	23	61%
5:	**Achieve gender equality and empower all women and girls**	**(3)**	**34**	**89%**
6:	Ensure availability and sustainable management of water and sanitation for all	(2)	12	32%
7:	**Ensure access to affordable, reliable, sustainable and modern energy for all**	**(3)**	**21**	**55%**
8:	**Promote sustained, inclusive and sustainable economic growth, full and productive employment and decent work for all**	**(3)**	**36**	**95%**
9:	**Build resilient infrastructure, promote inclusive and sustainable industrialization and foster innovation**	**(3)**	**24**	**63%**
10:	**Reduce inequality within and among countries**	**(3)**	**18**	**47%**
11:	Make cities and human settlements inclusive, safe, resilient and sustainable	(2)	14	37%
12:	**Ensure sustainable consumption and production patterns**	**(3)**	**32**	**84%**
13:	**Take urgent action to combat climate change and its impacts**	**(3)**	**37**	**97%**
14:	Conserve and sustainably use the oceans, seas and marine resources for sustainable development	(0)	10	26%
15:	Protect, restore and promote sustainable use of terrestrial ecosystems, sustainably manage forests, combat desertification, and halt and reverse land degradation and halt biodiversity loss	(1)	16	42%
16:	Promote peaceful and inclusive societies for sustainable development, provide access to justice for all and build effective, accountable and inclusive institutions at all levels	(0)	11	29%
17:	Strengthen the means of implementation and revitalize the global partnership for sustainable development	(2)	19	50%

Funding. This work was supported by the National Science Centre (Poland) under grant [UMO-2019/35/N/HS4/04367].

References

1. Brown, H.S., de Jong, M., Levy, D.L.: Building institutions based on information disclosure: lessons from GRI's sustainability reporting. J. Clean. Prod. **17**, 571–580 (2009)

2. GRI and UNGC: Integrating the SDGs into Corporate Reporting: A Practical Guide (2018)
3. GRI: Consolidated Set of the GRI Standards (2022)
4. Zieba, M., Johansson, E.: Sustainability reporting in the airline industry: current literature and future research avenues. Transp. Res. Part D Transp. Environ. **102**(103133) (2022)
5. ATAG Facts & figures. https://www.atag.org/facts-figures. Accessed 3 March 2023
6. Zhang, X.: Communicating social responsibilities through CSR reports: comparative study of top European and Asia-Pacific airlines. PLoS One **16**(10) (2021)
7. Yang, L., Ngai, C.S.B., Lu, W.: Changing trends of corporate social responsibility reporting in the world-leading airlines. PLoS One **15**(6) (2020)
8. Porter, M.E., Kramer, M.R.: Strategy and society: the link between competitive advantage and corporate social responsibility. Harv. Bus. Rev. **84**, 2–17 (2006)
9. KPMG: The time has come: the KPMG Survey of Sustainability Reporting 2020. KPMG Impact (2020)
10. Rosati, F., Faria, L.G.D.: Addressing the SDGs in sustainability reports: the relationship with institutional factors. J. Clean. Prod. **215**, 1312–1326 (2019)
11. Antolín-López, R., Delgado-Ceballos, J., Montiel, I.: Deconstructing corporate sustainability: a comparison of different stakeholder metrics. J. Clean. Prod. **136**, 5–17 (2016)
12. GRI: GRI 101 Foundation 2016. GRI, Amsterdam (2016)
13. United Nations: General Assembly Resolution A/RES/70/1. Transforming Our World, the 2030 Agenda for Sustainable Development. New York (2015)
14. IATA: Airline Sustainability Reporting Handbook, 1st edn. IATA, Geneva (2020)
15. ATAG: Flying in formation, ATAG, Geneva (2017)
16. Johansson, E.: The use of sustainability reporting instruments in the global airline industry. unpublished (2023)
17. Kondracki, N.L., Wellman, N.S., Amundson, D.R.: Content analysis: review of methods and their applications in nutrition education. J. Nutr. Educ. Behav. **34**(4), 224–230 (2002)
18. GRI: GRI Universal Standards 2021 Frequently Asked Questions (FAQs). GRI, Amsterdam (2022)
19. Pallant, J.: SPSS Survival Manual, 6th edn. Open University Press, Maidenhead (2016)
20. Roca, L.C., Searcy, C.: An analysis of indicators disclosed in corporate sustainability reports. J. Clean. Prod. **20**(1), 103–118 (2012)
21. Hess, D.: The transparency trap: non-financial disclosure and the responsibility of business to respect human rights. Am. Bus. Law J. **56**(1), 5–53 (2019)
22. Rosati, F., Faria, L.G.D.: Business contribution to the Sustainable Development Agenda: organizational factors related to early adoption of SDG reporting. Corp. Soc. Responsib. Environ. Manag. **26**(3), 588–597 (2019)

The Importance of Responsible Investment

António Augusto Baptista Rodrigues(✉)

ISG- Business and Economics School, Lisboa, Portugal
antonio.rodrigues@isg.pt

Abstract. Responsible investment (IR) has become increasingly relevant in the investment and asset management industry. According to the Global Sustainable Investment Alliance, assets incorporating IR strategies represent $22.9Tr globally, representing 26% of total professionally managed assets. ESG (Environment, Social and Governance) aspects are considered important areas when assessing the merits and risks of a company or an investment. The IR universe presents a high level of maturity and technical sophistication, typical of the financial sector. To achieve the proposed objectives, a qualitative methodology was adopted, using the analysis of norms and the literature review in order to analyze and theoretically understand the different concepts of Socially Responsible Investment and what ESG metrics are.

Keywords: Responsible investment · ESG · SRI Strategies

1 Introduction

The need for comparable, relevant and consistent information has become a requirement for asset managers when making investment decisions.That is why the sustainability financial services market has developed: ESG ratings, rankings, sustainability indices and (socially) responsible funds. These services provide asset managers of IR portfolios with the analytical information they need to assess the ESG performance, risks and opportunities of managed assets.

Most indices are set by ratings agencies, which use ESG factors in their research methodology to select companies to be included in the indices. Agencies also often use a traditional index provider to create a sustainability index, e.g. the DJSI Index published by S&P Dow Jones Indices and RobecoSAM, the FTSE Index for FTSE4GOOD, the MSCI ESG Index. These indices serve to establish benchmarking, so they play an important role for companies to influence their ESG strategy, since they want to be part of these indices. In addition, they define standards of responsible behavior. From the point of view of the responsible investor, they provide a potential investment universe.

According to data published by the Global Sustainable Investment (2018) Alliance, between 2016 and 2018 sustainable investment grew in absolute and relative values in most regions of the world. Thus, sustainable investing represents more than 50% of total assets under management in Canada, Australia and New Zealand, 48% in Europe, 26% in the US and 18% in Japan. The study states that this growth reflects the growing

C. Montenegro et al. (Eds.): ICMTT 2023, LNNS 774, pp. 185–194, 2024.
https://doi.org/10.1007/978-3-031-43733-5_17

awareness of the financial value associated with this type of fund, with many investors stating that the main motivations for investing in this type of fund are:

- desire to minimize risk and improve financial performance over time;
- the desire that the investments also have a positive impact, so that they can generate social and environmental benefits or fulfill the mission of their companies;
- help combat climate change and help implement the UN Sustainable Development Goals.

2 Theoretical Framework

According to the Global Impact Investing Network (GIIN), impact investing can be defined as "investments made with the intention of generating positive and measurable social and environmental impact along with financial return" [1]. Normally, in order for it to fit this definition, three common characteristics of impact investment are also assumed: the first is obviously the intentionality of the impact, the second, like any financial investment, has to do with the expectation of obtaining returns. Positive financial, and finally there must be a measurable impact along the non-financial dimensions, in particular; in the social, environmental or governance spheres of the companies in which it is invested.

Subsequently, in 2005 the so-called "United Nations Principles for Responsible Investment" (UNPRI) were created, whose objective was the introduction of environmental, social and corporate governance factors that had to be taken into account when make investment decisions, in order to more efficiently manage those extra-financial risks that result from investments and in order to specify what these factors had to be in order to improve and facilitate their measurement.

But, a key date in the process of raising awareness about the progress of Socially Responsible Investment (SRI), 2015, because the "2030 Agenda for Sustainable Development" was launched by the UN General Assembly, which together with the 17 Sustainable Development Goals (SDG) [2] they implied a universal commitment to the improvement of the planet and the prosperity of people [3]. SRI can be defined as an investment strategy that overweights environmental, social and governance criteria in the process of studying, analyzing and selecting securities for an investment portfolio.

The strong potential impact and the high probability of occurrence of environmental risks have been recognized in the Global Report on Global Risks prepared annually by the World Economic Forum, going from not including environmental risks in 2011 to being those with the highest probability of occurrence in 2021 [4].

Since Stern [5] published one of the first analyzes on the economic impact of climate change in 2005, estimating the cost of inaction at 5% of global GDP, many studies have been published, giving dimension to a very substantial risk for the world economy. Climate risks are considered from two points of view:

- Physical risks: They refer to impacts of climate change that can damage properties or assets. For example, extreme weather events, such as floods or storms that can cause property damage or disrupt business. In the longer term, they can even affect production capacity.

- Transition risks: These are risks that derive from changes in policies, responsibilities, market preferences or changes in technology to move towards a green economy. This transition may result in changes in asset valuations. For example, the loss of value of assets based on fossil fuels.

The transition to a green economy requires heavy investment, which IFC estimates at more than $10 trillion in emerging markets alone over the next decade. Obviously, this volume of investment cannot come exclusively from public sources. The private sector plays a leading role in this economic transformation, and this also implies an important opportunity derived from climate change for companies and investors. Some of the sectors in which this opportunity is most clearly materialized are: renewable energy and energy efficiency, infrastructure and buildings, agriculture and forestry, insurance, tourism and sustainable mobility.

In order to correct some of the inertia that the market economy has been creating in recent decades, from the perspective of sustainability, increasing attention has been paid to aspects such as diversity and gender, human rights, protection of data, or labor standards. Additionally, companies have come to give much greater importance to caring for their relationship with all the interest groups around them, not only with shareholders, but also with customers, employees, communities, suppliers, etc.

If the 2008 financial crisis increased the focus on some of these problems, the crisis caused by Covid-19 has catalyzed solidarity movements and further accelerated concern for social issues, which has been reflected in the incipient market for new investment instruments, such as social impact bonds, which allow the financing of activities exclusively guided by social purposes that seek commitments of additional social benefits and tangible positive impacts.

The third pillar that makes up the concept of traditional sustainability in investment refers to the corporate governance of companies. On a theoretical level, it is described by the agency problem and fiduciary duty, in which the managers (agent) have to make decisions on behalf of the interest of the shareholders (principal), which can generate different situations of conflict of interest. Additionally, various cases of fraud or irresponsible actions by managers have increased the scrutiny of company management by shareholders and other interest groups. Thus, the concept of investor activism has emerged, as a strategy in which the investor participates through the governing bodies and through other channels, with the aim of improving the governance of a company and generating value as a result of it.

With the aim of minimizing conflicts of interest and increasing transparency in actions, good corporate governance practices have been articulated, which try to define the set of processes, policies and relationships by which companies are controlled and managed.

Among the aspects included in the rules and codes of good governance, the definition of the responsibilities and mode of operation of the general meeting of shareholders and the board of directors stands out, with special emphasis in the latter case on the profile that directors must have, as well as the need to set up a series of committees reporting directly to the board (such as the audit committee, the appointments and remuneration committee, and the risk committee in the case of banking entities). Finally, another aspect to which special attention has been devoted is the establishment of a remuneration system that makes it possible to align the interests of shareholders, without falling into the short-term temptations that managers can often have.

3 Research Objectives

The vision and appearance of sustainable finance have been in the world for more than two decades, having a strong momentum in recent years, as part of its commitment to the adoption of goals associated with the 17 sustainable development goals, in line with this global trend. One way to be more tangible in the above is the adoption by companies in the financial sector of investment decisions, integrating financial aspects, ESG criteria.The main objective is to know how non-financial variables and their practices add value and sustainability over time, encouraging greater economic benefits.

4 Methodology

To achieve the proposed objective, a qualitative and interpretative methodology was adopted, using the literature review and analysis of norms/legislation. The methodology is suitable for collecting information within the framework of the study of subjective phenomena, which in the opinion of Yin [6] favors a more critical position of the researcher.

5 SRI Strategies

According to The European Sustainable Investment Forum (Eurosif) [19] there are seven different categories of ESG strategies, which can be applied exclusively or jointly and are valid, except for dialogue with companies, both for companies and for public entities (debt). Said categorization can be segmented into two large blocks, referenced as negative criteria or positive evaluation criteria.

5.1 Negative Criteria

5.1.1 Exclusion of Activities

The exclusion includes the elimination of certain companies or sectors from the investment universe insofar as part or all of their source of business is linked to activities considered unsustainable or that imply high reputational risks. Among them, the most common are usually weapons, tobacco, nuclear energy and pornography.

5.1.2 Standards-Based Assessment

Through this strategy, investors exclude investments in companies whose income or profits, in whole or in part, go against the spirit of international standards and fundamental rights that generate a broad consensus, such as: OECD guidelines for multinational companies, declaration ILO Tripartite Principles on Multinational Enterprises and Social Policy or the United Nations Global Compact.

5.2 Positive Criteria

5.2.1 Thematic Investments of Sustainability

This strategy is aimed at investing in specific sectors with positive externalities in social and/or environmental aspects, such as renewable energy, sustainable infrastructure, water management or waste management, among others.

5.2.2 Outstanding Ratings (Best-in-Class)

Through which investors select the companies or issuers with the best ESG metrics or ratings in a particular sector. Unlike other similar positive approaches, such as Best-in Universe or Best-Effort, a priori any sector or type of asset can be present in said strategy, so that only the best ones will be selected.

5.2.3 Dialogue with Companies or Shareholding Engagement

It consists of improving corporate governance through dialogue processes with companies, such as through meetings with management teams in which potential aspects of improvement in corporate management can be conveyed. It has a very strong link to fiduciary duty, as it is driven in large part by the current of opinion that asset managers are asset managers who are accountable to their investors.

5.2.4 Impact Investments

These are investments that take into account, in addition to economic indicators, the company's social and/or environmental impact, as well as being implemented in both developing and developed countries. One of the main catalysts for this strategy is the contribution to sustainable development, a fact that partly explains the success in the development of the Green Bonds market as one of the most successful initiatives among the investment community in recent years.

Unlike other strategies, impact investing has a larger number of investment vehicles. Along with investing in listed companies, this strategy uses other assets such as loans, direct participation in the capital of a company, or hybrid products [7].

5.2.5 SG Integration

It is considered the most advanced strategy and that maximizes the spirit of SRI through the incorporation of ESG criteria in traditional financial analysis.

This strategy consists of the "explicit and systematic integration, by managers, of ESG factors in financial analysis" [8] that precedes the selection of assets for an investment portfolio. Said integration requires, in particular, the identification of the materiality elements within the ESG scope for each of the sectors of the companies in which the fund wants to invest, in addition to the analysis of the potential impact on the performance of each company.

Investments that are made globally with sustainability criteria have experienced strong growth rates in recent years, already exceeding 30 trillion dollars in 2018, according to the latest report published by GSIA [8], with a special role for Europe as the region that has invested the most in the application of these investment criteria.

Although in its origins socially responsible investment was mainly focused on certain ethical funds that applied certain exclusions to sectors or companies that carried out practices contrary to certain modal values, today, investors have also incorporated other more diverse and complex strategies.

Among these strategies, the following stand out: the selection of companies based on the best in class criteria (that is, investing within a sector in the company with the best behavior in terms of ESG); the use of the best in universe criterion (which invests in the best valued companies from the ESG perspective, regardless of their sector); the use of the best efforts criterion (which gives priority to those issuers that have improved the most in their ESG practices); investment that integrates ESG criteria into the analysis; thematic investments (for example, in environmental or social issues); those that pursue impact; or the so-called engagement and voting, which involves active management in which the investor begins to exercise their rights as a shareholder to, through the governing bodies, change the practices of the company and align them with ESG criteria.

6 Responsible Investment Products

The use of the term "Socially Responsible Investment Fund" refers to any investment scheme that explicitly refers to profit, based on social or environmental directives, with a view to building investment portfolios [9]. The ISR originated in 1920, when certain religious institutions refused to invest in companies linked to areas of activity related to the alcohol and tobacco industry, thus avoiding investing in the so-called sin actions [10].

Environment-related funds emerged later in the 1970s and 1980s, reaching enormous prominence in the 1990s [11]. The SRI concept supports the channeling of funds in a way that combines the financial objectives of investors with a commitment to social concerns such as social justice, economic development, peace, and a healthy environment [9]. SRI funds have a double objective, which is to achieve financial and social returns. Its investment strategy is guided by principles of social and environmental responsibility and will only be made in the assets of companies that also adopt this type of practice [11].

SRI fund managers argue that investing in these types of assets promotes the performance of normatively desirable activities. Similarly, the management of these funds has implications for business change, as good companies reduce their capital costs relative to bad ones [9].

When an SRI fund selects a good company (positive screen) to invest in, the additional capital resulting from that choice will lower its cost of capital. Holding the investment will allow the company to provide capital, giving it control over its capital costs. Most SRI funds maintain their profiles socially responsible through the use of positive or negative filters. Positive selection identifies companies with superior social or environmental performance, while negative selection identifies companies with unwanted activities.

Conversely, not investing in a "bad" company (negative screen) decreases the supply of capital for that company, thus increasing its costs of capital. The profitability of investing in shares will be affected by the decline in share projects and will lead to the consequent drop in the value of shares.

In this way, companies will modify their operations to meet the requirements of the ISR funds.SRI fund managers invest in companies that, being socially responsible, proactively anticipate and avoid the costs of government regulation [9]. These behaviors are also those of the business managers reported in the study by Corten and Vermeir [18]. Fund managers, by acting in this way, will put companies in a position to take advantage of opportunities derived from the negative or positive assessment of certain external factors, thus creating a competitive advantage over less socially responsible competitors, for which a performance above the market [9].

7 Green Bonds

The Paris agreement, in 2015, reinforced the need to mobilize significant financial resources to allow the investments necessary to make the world economy carbon neutral during the second half of the 21st century. This global commitment thus reinforced what was already evident after the assumptions of the Vision for 2050 developed by the World Business Council for Sustainable Development and the Goals for Sustainable Development that today constitute the global agenda for 2030 [12]. Initially introduced by the European Investment Bank in 2007, Green Bonds are still far from having a relevant position within the framework of financial instruments available to countries and public and private entities. However, the growth of the Green Bond market is an excellent sign to move towards the scale needed under the pressure of climate change and other environmental pressures.

There is no uniformly accepted definition for these financial instruments, although the one transcribed below is quite clear: "Green Bonds are any type of bond instrument where the proceeds will be exclusively applied to finance or re-finance in part or in full new and/or existing eligible Green Projects and which are aligned with the four core components of the Green Bonds Principles."

Therefore, Green Bonds are financial instruments intended for the execution of specific and eligible projects, and the entire process of obtaining and using financing is subject to its own scrutiny and reporting procedures. Its relevance in recent years and many experts believe that it will have a growing and significant evolution in the coming years given the growing importance of socially responsible investments.

Until the very recent arrival of the European Union sustainable investment taxonomy in 2020, there has not seemed to be a consensus on what is considered a green bond and what is not [13]. Although the Association International Capital Markets (ICMA) developed the first green bond standard in 2013 [14], and this was mostly taken up by private companies, sovereign states and other industry players, numerous countries issued their "variations" on green bond standards, which has generated some confusion. in the sector.

An important milestone when it comes to aligning the criteria under which this type of bond should be issued was the publication in 2014, and its subsequent update in 2018, of the Green Bond Principles (GBP), formulated by the ICMA. These establish voluntary adoption guidelines that recommend transparency and disclosure, and promote integrity in the development of the green bond market, establishing a framework for their issuance. Finally, and in order to give investors greater guarantees regarding the alignment of green bonds with the GBP, it is considered good practice for the issuer of this type of instrument to request external verification by a rating agency or independent entity. Internationally recognized.

A second type of bonds, which has had a growing development, are social bonds. Among the eligible categories for this type of issuance, according to the Social Bond Principles updated by ICMA in 2020, are access to essential services (such as health or education), access to affordable housing, basic infrastructure, or the employment creation. Social bonds have shown particular dynamism in 2020 as a response to finance some of the pressing problems derived from Covid-19. Additionally, other instruments have emerged on the market that allow investing in projects linked to sustainability criteria. These include: sustainable bonds (whose funds are applied to finance a mix of eligible green and social projects); "sustainability-linked" bonds (which link the coupon to compliance with certain sustainability indicators); "SDG-linked" bonds (which, more specifically, link the coupon to compliance with the SDGs); gender bonds (whose use of funds is linked to investments linked to promoting gender equity initiatives, such as the development of SMEs led by women); blue bonds (whose use of funds is linked to investments linked to promoting the conservation of ocean life); transition bonds (the use of which is intended to provide financial resources to those companies that are "brown" today, but have the ambition to transition to green in the future); or green loans (which encompass any type of loan instrument available exclusively to finance or refinance eligible green project.

On the other hand, funds can be managed actively or passively. Active management is characterized by the fact that the manager decides what to buy or sell, believing that those companies in which they have decided to invest will be the ones that will achieve the highest profitability in the future. On the contrary, in passive management the manager manages index funds that follow stock market indices, its main objective is to replicate those indices.

8 Conclusions

Investors and savers are increasingly aware of sustainability issues, and want to know what activities are financed with their savings, and in many cases invest in instruments that allow them to achieve certain social or environmental objectives.

Multiple analyzes have been published to defend the hypothesis that investments that take sustainability criteria into account generate better financial results than investments that do not comply with these criteria.

More specifically, some studies, such as the one prepared by MSCI [15], suggest for example, that sustainability is associated with higher margins and less volatile profits. Others, such as the one prepared by Forética [16], document the losses generated to investors as a consequence of the ESG shocks that have occurred in the last decade. Other studies point in a similar direction, such as the one carried out by Bank of America

Merril Lynch [17], which shows how better sustainability behavior is correlated with lower risk, reducing the volatility of profits and improving cash flows.

In summary, it is possible to conclude that the local market is in the initial stages of incorporating sustainability into its investment decisions, requiring collaborative work between the various actors in the financial ecosystem in order to advance in the dissemination and understanding of the sustainable investment, definition of financially material metrics and standards.

In the construction process of the respective portfolio, these funds meet, in addition to traditional financial criteria (for example, risk and return), non-financial criteria (environmental, social and best corporate governance practices - ESG), seeking financial return as well as social good. To ensure that the investment is made in shares of companies that stand out for the best practices in terms of sustainability, the funds use stock indices or bonds.

The strategy of excluding specific industries or sectors from portfolios remains the most prevalent. This is followed by the strategy of systematically and explicitly including ESG factors in the financial analysis of the project and the strategy where shareholders influence strategic decisions with ESG objectives.

One of the actions that the European Commission is carrying out within the scope of the Action Plan for Sustainable Finance is related to the dissemination of information on how ESG issues were incorporated or not in the creation of a fund. This EC initiative is called "Duties of institutional investors and asset managers in terms of sustainability" [13] and there has already been a proposal for a regulation on the information to be disclosed in relation to sustainable investments and ESG risks, which had a period of public consultation, pending decisions on the final text.

ESG funds will become more mainstream as there will be regulation requiring you to communicate how ESG themes have been incorporated into asset selection. Funds may continue to be built without ESG criteria, but entities will be required to expressly report that said fund did not include ESG criteria or an ESG risk analysis when selecting assets in their portfolio.

Impact investing is "investment in business models behind which there are entrepreneurs who directly intend to generate a positive social or environmental impact as well as attractive financial returns for the investor", as well as emphasizes its relevance due to the potential it has to solve "the most pressing problems we face as a society."

As trends and challenges with this investigation it is suggested to break with certain stereotypes regarding its profitability and above all to analyze and demonstrate that there are many efforts that are being made to promote SRI at the institutional level, so we must trust and begin to be more interested for this alternative, to continue promoting changes, improvements and unifications by the system.

References

1. Mudaliar, A., Bass, R., Dithrich, H., Nova, N.: Annual Impact Investor Survey. Global Impact Investing Network, New York (2019)
2. United Nations, The 17 Goals. https://sustainabledevelopment.un.org/?menu=1300
3. Analistas Financieros Internacionales; Divulgación en Sostenibilidad: El GAR (Green Asset Ratio) como indicador clave. AFI, Madrid (2021)

4. The Global Risks Report 2021. http://www3.weforum.org/docs/WEF_The_Global_Risks_Report_2021.pdf
5. Stern,N.H.: The Economics of Climate Change: The Stern Review. Cambridge University Press.Cambridge, United Kingdom (2007)
6. Yin, R.K.: Qualitative Research from Start to Finish, 2nd edn. Guilford Publications, New York, NY (2016)
7. Fontrodona, J., Levy, D., Sanz, P., Casas, M.: La apuesta del impact investing. IESE Business School (Center for Business in Society) y DiverInvest, Invertir contribuyendo a la mejora de la sociedad. Barcelona (2016)
8. GSIA. http://www.gsi-alliance.org
9. Haigh, M., James, H.: Financial Markets: A tool for social responsability?, Journal of Business Ethics, **52** (1), (2004)
10. Skillius, A., Wennberg, U.: Continuity, credibility and comparability: key challenges for corporate environmental performance measurement and communication. The International Institute for Industrial Environmental Economics at Lund University (1988)
11. Rezende, I.C.: Análise da rentabilidade dos fundos de investimento socialmente responsáveis (2005)
12. WBCSD. www.wbcsd.org/Overview/About-us/Vision2050
13. Ec.Europa. https://ec.europa.eu/info/business-economyeuro/banking-and-finance/sustainable-finance/eu-taxonomy-sustainable-activities_en
14. Icmagroup. https://www.icmagroup.org/sustainable-finance/theprinciples-guidelines-and-handbooks/green-bond-principles-gbp/
15. Foundations of ESG investing: How ESG Affects Equity Valuation, Risk and Performance. MSCI (2017)
16. El valor de la gobernanza: Incrementando el retorno a través de la gestión extra financiera. Forética (2018)
17. ESG Matters – US. 10 reasons you should care about ESG. Bank of America Merrill Lynch (2019)
18. Corten, F., Vermeir, W.: Sustainable Investment: the complex relationship between sustainability and return. Banken Financiewezen (2001)
19. The Sustainable Investment Forums. https://www.eurosif.org/

Organizational Culture and Social Responsibility, A Look from Latin American SMEs

Linda Núñez[1](✉) and Ronaldo Castro[2]

[1] Universidad Estatal Península de Santa Elena, Santa Elena, SE 240350, Ecuador
lnunez@upse.edu.ec
[2] Escuela Superior Politécnica del Litoral, Guayas ECO90211, Ecuador
ronacast@espol.edu.ec

Abstract. Latin America is a region with many cultural riches and ethnic diversity, with a diverse and sometimes unstable economy, with extreme political ideologies where its rulers try to implement solutions to deep-rooted problems. Under these conditions, citizens seek mechanisms to generate sources of employment, giving way to SMEs, which, in turn, incorporate business practices to stay in the competitive market. The present investigation is developed with the objective of identifying the conditions of Latin American SMEs from the approach of Organizational Culture and Corporate Social Responsibility. As the theoretical background was elaborated, experiences from different countries were randomly established, how they have put the variables under analysis into practice. The temporality of the study was defined by the consulted scientific productions themselves; each variable took into account contextualizations, approaches, models and dimensions. A meta-study was carried out based on the review of various articles searched in Latindex and high-impact databases to later characterize them, analyze their findings, and identify possible interrelationships between variables. The results effectively conclude in the identification of SME realities in various Latin American countries, conditions of the region, the imprint generated by the application of the variables studied; in addition to the selection, as a proposal, of the Denison Organizational Culture model for future comparative studies with CSR. Each variable took into account contextualizations, approaches, models and dimensions.

Keywords: Organizational culture · CSR · SMEs in Latin America

1 Introduction

By fixing your gaze on Latin American SMEs, you can learn many particularities of the countries that comprise it, but at the same time similarities that make them unique compared to the rest of the world. The studies, investigations, reports in these territories capture the attention of world organizations due to the resources it possesses, culture, tradition, customs, etc. They are sectors where international contributions are waiting to

C. Montenegro et al. (Eds.): ICMTT 2023, LNNS 774, pp. 195–205, 2024.
https://doi.org/10.1007/978-3-031-43733-5_18

intervene in these emerging economies. Another particularity in terms of SMEs and the attention of international communities, whether academic, social, economic or environmental, is due to the presentation of sustainable and sustainable services or products, where for many large companies these products are the raw material in their production chain.

Regarding the variables subject to study and as part of the methodology, a journey was made from the first definitions of Organizational Culture, the effects that it has exerted on workers and companies, characteristics in the construction of a social collective, are also addressed the elements that each model contemplates, from Denison in 1996, Hofstede in 1997, Cameron and Quinn in 1999, Schein in 2010. In the case of Corporate Social Responsibility (CSR), definitions, approaches, trends from different countries of Latin America and various moments, among these realities were found SME cases in Peru, Colombia, Mexico, Argentina, Ecuador and Venezuela.

The objective of this study was to identify the conditions of Latin American SMEs from the approach of Organizational Culture and Corporate Social Responsibility; consequently, for each variable its contextualization, models and dimensions were taken into account. For a better synthesis and understanding, the findings section was structured into: Realities of Latin American SMEs, References of studies on Latin American SMEs cases, Approach to Organizational Culture and CSR in Latin American SMEs. Within the conclusions, some aspects were raised, but in particular a recommendation is made for future studies between the two variables. Based on the models and dimensions reviewed in this paper, indicators were selected to facilitate their comparison.

2 Methodology

The study on Organizational Culture and Corporate Social Responsibility from SMEs in Latin America has allowed us to know a little more about the realities that arise in various countries that comprise it, it was then necessary to make a bibliographical tour of various relevant investigations, be these scientific articles, books, reports, fourth level theses and others where various approaches, approaches, findings, enriching conclusions were found, the same ones that visualized a horizon to assemble the work route of the present study.

A first review of the scientific productions was carried out, making literal readings for the recognition of details, main ideas and identifying relationships or even discarding what did not correspond. Subsequently, the readings were more inferential and critical so that the data is relevant and linked to the objective of the study. This review was not limited to choosing countries that are in a better economic position, it was simply enough that they are from the region and among these realities, cases of SMEs were known in Peru, Colombia, Mexico, Argentina, Ecuador and Venezuela.

From the first moment of the search, documentation from various scientific databases, whether Latindex, Dialnet or high-impact, was considered; regardless of the type, scope, level or focus of the investigation, since what was significant was finding the relevant realities of Latin America from the case of SMEs; Quantitative and correlational scope types of research greatly facilitated the definition of findings and conclusions.

3 Theoretical/Conceptual Background

3.1 Contextualization of Organizational Culture

Starting from the identification of conceptual aspects of organizational culture, the following is a summary of some definitions from the precursors of this theory (Table 1):

Table 1. First definitions of Organizational Culture.

Year	Author	Definition
1981	Ouchi William	It constitutes symbols, ceremonies and myths that communicate to the staff their most deeply rooted values and beliefs and are gradually put into practice on a voluntary basis
2010	Schein Edgar	Basic model created by people as they learn about the problems and later become institutional patterns
1997	Hofstede Geert	Collective mental program that distinguishes members of a group or category
1992	Allaire Ivan and Firsirotu Mihaela	Composed of a socio-structural system that has to do with strategies, policies, formal processes; system of significant symbols that groups expressive and affective aspects; and, the staff, as an organizational reality with talents, experience and personality

Organizational culture as a set of values and beliefs has been in force since the eighties until it became a vital system within companies, and in its evolution it ceases to be just an element in organizations to become a component of relevant strategic and differentiator for companies seeking a competitive advantage (Barney 1986; Porter 1980). In several organizations, the culture is so strong that you can see how people's behavior changes the moment they walk through the doors of the company, transforming the way they operate (Ruiz & Naranjo 2013).

The study of culture has gained strength in recent years, thanks to being considered a fundamental element in organizations. (Garcia Vargas 2011). In accordance with (Huang 2005), affirms that the organizational culture allows companies to know their positive and negative aspects and manages to develop strategies that help them identify strengths and weaknesses, which have an impact on performance, thus achieving permanence within the market; allowing to increase productivity (Salazar Estrada et al. 2009).

3.2 Organizational Culture Models

There are some authors who present their perspectives on the elements that a company should have in its organizational culture. They are summarized below (Table 2):

Table 2. Organizational Culture Models

Author	Component elements of each Model
Daniel R. Denison (1996)	It highlights 4 attributes in an organization with a central axis of Beliefs and Assumptions, deep aspects of institutional identity, often difficult to access. Mission, Adaptability, Involvement, and Consistency measure behaviors guided by these beliefs and assumptions that create an organization's culture
Geert Hofstede (1997)	It encompasses 5 dimensions: Individualism, the extent to which people feel independent as a member of a larger whole. Power distance, less powerful members of the organization accept and expect power to be distributed unequally. Masculinity/femininity, the extent to which the use of force is socially endorsed. Uncertainty avoidance, society's tolerance for uncertainty and ambiguity. Long-term orientation, the world is constantly changing so prepare for the future
Cameron and Quinn (1999)	Model covered in 2 dimensions. One: Stability (control and order) and Flexibility (innovation and dynamism). Other: orientation of the company and can be external (customers, competitors, environment, etc.) and internal (workers, products, processes, etc.). Here 4 types of cultures arise: clan, adhocratic, market, hierarchical

3.3 Contextualization of Corporate Social Responsibility

It should be noted that CSR practices vary from country to country and from culture to culture, and have a different role in each reality. In particular, they must be situated in contexts of public policies and the institutional development of the country. Thus, practices must be adapted to the environment in which it operates and the definition of what is the responsibility of the company varies in each situation (Vives and Star 2011).

In Colombia, a large percentage of companies belong to SMEs, their contribution to national production is high. Most of them do not have formalized CSR activities and policies, often due to lack of knowledge and other times because they focus interest on activities specific to their economy: assuming expenses, investing in capital, among others. (Sepúlveda Romero et al. 2015). For his part, (Costin 1999), when reflecting on CSR in SMEs, concludes that it can be considered as basic expectations that the company can meet in terms of environmental protection, health and public safety, closely related to the leadership exercised by the manager or owner.

Vives, Corral and Isusi (2005), in their work Tendencies of Social Responsibility in Argentina, indicate that:

> "This process begins through responsible practices in social assistance programs, fundamentally as charitable works carried out by the Catholic Church and financed through private donations and requirements of a fundamentally ethical or religious order, of an absolutely individual nature. From the first decade of the 20th century, the issue of SR is interpreted as philanthropic actions carried out by businessmen,

in some cases in a timely manner, being able to find important cases closely linked to the current vision of developing programs related to the internal dimension of the SR, oriented to the well-being of the worker, as it happens in Venezuela and Colombia. (p.18)"

Montoya and Martínez (2012), professors at the University of La Salle, Mexico, in their work entitled "Corporate Social Responsibility: An ethical response to global challenges", indicate that:

"It must be stated that CSR cannot and should not be seen as a complicated network of obligations. Neither is it about the development of a set of solidarity actions, which are more like market positioning measures. It is simply a contribution of analysis and knowledge of the company, which starts primarily from internal reflection to emerge, define and reaffirm the objectives and principles with which the business project is carried out, taking into account the ways of interaction with its workers, relationships with suppliers, communication and search for customer satisfaction, the remuneration to shareholders if any."

3.4 CSR Dimensions

Based on the Green Book of the Commission of the European Communities (2001), the dimensions on which CSR is practiced are presented below (Table 3):

Table 3. CSR Dimensions.

Dimensions	Indicator	Description
Internal Dimension	Human resources management	Attract qualified workers, maintaining a retention program, practicing inclusion, promoting and valuing learning
	Health and safety in the workplace	Due to subcontracting, companies are not exempt from supporting this indicator, they become voluntary and complementary actions to regulations
	Adaptation to change	Balance and take into account the interests and concerns of all those affected by the changes and decisions

(*continued*)

Table 3. (*continued*)

Dimensions	Indicator	Description
External Dimension	Local communities	Development of communities by providing jobs, wages and benefits, and tax revenue. Clean environments and environmental education
	Business partners, suppliers and consumers	Efficient, ethical and ecological offer of products and services that consumers need and want. Mutual aid between partners and suppliers
	Human rights	Linked by international activities and global supply chains. Take care and guarantee respect for labor standards and protection of the environment and human rights

4 Results

The business, economic, social and environmental conditions experienced by SMEs in Latin America are very particular to those that companies from other environments or world powers may have. Some random but relevant findings that emerged from the inquiry in this study and structured as follows are presented below:

1. Realities of Latin American SMEs

For this section, two studies applied at different times are taken into account but that despite the years that have elapsed are realities that are still maintained in Latin American SMEs (Table 4):

The conditions presented as a synthesis at two different moments, one 16 years ago and the other 10 years ago, leave many musings about the context of SMEs in Latin America. Is this region condemned to integrate only emerging countries? Regional or sectorial are not competent or sufficient to supply a demanding demand? Will the companies that emerge as SMEs be limited to this, without being able to grow their economies? Do the governments on duty know definitively about the realities of their countries? Do the implemented regulations correspond to the realities of each country? Could neighboring countries create alliances/economic groups to meet individual needs? Among other questions.

2. References of case studies of Latin American SMEs

A study carried out in Colombia in 2013 on the perception and practices of CSR in SMEs, where its data was collected through an online questionnaire, sought to identify the understanding and practices of CSR, in addition to demographic issues. The companies had various activities from manufacturing, service, trade, technology, construction, banking, etc. Synthesizing the results it is mentioned:

Table 4. Realities of Latin American SMEs.

Author/Year	Findings
Vives, Corral & Isusi (2005)	Need to find solutions to poverty and exclusion based on the creation of economic opportunities for groups that supply goods and services, support these groups and improve aspects such as supply from nearby places, improvement of the quality of products and services There are companies in countries like Brazil and Colombia that contribute to sustainable development focusing on environmental conservation and biodiversity, use natural products, buying part of their inputs under fair trade conditions and organic origin For SMEs, the incentive to be more responsible often comes through the value chain to which they belong; Generally, the pressure is exerted by their clients, who, in many cases, are the largest companies In terms of markets, most SMEs tend to serve local markets and are not exposed to international pressures or incentives
Vives and Peinado-Varas (2011)	More than 60% of companies are SMEs and they provide between 40% and 50% of employment, even though these statistics vary from country to country and are difficult to consolidate, since countries use different definitions of size and capture them at different dates There are many difficulties that make CSR evolve more slowly than in other countries in the world, such as less institutional capacity, weaker corporate governance, less favorable business climate and smaller business scale. Since the late 1990s, numerous organizations have emerged that are advancing in the promotion of CSR, such as: World Business Council for Sustainable Development, with national chapters in almost all Latin American countries, Business for Social Responsibility - BSR, focused in the United States and one of the promoters of the creation of the umbrella organization in the Americas called Foro Empresa, More than 60% of companies are SMEs and they provide between 40% and 50% of employment, even though these statistics vary from country to country and are difficult to consolidate, since countries use different definitions of size and capture them at different dates

– SMEs apply several activities on the practice of social responsibility, those that are between 40% and 80% are mentioned: employee welfare, training and development, adoption of charities or other non-profit causes, charitable donations, environmental projects, code of ethics, sports, health care.

- Another resulting aspect was the important elements to motivate companies to be socially responsible and they are mentioned in an order of priority: improving community trust and support, long-term sustainability, improving public welfare, company tradition, recruit/retain employees, enhance reputation, attract investors, charitable donations, increase profits, and promote transactions/partnerships.
- Among the benefits that SMEs have obtained when considering CSR practices, those that are between 46% and 80% are mentioned: Improvement of organizational culture, attracting and keeping the best employees, improved image and reputation, loyalty of improved clients, improvement of the quality of management.

Another study applied in 2019 focused its analysis on CSR and Organizational Climate aimed at SMEs in Ecuador with a sample of 384 employees from a mediumhigher hierarchical level and from different economic sectors. A questionnaire of 28 statements on a Likert scale was used, synthesizing the following:

- From the use of statistical tools, a series of analyzes were carried out that conclude in the existence of a moderate positive link between CSR and Organizational Climate.
- It is necessary to promote CSR actions based on human capital and processes to cause a positive effect on the business climate.
- A weak positive relationship between the internal dimension of CSR and CO was confirmed, which reinforces that CSR actions help strengthen the company's reputation based on a good work environment and environmental practices.
- Both private and public companies must develop action programs and, in order to carry out future studies on the evolution of CSR in SMEs, recognize its impact on the organizational culture to evaluate behavioral aspects.

5 Approach to Organizational Culture and CSR in Latin American SMEs

The last aspect considered to identify the conditions of Latin American SMEs is to randomly highlight research that refers to the approach to the two variables subject to study, regardless of approaches, methods and levels of research; or, if they were channeled independently or jointly. They are summarized below (Table 5):

With this section, despite the fact that the findings are not literally the same, its essence leads to meeting points even when the studies belong to economies of very divergent countries at the present time; It is recognized then that both for the application of CSR or Organizational Culture in the daily life of SMEs these represent: an improvement in production processes that imply the incorporation of technologies, motivation and loyalty of workers, improvement of the business image, recognition government entities and customer loyalty, increased sales and profitability.

Table 5. Studies on Organizational Culture and CSR in Latin American SMEs.

Author/Year	Origin	Findings
Aristimuno, M.; Velásquez, Y. & Rodríguez C (2010)	Venezuela	– Improvement of products and/or production processes, resulting in greater customer satisfaction and loyalty – Greater motivation and loyalty of workers, which increases their creativity and innovation – Better public image, due to awards and/or greater awareness of the company in the community – Better position in the labor market and better interrelation with other business partners and authorities
Bust, L (2018)	Argentina	– When the members of the organization understand the mission, strategy and identify with the objectives of the company, they can be involved in the development of the strategic plan, despite the fact that at the SME level very few have a planning – Many times customers buy products or services because of the brand because they identify with the values of the company, but other times, being products or services of excellent quality, not having a strong image, the customer discards the purchase – SMEs generally have low profitability or are in equilibrium at most, and need to increase sales volume to survive and prevent debt from growing, so it is necessary for them to have loyal customers to maintain their profitability

6 Conclusions

Based on the theoretical background of the two variables studied, an articulation between models for the Organizational Culture variable and dimensions for Corporate Social Responsibility is conceived. Therefore, it is ideal to work under the proposal of the Daniel Denison model for comparative studies with the CSR variable, taking into account that its attributes are more explicit, detailed and with an internal and external perspective.

For the Organizational Culture variable, it is evident that its definition has evolved, ceasing to be interpreted only as symbols, ceremonies and myths transmitted to collaborators, but rather it has become a structured, integrated system that shares with human talent involved and motivated to achieve of institutional objectives. There are companies that become a benchmark in the environments where they settle, making them attractive not only for consumers who would like to have the products or services they offer; but, in addition, it motivates many applicants to be part of their payrolls.

Both the organizational culture variable and social responsibility are business conditions that despite having their own elements, dimensions, regulations; they are adapted to the contexts where the institutions are established, together with the requirements of the stakeholders. Analyzing the variables independently or looking for causalities between them, it is inferred that the organizational culture is the fundamental pillar for the practical application of CSR since it involves, among other elements, values, beliefs, lifestyles that, internalized throughout the human talent, daily actions become innate processes in each entity.

A weakness detected in the present investigation is the scarce study between both variables, mostly the information was obtained independently, so it is necessary to consider inquiring about them, since one is a complement to the other within the business field, either for large-scale organizations or for the so-called SMEs, particularly the latter, the findings that can be obtained would allow them to strengthen their economies and change the history of many Latin American countries that, apparently due to their results, have stopped at the time.

References

Aristimuno, M.; Velásquez, Y., Rodríguez, C.: Corporate social responsibility (CSR), as an organizational value in Small and Medium Industry (SME). Eighth LACCEI Latin American and Caribbean Conference for Engineering and Technology. Arequipa, Peru (2010)

Barney, J.B.: Organizational culture: can it be a source of sustained competitive advantage? Acad. Manag. Rev. **11**(3), 656 (1986). https://doi.org/10.2307/258317

Bust, L.M.: Master's Thesis: Transmission of Organizational Culture, a challenge for today's SMEs. National Technological University (2018). https://ria.utn.edu.ar/bitstream/handle/20.500.12272/3274/BUSTO.pdf?sequence=1&isAllowed=y

Cameron, W., Quinn, R.E.: Diagnosing and Changing Organizational Culture: Based on the Competing Values Framework, 3rd edn. Jossey-Bass (2011)

European Comission. GREEN PAPER - Promote a European framework for corporate social responsibility (2001). https://ec.europa.eu/transparency/documentsregister/detail?ref=COM(2001)366&lang=es

Costin, H. Strategies for Quality Improvement: Tqm, Reengineering, and ISO 9000, 2nd edn. The Dryden Press (1999)

Denison, D.R.: What is the difference between organizational culture and organizational climate? A native's point of view on a decade of paradigm wars. Acad. Manag. Rev. **21**(3), 619–654 (1996). https://doi.org/10.5465/amr.1996.9702100310

Garcia Vargas, O.: Human culture and its interpretation from the perspective of organizational culture. Thought and Manage. **22**, 143–154 (2011). https://rcientificas.uninorte.edu.co/index.php/pensamiento/article/view/3531

Hofstede, G.: Cultures and Organizations: Software of the Mind Intercultural Cooperation and Its Importance for Survival (Revised ed.). McGraw-Hill (1997)

Huang, H.: Investigation of the fit among current and preferred organizational cultures, personality styles, and job attitudes in employees of international tourist hotels in Taiwan. In Idaho State University. ProQuest Dissertations Publishing (2005). https://www.proquest.com/openview/be20260affffba193670e2087261a53/1?pqorigsite=gscholar&cbl=18750&diss=y

Miron, E., Erez, M., Naveh, E.: Do personal characteristics and cultural values that promote innovation, quality, and efficiency compete or complement each other? J. Organ. Behav. **25**(2), 175–199. https://doi.org/10.1002/job.237

Ouchi, W.G.: Theory Z: How American Business Can Meet the Japanese Challenge. Addison-Wesley Pub (Sd) (1981)

Porter, M.: Competitive strategy: techniques for analyzing industries and competitors (Free Press) (1980)

Porter, M., Kramer, M.: Strategy and society. The link between competitive advantage and corporate social responsibility. Harvard Bus. Rev. **84**(12), 42–57 (2006). https://www.iarse.org/uploads/Porter_y_Kramer_Estrategia_y_Society_HBR_dic_2006.pdf

Ruiz, Y.B., Naranjo, J.C.: Research on organizational culture in Colombia: a look from the diffusion in scientific journals. Diversitas **8**(2), 285 (2013). https://doi.org/10.15332/s17949998.2012.0002.06

Salazar, J.G., Guerrero, J.C., Machado, Y.B., Cañedo, R.: Organizational climate and culture: two essential components in labor productivity. ACIMED **20**(4), 67–75 (2009). http://scielo.sld.cu/scielo.php?script=sci_abstract&pid=s1024-94352009001000004

Sepúlveda, M.E., Moreno, E.P., Tovar, J.E., Franco, J.A., Villarraga, C.L.: Social responsibility of SMEs in the post-conflict framework, as a support for the labor inclusion of the reinserted population. Commun. Culture Politics **6**(1), 45–62 (2015). https://journal.universidadean.edu.co/index.php/revistai/article/view/1335. (2015)

Schein, E.H., Schneider, B., Mason, R.O., Mitroff, I.: Organizational Culture and Leadership: A Dynamic View (The Jossey-Bass Business & Management Series), 1st edn. Jossey-Bass (2010)

Technical Economy Management

Inflation in Different Economies: A Brief Analysis

Andrea Suárez López[1], Priscilla Massa-Sánchez[2], Nelson Becerra Arevalo[1], and Héctor F. Gómez A[3(✉)]

[1] Docente de la Carrera de Medicina de la Universidad Regional Autónoma de los Andes (UNIANDES), Ibarra, Ecuador
{ua.andreasl01,up.nelsonbecerra}@uniandes.edu.ec
[2] Departamento de Economía, Universidad Técnica Particular de Loja, Loja, Ecuador
pmassa@utpl.edu.ec
[3] Centro de Posgrados, Universidad Tecnica de Ambato, Ambato, Ecuador
Hf.gomez@uta.edu.ec

Abstract. The steady rise in the average price of goods and services over time in an economy is known as inflation, and it is a key economic indicator. It happens frequently and regularly in the economy, and both individuals and the economy as a whole may be significantly impacted by its repercussions. We must gather information on both variables Inflation Ecuador-Chile, Ecuador-Venezuela in order to perform a linear regression analysis to look at the relationship between inflation and the economy, using Linear Regression, Kmeans and Hierarchy Cluster. The analysis' findings can offer important new perspectives on the variables that affect inflation and the economic circumstances that cause changes in inflation. Making informed judgments about the economy and fostering economic stability require an understanding of the connection between inflation and the overall economy.

Keywords: Inflation · Economy · Analysis

1 Introduction

Inflation is a crucial economic indicator that refers to the sustained increase in the general price level of goods and services in an economy over a period of time. It is a common and normal occurrence in the economy, and its effects can have far-reaching impacts on both individuals and the economy as a whole (Zheng, Wan, & Huang 2023).

Inflation can be caused by various factors, including an increase in demand for goods and services, an increase in production costs, or a decrease in the supply of money. Some of the most common causes of inflation include (Eti, Dinçer, Gökalp, Yüksel, & Kararoğlu 2023):

- Demand-pull inflation: This occurs when the demand for goods and services in an economy exceeds the supply, causing prices to rise.
- Cost-push inflation: This occurs when the cost of producing goods and services increases, leading to higher prices.

C. Montenegro et al. (Eds.): ICMTT 2023, LNNS 774, pp. 209–217, 2024.
https://doi.org/10.1007/978-3-031-43733-5_19

- Monetary inflation: This occurs when there is an increase in the money supply, causing prices to rise.
- Inflation can have both positive and negative effects on the economy. Some of the most significant effects include:
- Increased economic activity: Inflation can lead to increased economic activity as consumers and businesses respond to higher prices by increasing their spending.
- Increased borrowing costs: Inflation can lead to higher interest rates, making borrowing more expensive and potentially slowing down the economy.
- Decreased purchasing power: Inflation can decrease the purchasing power of consumers as the value of their money decreases.
- Increased uncertainty: Inflation can increase uncertainty in the economy, making it difficult for businesses and consumers to make plans and investments.

The most widely used measure of inflation is the Consumer Price Index (CPI), which measures the change in the price level of a basket of goods and services consumed by households. Other measures of inflation include the Producer Price Index (PPI) and the Gross Domestic Product deflator (GDP deflator) (Karanfil & Omgba 2023).

Ecuador is a South American country located on the Pacific coast. Despite being rich in natural resources, the country has faced persistent economic challenges, including high levels of inflation. Inflation in Ecuador has been a significant concern for policymakers and citizens alike, as it affects the purchasing power of the population, hinders economic growth, and creates uncertainty in the economy (Eti et al. 2023). Ecuador's inflation has been driven by a variety of factors, including:

- Monetary expansion: The central bank of Ecuador has pursued expansionary monetary policies in an effort to stimulate the economy, which has resulted in an increase in the money supply and higher inflation.
- Fiscal policy: The government has increased spending to fund various social programs and infrastructure projects, which has led to higher prices and inflation.
- Currency devaluation: Ecuador has experienced significant currency devaluation in recent years, which has led to higher import prices and inflation.
- External factors: External factors, such as changes in commodity prices, can also affect inflation in Ecuador.
- The high levels of inflation in Ecuador have had various effects on the economy, including:
- Decreased purchasing power: Inflation has reduced the purchasing power of consumers, as the value of their money has decreased.
- Increased uncertainty: High levels of inflation have created uncertainty in the economy, making it difficult for businesses and consumers to make plans and investments.
- Reduced investment: The high levels of inflation have made investment in the country less attractive, as returns on investments may be lower due to inflation.
- Widening income disparities: Inflation can have a disproportionate effect on the poor, as they tend to spend a larger share of their income on basic necessities, which become more expensive during periods of inflation.

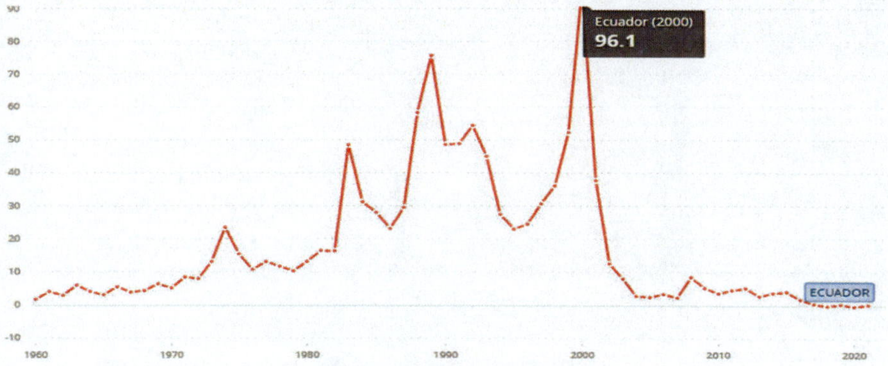

Fig. 1. Ecuador inflation[1]

In Ecuador, the National Institute of Statistics and Census (INEC) is responsible for measuring inflation using the Consumer Price Index (CPI). The CPI measures the change in the price level of a basket of goods and services consumed by households (Fig. 1).

Ecuador's annual inflation rate increased from a six-month low of 3.64% in November 2022 to 3.74% in December 2022. Some CPI items, including food and non-alcoholic drinks (7.60% versus 7.57% in November), education (5.04% vs 4.94%), and restaurants and hotels (3.70% vs 3.44%), exerted upward pressure. Consumer prices increased by 0.16% on a monthly basis after being nearly flat the previous month (Cajas Barba 2023; del Cioppo Morstadt 2023; Flores, Ulcuango, & Núñez 2023).

Food and non-alcoholic beverages make up 25% of the consumer price index in Ecuador, followed by transportation (14%), housing, water, electricity, gas, and other fuels (10%), clothing and footwear, and transportation (9%). 7% goes to restaurants and hotels, 7% to education, 6% to furniture and household goods, and 6% to health care. Other sectors include communications (4.4%), alcoholic beverages, tobacco, and illegal drugs (4.9%), leisure and culture (5.9%), other goods and services (5.2%), and other goods and services (1%). As of 2004, the national index's base value was 100. The CPI reflects the prices and spending trends recorded in eight sample cities, accounting for 67.44% of the total: Quito, Guayaquil, Cuenca, Machala, Esmeraldas, Manta, Ambato, and Loja (Shina 2023; Torres-Quezada, Sánchez-Quezada, & Vélez-Romero 2023).

Ecuador's annual inflation rate increased from a six-month low of 3.64% in November 2022 to 3.74% in December 2022. Some CPI items, including food and non-alcoholic drinks (7.60% versus 7.57% in November), education (5.04% vs 4.94%), and restaurants and hotels (3.70% vs 3.44%), exerted upward pressure. Consumer prices increased by 0.16% on a monthly basis after being practically flat the previous month.

For the third consecutive month, Ecuador's annual inflation rate decreased to 3.64% in November 2022 from 4.02% in October 2022. Due in large part to the decline in transportation costs (2.89% vs. 5.11% in May), it was the lowest reading since that month (Fig. 2).

[1] https://data.worldbank.org/indicator/FP.CPI.TOTL.ZG?end=2021&locations=EC&start=1960&view=chart.

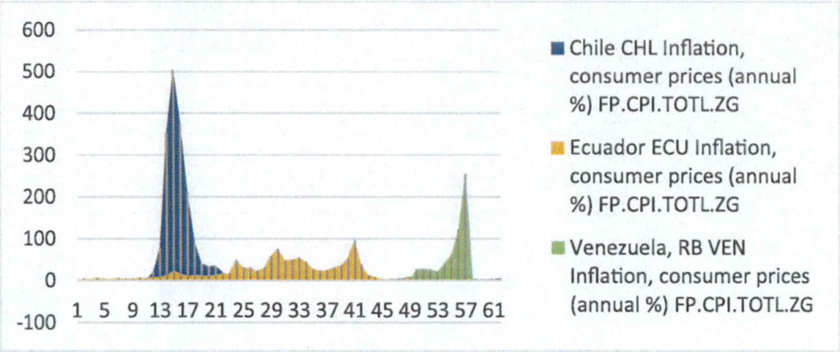

Fig. 2. Inflation Ecuador, Chile and Venezuela

Despite this, the country has experienced periods of high inflation, particularly in the 1970s and 1980s. In recent years, inflation in Chile has been relatively low, and the country has implemented various monetary and fiscal policies to maintain low levels of inflation and promote economic stability (Piscopo & Siavelis 2023).

Venezuela is a country located in the northern coast of South America. It is widely considered to be one of the most oil-rich countries in the world. Despite this, the country has experienced significant economic challenges, including hyperinflation, in recent years. The high levels of inflation in Venezuela have been driven by various factors, including monetary expansion, fiscal policy, currency devaluation, and external factors. The high levels of inflation have reduced the purchasing power of consumers, created uncertainty in the economy, and made investment less attractive (Mendoza 2023; Piscopo & Siavelis 2023).

To examine the relationship between inflation and the economy using linear regression analysis, we need to collect data on both variables. The data can be collected from Fig. 1 and Fig. 2 resources. Once the data has been collected, it can be analyzed using excel. Excel calculate the regression equation and provide various statistical measures, including the R-squared value, which measures the strength of the relationship between the two variables (Abbas, Kazeem, Abdulmalik, & Abdussalam 2023).

K-means is a popular machine learning technique used for clustering and classification. In this paper, we will examine the relationship between inflation and the economy using K-means clustering analysis (Shafi 2023; Wardhani, Rukmi, & Ardamayanti 2023).

The results of the K-means clustering analysis can be interpreted by examining the resulting clusters and the patterns in the data. The clusters can provide valuable insights into the relationships between inflation and the economy, such as the economic conditions that lead to changes in inflation. The patterns in the data can also provide valuable insights into the factors that influence inflation, such as monetary policy and economic growth.

Hierarchical clustering is a machine learning technique used for clustering data into a hierarchical structure. In this paper, we will examine the relationship between inflation and the economy using hierarchical clustering analysis (Liu, Ye, & Jianwei 2023; Ngo, Van Nguyen, Nguyen, Tram, & Hoang 2023).

2 Methodology

Linear regression analysis can be used to examine the relationship between inflation and the economy by modeling the relationship as a linear equation. The regression equation can be written as follows:

$$Y = \beta 0 + \beta 1X \quad (1)$$

where Y is the dependent variable (inflation), X is the independent variable (the economy), β0 is the intercept, and β1 is the slope of the regression line. The slope represents the change in the dependent variable (inflation) for a unit change in the independent variable (the economy).

K-means is a clustering algorithm that partitions a set of data into k clusters, where k is a user-defined parameter. The algorithm works by iteratively assigning each data point to the nearest cluster centroid and updating the cluster centroids based on the mean of the data points assigned to the cluster. The algorithm repeats this process until the cluster centroids no longer change.

Hierarchical clustering is a clustering algorithm that creates a hierarchical structure of clusters by iteratively merging smaller clusters into larger clusters. The algorithm starts with each data point as a single cluster and merges the closest clusters until all data points are part of a single cluster or a stopping criterion is met. The resulting hierarchy of clusters can be represented as a dendrogram, which visualizes the relationships between the clusters.

3 Experimentation

The results of the linear regression analysis can be interpreted by examining the slope and intercept of the regression equation, as well as the R-squared value. The slope represents the change in inflation for a unit change in the economy, while the intercept represents the level of inflation when the economy is at zero. The R-squared value measures the strength of the relationship between inflation and the economy, with higher values indicating a stronger relationship (Tables 1 and 2).

Table 1. Regression: Ecuador-Chile

Mult. Coef. Correl.	0,285888898
R^2	0,081732462
R^2 Tigth.	-1,25
Typical Error	1,060158499

Table 2. Regression: Ecuador-Venezuela

Mult. Coef. Correl.	0,46974545
R^2	0,22066
R^2 Tigth.	1,33 333
Typical Error	1,45557

Table 3. Cluster inflation: Ecuador-Chile

		Cluster #	
Attribute	**Full Data**	**0,00**	**1,00**
	(51.0)	**(3.0)**	**(48.0)**
Ecuador	41,36	410,76	18,27
Chile	20,53	17,23	20,74

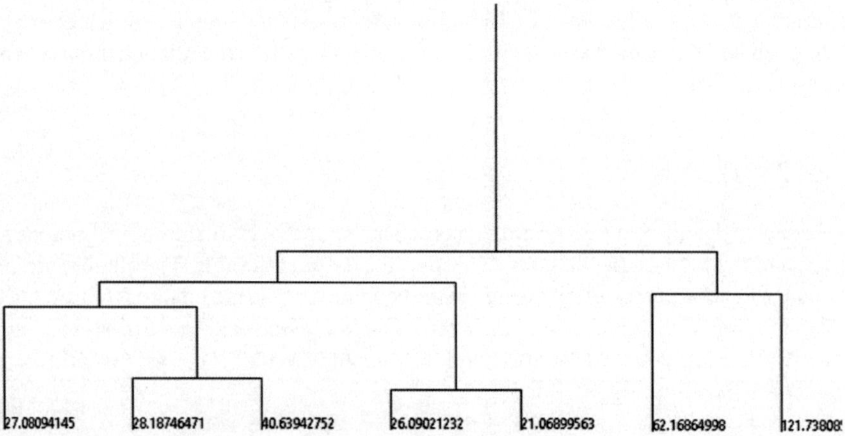

Fig. 3. Hierarchy cluster: Ecuador-Venezuela

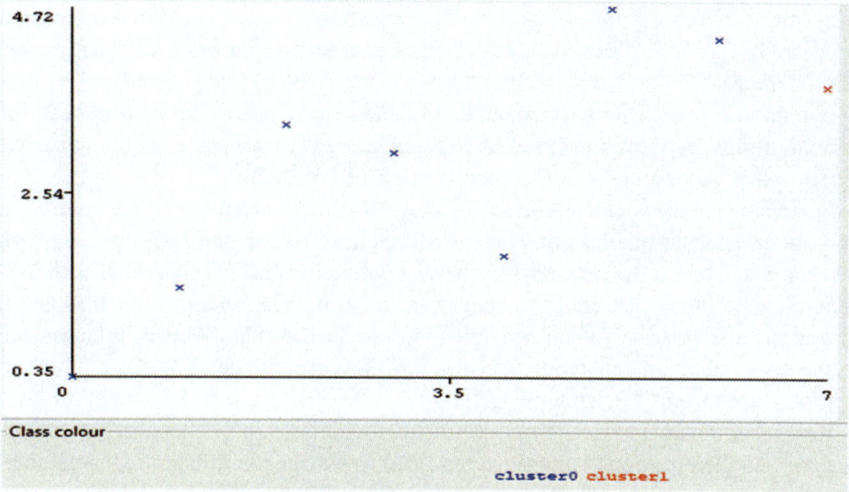

Fig. 4. Assignment cluster: Ecuador-Venezuela

4 Conclusion

Inflation is a normal and crucial economic indicator that can have far-reaching impacts on both individuals and the economy as a whole. Understanding the causes and effects of inflation, as well as the methods used to measure it, is essential for making informed economic decisions.

Inflation has been a persistent challenge for Ecuador's economy, affecting the purchasing power of the population, hindering economic growth, and creating uncertainty in the economy. Policymakers and citizens alike must be aware of the causes and effects of inflation, as well as the methods used to measure it, in order to make informed decisions and promote economic stability.

Inflation is a critical economic indicator that affects countries in different ways. Ecuador, Chile, and Venezuela are three South American countries with diverse economic and political systems that have experienced different levels and causes of inflation. Understanding the causes and effects of inflation, as well as the methods used to measure it, is essential for making informed economic decisions and promoting economic stability.

Linear regression analysis and Kmeans can be used to examine the relationship between inflation and the economy. The results of the analysis can provide valuable insights into the factors that influence inflation and the economic conditions that lead to changes in inflation.

Understanding the relationship between inflation and the economy is essential for making informed economic decisions and promoting economic stability. For the present case, no relationship is established between the countries under analysis with linear regression.

Although it calls the case of a certain closeness between Ecuador and Venezuela (0.46), it should be noted that the small amount of existing data on the second country can result in this data (Tables 1 and 2). The variation r2 is not explained between the countries, in fact it is close to 0, which implies a low explanation of what happens between them.

In inflation clusters, it may turn out that Ecuador and Chile at certain times can form regional groups, but not in the case of Venezuela, which marches at its own pace with respect to this variable in the hierarchical cluster (Table 3, Fig. 3).

Results of the hierarchical clustering analysis can be interpreted by examining the dendrogram and the relationships between the clusters. The dendrogram can provide valuable insights into the relationships between inflation and the economy, such as the economic conditions that lead to changes in inflation. The relationships between the clusters can also provide valuable insights into the factors that influence inflation, such as monetary policy and economic growth. For this case, the dendrogram only inherits inflation values from Venezuela (Fig. 4).

Results of the analysis can provide valuable insights into the factors that influence inflation and the economic conditions that lead to changes in inflation. Understanding the relationship between inflation and the economy is essential for making informed economic decisions and promoting economic stability.

References

Abbas, U., Kazeem, E., Abdulmalik, M., Abdussalam, M.: The impact of using robust mm-estimator in regression models: an application of nigerian inflation on the economic growth. Int. J. Sci. Appl. Res. (ISSN: 2504–9070) **5**(2), 66–74 (2023)

Cajas Barba, M.S.: El tamaño óptimo del gasto público y su incidencia en el crecimiento económico en el Ecuador, período 2000–2021. Universidad Nacional de Chimborazo, Riobamba (2023)

del Cioppo Morstadt, J.: Impacto económico del SARS-COV2 en las exportaciones del banano ecuatoriano. ECONÓMICAS CUC **44**(1), 31–46 (2023)

Eti, S., Dinçer, H., Gökalp, Y., Yüksel, S., Kararoğlu, D.: identifying key issues to handle the inflation problem in the healthcare industry caused by energy prices: an evaluation with decision-making models. Managing Inflation and Supply Chain Disruptions in the Global Economy 162–178 (2023)

Flores, C.O.G., Ulcuango, O.M.R., Núñez, I.M.: The behavior of spending and its financial impact on the consumer of basic basket products in Ecuador. J. Positive Psychol. Wellbeing 251–273 (2023)

Karanfil, F., Omgba, L.D.: The energy transition and export diversification in oil-dependent countries: the role of structural factors. Ecol. Econ. **204**, 107681 (2023)

Liu, J., Ye, J., Jianwei, E.: A multi-scale forecasting model for CPI based on independent component analysis and non-linear autoregressive neural network. Phys. A **609**, 128369 (2023)

Mendoza, E.G.: Monetary Policy Frameworks in Latin America: Evolution, Resilience and Future Challenges: A Discussion. Paper presented at the MACRO-FINANCIAL STABILITY POLICY IN A GLOBALISED WORLD: LESSONS FROM INTERNATIONAL EXPERIENCE: Selected Papers from the Asian Monetary Policy Forum 2021 Special Edition and MAS-BIS Conference (2023)

Ngo, V.M., Van Nguyen, P., Nguyen, H.H., Tram, H.X.T., Hoang, L.C.: Governance and monetary policy impacts on public acceptance of CBDC adoption. Res. Int. Bus. Financ. **64**, 101865 (2023)

Piscopo, J.M., Siavelis, P.M.: Chile's constitutional chaos. J. Democr. **34**(1), 141–155 (2023)

Shafi, M.A.: K-means clustering analysis and multiple linear regression model on household income in Malaysia. Int. J. Artif. Intell. **12**(2), 731–738 (2023)

Shina, B.: Economic Precariousness and Political Ownership of Media in Nigeria: Implications for Democratic Consolidation in Nigeria Media Ownership in Africa in the Digital Age, pp. 119–142. Routledge (2023).

Torres-Quezada, J.E., Sánchez-Quezada, T., Vélez-Romero, G.: Construction Development, Economic Evolution, and Environmental Impact in Ecuador. In: Torres-Quezada, J.E. (eds.) Energetic Characterization of Building Evolution. Green Energy and Technology, pp. 79–100. Springer, Cham (2023). https://doi.org/10.1007/978-3-031-21598-8_3

Wardhani, L.P., Rukmi, A.M., Ardamayanti, T.F.: Investigating the Association Between Carbon Dioxide (CO2) emissions and energy consumption on the economic growth of clusterized countries in the world. Euro. J. Pure Appl. Math. **16**(1), 657–669 (2023)

Zheng, Z., Wan, X., Huang, C.-Y.: Inflation and income inequality in a Schumpeterian economy with heterogeneous wealth and skills. Econ. Model. 106193 (2023)

Microcredit, Gender and Geography. Case of Banecuador

Aurora Samaniego-Namicela(✉) and Reinaldo Armas-Herrera

Universidad Técnica Particular de Loja, Loja, Ecuador
{afsamaniego3,ahreinaldo}@utpl.edu.ec

Abstract. This article investigates the factors to access to credit of the largest public bank in Ecuador. The methodology includes a logistic regression model to a database corresponding to loan applications from associations of the Popular and Solidarity Economy of Ecuador during the period 2016–2019. The estimates show that the unsecured guarantee, belonging to the tertiary sector, being located in a rural environment and the sex of the loan officer influence the access to financing. In the first place, it was found that the businesses that belong to the tertiary sector show a higher statistically significant probability for the granting of credit, this is in accordance with the business structure of Ecuador where most of the companies are part of this sector.

Keywords: Credit granting · gender · geographical area

1 Introduction

Microfinances are intended to serve clients in situations of poverty, geographically dispersed and historically excluded from traditional banking [3, 10, 16, 26].

Rural microfinance programs focus their objectives mainly on women with the use of various alternatives in terms of guarantees, requirements and mobility [14]. In addition, women manage their income better than men; that is, the income of women is low but permanent, unlike men whose income is higher, but not frequent [15], which allows women to have greater access to financial products [28, 30].

This characteristic makes women obtain low amounts of credit for agricultural activities, considering that in rural environments women face subordination and are engaged in agriculture and household chores [20, 27]. Women's unpaid domestic work limits their achievement of economic freedom.

Access to microcredit for rural women has a positive influence on income, savings, job creation, which allows them to meet their needs and those of their families, improving their living conditions [1, 2].

There are studies that indicate that microfinance programs are oriented towards the empowerment of women, transferring autonomy to decide, build trust, reduce gender gaps, fight poverty and generate social capital [16, 24, 26].

Within the credit evaluation, management and monitoring process, loan officers show aversion to risk, with women loan officers being the ones who show greater risk aversion

C. Montenegro et al. (Eds.): ICMTT 2023, LNNS 774, pp. 218–228, 2024.
https://doi.org/10.1007/978-3-031-43733-5_20

and less confidence. This position mainly affects new clients who have limited financial information [4, 6, 25].

The purpose of this work is to analyze the factors that affect access to credit, among the factors are the amount, type of guarantee, economic sector, belonging to the rural sector, sex of the borrower, sex of the territorial manager and sex of the loan officer.

The document is structured as follows: Sect. 2 presents the literature review, Sect. 3 describes data, variables and methodology, Sect. 4 details the results and Sect. 5 establishes conclusions.

2 Literature Review

Microfinance clients do not have financial information on their businesses, so microfinance institutions have implemented methodologies to obtain this information with on-site visits [11]. These visits make it possible to generate an intense relationship between the client and the loan officer [7, 10, 12]. This interaction allows a certain relationship to be built that affects the granting of credit. This panorama is more evident in clients from the rural sectors mainly.

In the process of interaction, loan officers generate a certain affinity with clients that influences the decision to grant credit. The relationship is usually given by social, cultural factors, risk attitude, human capital, gender, among others [4, 6, 8, 17].

For [8], loan officers with higher human capital in terms of education, banking experience and credit experience are attracted to clients with a high level of human capital. For their part, [17] point out that the clients that are favored by loan officers are those that belong to the same social network.

Loan officers perform better when transacting for clients of the same gender [5]. The loan officer must integrate the hard and soft information for the evaluation of the credit application [24]. When loan officers are women, they tend to select women clients, offer larger loans, longer terms, fewer guarantees, and lower interest rates. [4, 5, 9].

The determinants of access to microcredit are the client's sex [25, 28, 30, 31], geographic sector, economic sector and guarantees [21, 22]. As [13] points out, the amount of credit directed to women is lower compared to male clients and this does not allow them to venture into businesses that generate high income. The women who received loans mostly used them for businesses in the trade sector. For microenterprises in the manufacturing sector, they are the ones with the greatest access to credit and if a young man is managed, it is much better, [19].

3 Data, Variables and Methodology

The data for this work was provided by BanEcuador, which is a public banking in the Republic of Ecuador. The data corresponds to 3,290 credit operations of the Popular and Solidarity Economy associations during the 2016–2019 period.

The variables used in this work are:

Credit (Cred): It is a dichotomous variable that takes the value of 1 if the credit is granted and 0 if not. It is the variable to be explained.

Amount: It is the amount or value of the credit.
Guarantee class (C_garan): It is a discrete variable that takes the value of 1 if the guarantee is mortgage, 2 if it is unsecured and 3 if it is personal.
Sector (Sec): It is a discrete variable that takes the value of 1 if the association belongs to the primary sector, 2 to the secondary and 3 to the tertiary.
Rurality (Rural): It is a dichotomous variable that has the value of 1 if the place where the association is located is rural and 0 if it is urban.
Borrower sex (S_p): It takes the value of one if it is female and 0 if it is male.
Manager sex (S_g): It takes the value of one if it is female and 0 if it is male.
Official sex (S_o): It takes the value of one if it is female and 0 if it is male.

The hypotheses to be tested are the following:

H_1: The amount is associated with a greater probability of receiving the credit from BanEcuador.
H_2: The type of guarantee is associated with a greater probability of receiving the credit.
H_3: The sector to which the association belongs influences the granting of credit by BanEcuador.
H_4: The association's belonging to the rural sector influences the granting of credit from the public entity.
H_5: The sex of the borrower influences when granting credit in the public bank.
H_6: The sex of the territorial manager influences when granting credit in the public bank.
H_7: The sex of the loan officer influences when granting credit in the public bank.

The econometric equation to be estimated is the following:

$$\begin{aligned} Cred_i = \beta_0 + \beta_1 Amount_i + \beta_2 C\_garan_i + \beta_3 Sec_i + \beta_4 Rural_i + \beta_5 S\_p_i + \\ \beta_6 S\_g_i + \beta_7 S\_o_i + \varepsilon_i \end{aligned} \tag{1}$$

Where i represents the ith loan and ε_i the estimation error.

Due to the nature of the dependent variable, a logistic regression such as [18, 23, 29], among others, will be used. To make the estimates, the statistical software JASP 0.17.1 from the University of Amsterdam was used.

4 Results

Figure 1 indicates that the bank grants 51 percent of the credits requested in the 2016–2019 period.

Table 1 shows the correlations of the study variables, there being a certain correlation between the economic sector and credit, due to the fact that the productive activities to which the credit is intended have different risks that influence the granting of the loan. The granting of credit has a positive correlation with the sex of the client, since microfinances aim to provide loans especially to women. The granting of credit has a negative correlation with the rural sector, since fewer credits tend to be granted to clients in the rural sector.

There is a negative relationship between the amount and the guarantee, considering that microfinance clients who do not have accounting and financial information on their

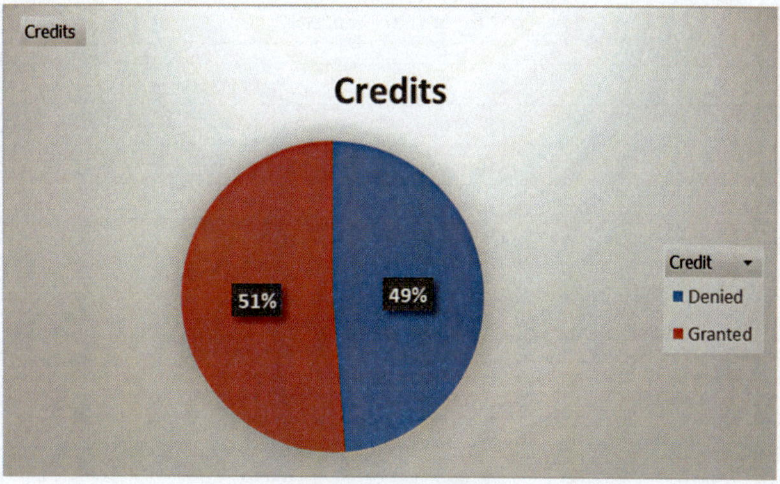

Fig. 1. Granted credits.

businesses represent a high risk for the institution, therefore, the requirements for obtaining the first credit are more rigorous, the requirements are decreasing for the following credits, when the customer's payment history has been generated.

The guarantee and the sex of the client are related, a greater guarantee is required to grant credit due to the risk that certain businesses managed by women represent, usually women are engaged in businesses that generate less income than men.

Table 1. Correlations.

			Pearson's r	
Cred	-	Amount	–0.064	***
Cred	-	C_garan	0.017	
Cred	-	Sec	0.205	***
Cred	-	S_p	0.112	***
Cred	-	S_g	–0.020	
Cred	-	S_o	0.089	***
Cred	-	Rural	–0.104	***
Amount	-	C_garan	–0.169	***
Amount	-	Sec	–0.148	***
Amount	-	S_p	–0.108	***
Amount	-	S_g	0.005	

(*continued*)

Table 1. (*continued*)

			Pearson's r	
Amount	-	S_o	−0.054	*
Amount	-	Rural	0.060	***
C_garan	-	Sec	0.104	***
C_garan	-	S_p	0.158	***
C_garan	-	S_g	−0.002	
C_garan	-	S_o	0.110	***
C_garan	-	Rural	−0.051	**
Sec	-	S_p	0.273	***
Sec	-	S_g	0.063	***
Sec	-	S_o	0.075	**
Sec	-	Rural	−0.252	***
S_p	-	S_g	0.032	
S_p	-	S_o	0.107	***
S_p	-	Rural	−0.097	***
S_g	-	S_o	0.045	
S_g	-	Rural	−0.021	
S_g	-	Rural	−0.016	

* $p < .05$, ** $p < .01$, *** $p < .001$

Table 2 shows the global regression without considering the geographical area where the credit transaction was generated. Next, the hypotheses are tested at the country level, the variables that influence the granting of credit are: unsecured guarantee, being part of the tertiary sector, belonging to a rural environment and the sex of the loan officer.

Hypothesis 1 is rejected because the amount is not significant, hypothesis 2 is not rejected because the coefficient of the unsecured guarantee variable is significant. Although the unsecured guarantee ensures payment of the debt only with the borrower's signature, this means that the institution trusts its clients, a fundamental principle in microfinance, where by the very nature of the credit granting process it favors to establish relationships of trust between the institution and clients. Most of the companies in the country belong to the tertiary sector, in addition to having few barriers to entry, therefore, this sector has an influence on the granting of credit, so hypothesis 3 is not rejected.

Hypothesis 4 is not rejected. The bank's objective clients are natural and legal persons from the marginal rural and urban sectors of the country, which is in relation to the postulates of microfinance where microcredit is granted to vulnerable people mainly from rural sectors who they do not have a guarantee and have barriers to access traditional banking. Hypotheses 5 and 6 are rejected because the gender of the borrower and the gender of the manager are not relevant when granting credit.

Hypothesis 7 is not rejected, credit officers play a fundamental role in granting credit, since they are the ones who carry out on-site visits to generate in some cases financial information on clients, necessary in the credit application, in addition to generates a client-institution (loan officer) relationship.

Table 2. Global regression.

					Wald Test		
	Estimate	Robust Standard Error	Odds Ratio	z	Wald Statistics	df	p
(Constant)	−0.319	0.180	0.727	−1.777	3.566	1	0.076
Amount	−0.000	0.000	1.000	−0.318	0.160	1	0.750
C_garan (chirographical)	0.407	0.157	1.502	2.586	7.428	1	0.010
C_garan (personal)	0.365	0.217	1.441	1.687	3.049	1	0.092
Sec (secondary)	0.187	0.260	1.206	0.719	0.517	1	0.472
Sec (tertiary)	1.066	0.139	2.902	7.689	61.073	1	< .001
Rural	−0.279	0.131	0.757	−2.132	4.776	1	0.033
S_p	0.206	0.130	1.229	1.588	2.562	1	0.112
S_g	−0.034	0.126	0.967	−0.266	0.071	1	0.790
S_o	0.313	0.131	1.368	2.394	5.600	1	0.017

Note. Cred 'granted' level coded as class 1.

Table 3 shows the credit granting regression when the association is on the coast. Next, we test the hypotheses in the coastal region, the variables that influence the granting of credit are: being part of the tertiary sector, belonging to a rural environment and the gender of the territorial manager.

Hypotheses 1 and 2 are rejected because the amount and the guarantee are not relevant when granting the credit. Hypothesis 3 is not rejected, the sector to which the business belongs has a certain risk which influences the granting of the loan. Hypothesis 4 is not rejected, the clients are in the rural sector, who sometimes do not have the financial information of the business, and the difficulty of going to the financial institution in person.

Hypothesis 5 is rejected because the sex of the borrower does not determine the probability of granting credit in the coastal area. Hypothesis 6 is not rejected since the territorial managers play an important role because among their functions is the on-site visit to the potential client, during the visit they conduct a survey that collects organizational, productive, market and financial information, in addition to generate skills in business management in potential customers. Hypothesis 7 is rejected because the sex of the loan officer does not influence when granting credit.

Table 3. Coast regression.

					Wald Test		
	Estimate	Robust Standard Error	Odds Ratio	Z	Wald Statistics	df	p
(Constant)	0.213	0.229	1.237	0.930	0.928	1	0.353
Amount	−0.000	0.000	1.000	−0.424	0.305	1	0.672
C_garan (chirographical)	−0.252	0.222	0.777	−1.135	1.408	1	0.256
C_garan (personal)	−0.416	0.291	0.659	−1.431	2.120	1	0.152
Sec (secondary)	0.126	0.379	1.134	0.332	0.104	1	0.740
Sec (tertiary)	0.818	0.186	2.267	4.408	19.729	1	< .001
Rural	−0.366	0.183	0.694	−2.004	4.060	1	0.045
S_p	0.199	0.189	1.220	1.052	1.132	1	0.293
S_g	0.469	0.192	1.599	2.446	6.000	1	0.014
S_o	0.210	0.183	1.233	1.143	1.282	1	0.253

Note. Cred 'granted' level coded as class 1.

Table 4 displays the regression when the association belongs to the mountain range zone. Next, the hypotheses are tested in the mountain region, the variables that influence the granting of credit are: being part of the tertiary sector, the unsecured guarantee, personal guarantee, gender of the territorial manager. Hypothesis 1 is rejected because the amount does not determine the granting of credit. Hypothesis 2 is not rejected; both the unsecured guarantee and the personal guarantee have a trust component of the financial institution towards its clients. Hypothesis 3 is not rejected, a business belonging to the tertiary sector has a higher probability of granting credit. Hypothesis 4 is rejected because belonging to rural areas in the mountain range does not determine the granting of credit. The sex of the borrower is not relevant when receiving the credit (hypothesis 5 is rejected), although that of the manager is (hypothesis 6 is not rejected), but not that of the loan officer (hypothesis 7 is rejected).

Table 5 displays the regression of the eastern zone. In the eastern zone, the most significant variables are: amount, belonging to the tertiary sector. Hypothesis 1 is not rejected because the amount is relevant when granting a loan.

In the East, guarantees are not relevant when it comes to giving credit (hypothesis 2 is rejected), belonging to the tertiary sector is associated with a greater probability of credit (hypothesis 3 is not rejected). Belonging to the rural sector or the sex of the borrower, manager and officer are not relevant when granting a loan (hypotheses 4, 5, 6 and 7 are rejected).

The results are in line with the investigations of [4–7, 29], which indicate that the sex of the loan officer influences the granting of credit, this may be given by the risk tolerance levels, your desire to benefit the borrower, and your style in interacting with the

Table 4. Mountain range regression.

					Wald Test		
	Estimate	Robust Standard Error	Odds Ratio	z	Wald Statistics	df	p
(Constant)	−1.112	0.349	0.329	−3.187	10.442	1	0.001
Amount	−0.000	0.000	1.000	−0.266	0.037	1	0.790
C_garan (chirographical)	1.380	0.279	3.974	4.944	24.856	1	< .001
C_garan (personal)	1.874	0.434	6.512	4.322	20.136	1	< .001
Sec (secondary)	0.105	0.446	1.111	0.237	0.058	1	0.813
cod_sect (tertiary)	1.393	0.264	4.028	5.273	29.879	1	< .001
Rural	−0.260	0.237	0.771	−1.097	1.328	1	0.273
S_p	0.304	0.220	1.355	1.378	1.835	1	0.168
S_g	−0.531	0.213	0.588	−2.488	5.941	1	0.013
S_o	0.280	0.241	1.324	1.164	1.304	1	0.244

Note. Cred 'granted' level coded as class 1.

Table 5. East regression.

					Wald Test		
	Estimate	Robust Standard Error	Odds Ratio	z	Wald Statistics	df	p
(Constant)	−0.395	0.682	0.674	−0.579	0.365	1	0.563
Amount	−0.000	0.000	1.000	−2.403	4.802	1	0.016
C_garan (chirographical)	0.240	0.466	1.271	0.514	0.268	1	0.607
C_garan (personal)	−0.640	0.602	0.528	−1.063	0.947	1	0.288
Sec (secondary)	1.726	0.944	5.616	1.828	3.848	1	0.068
Sec (tertiary)	1.750	0.495	5.754	3.534	12.812	1	<.001
Rural	−0.202	0.433	0.817	−0.467	0.259	1	0.640
S_p	0.348	0.371	1.416	0.937	0.892	1	0.349
S_g	−0.493	0.385	0.611	−1.280	1.609	1	0.201
S_o	0.659	0.400	1.934	1.649	2.764	1	0.099

Note. Cred 'granted' level coded as class 1.

borrower. In addition, the results show that guarantees are a fundamental requirement when granting credit [20, 21]. In microfinance, guarantees include a trust component, which allows access to financing for people with a low economic level. Another outstanding result is that businesses belonging to the tertiary sector are more likely to access credit, this is due to the few entry barriers that certain activities belonging to the tertiary sector represent such as: commerce, small restaurants and beauty salons, among others, and finally that the business is located in the rural sector has a negative correlation in the granting of credit.

5 Results

The article found that the factors that influence access to credit in the largest public bank in the country are the guarantee, that the business belongs to the tertiary sector, that the business is located in a rural environment and the sex of the loan officer, considering that the unsecured guarantee is predominant, which is favoring access to financing for clients in vulnerable conditions. In addition, the business fabric of Ecuador is made up mostly of microenterprises from the tertiary sector, which means that this variable has significance. Another finding is that if the company is located in the rural sector, it has a negative correlation in granting credit, and the sex of the loan officer has a positive correlation in granting credit.

Differences are evident in terms of the coastal, highland and eastern areas. On the coast, the gender of the land manager who served the client before the loan and the rurality of the clients are of great importance; In the eastern region, the relevant variables in granting credit are the amount and that the clients belong to a rural environment, and in the mountain range, the variables that influence the granting of credit are the unsecured guarantee, the personal guarantee and the sex of the client.

The way is left open for future research that examines access to credit considering other variables.

References

1. Alnaa, S.: Microcredit to rural women, intra-household power play and employment creation in Northern Ghana. J. Interdiscip. Econ. **29**(2), 197–213 (2017)
2. Al-Shami, S., Razali, R., Rashid, N.: The effect of microcredit on women empowerment in welfare and decisions making in Malaysia. Soc. Indic. Res. **137**, 1073–1090 (2018)
3. Armendariz de Aghion, B., Morduch, J.: Microfinance beyond group lending. Econ. Transit. **8**(2), 401–420 (2000)
4. Beck, T., Behr, P., Guettler, A.: Gender and banking: are women better loan officers? Rev. Finan. **17**(4), 1279–1321 (2013)
5. Beck, T., Behr, P., Madestam, A.: Sex and credit: is there a gender bias in lending? SSRN Electron. J. **2013**, 1–49 (2012)
6. Bellucci, A., Borisov, A., Zazzaro, A.: Does gender matter in bank–firm relationships? Evidence from small business lending. J. Bank. Finance **34**(12), 2968–2984 (2010)
7. Blanco-Oliver, A., Reguera-Alvarado, N., Veronesi, G.: Credit risk in the microfinance industry: the role of gender affinity. J. Small Bus. Manage. **59**(2), 280–311 (2021)

8. Bruns, V., Holland, D., Shepherd, D., Wiklund, J.: The role of human capital in loan officers' decision policies. Entrep. Theory Pract. **32**(3), 485–506 (2008)
9. Buttner, E., Rosen, B.: Bank loan officers' perceptions of the characteristics of men, women, and successful entrepreneurs. J. Bus. Ventur. **3**(3), 249–258 (1988)
10. Canales, R., Greenberg, J.: A matter of (relational) style: loan officer consistency and exchange continuity in microfinance. Manage. Sci. **62**(4), 1202–1224 (2016)
11. Canales, R.: Rule bending, sociological citizenship, and organizational contestation in microfinance. Regul. Gov. **5**(1), 90–117 (2011)
12. Doering, L.: Risks, returns, and relational lending: personal ties in microfinance. Am. J. Sociol. **123**(5), 1341–1381 (2018)
13. Ganle, J.K., Afriyie, K., Segbefia, A.Y.: Microcredit: empowerment and disempowerment of rural women in Ghana. World Dev. **66**, 335–345 (2015)
14. Johnson, S., Nino-Zarazua, M.: Financial access and exclusion in Kenya and Uganda. J Devel. Stud. **47**(3), 475–496 (2011)
15. Johnson, S.: Gender norms in financial markets: evidence from Kenya. World Dev. **32**(8), 1355–1374 (2004)
16. Kah, J., Olds, D., Kah, M.: Microcredit, social capital, and politics: the case of a small rural town—Gossas, Sendgal. J. Microfinance/ESR Rev. **7**(1), 8 (2005)
17. Labie, M., Méon, P., Mersland, R., Szafarz, A.: Discrimination by microcredit officers: theory and evidence on disability in Uganda. Q. Rev. Econ. Finance **58**, 44–55 (2015)
18. Lalangui, P.S., Herrera, R.A., Jaramillo, S.E.R.: Female entrepreneurship and religious tourism in Ecuador. In: 2016 11th Iberian Conference on Information Systems and Technologies (CISTI), pp. 1–6. IEEE (2016, June)
19. León, J., Sánchez, M., Jopen, G.: Acceso y uso de microcréditos para la MYPE: la clave está en la diferenciación. Economía y sociedad **92**(6), 6–11 (2017)
20. Maclean, K.: Capitalizing on women's social capital? Women-targeted microfinance in Bolivia. Dev. Chang. **41**, 495–515 (2010)
21. Murcia Pabón, A.: Determinantes del acceso al crédito de los hogares colombianos. Ensayos sobre política económica **25**(55), 40–83 (2007)
22. Peñaloza, H.A.B.: Determinantes del acceso al crédito de las PYMES en Colombia. Ensayos de economía **25**(46), 135–156 (2015)
23. Peñarreta, M., Teijeiro, M., Armas, R.: Explorando las claves del éxito de las campañas de crowdfunding en Ecuador. Revista Ibérica de Sistemas e Tecnologias de Informação **E16**, 247–256 (2018)
24. Pérez, F.: Banca relacional y capital social en España: competencia y confianza. Fundacion BBVA/BBVA Foundation (2006)
25. Purmiyati, A., Setyowati, R., Rakhima, M.S.: The role of microcredit program and micro enterprises in poverty reduction. J. Ekonomi Malaysia **55**(2), 29–38 (2021)
26. Rad, A., Yazdanfar, D., Öhman, P.: Female and male risk aversion: an empirical study of loan officers' assessment of SME loan applications. Int. J. Gend. Entrepreneurship **6**(2), 121–141 (2014)
27. Rankin, K.: Social capital, microfinance, and the politics of development. Fem. Econ. **8**(1), 1–24 (2002)
28. Salas, E., Espinoza, V., Samaniego, A., Martínez-Fernández, V.: Inclusión financiera a través del crédito de desarrollo humano como herramienta para el fomento de capacidades y emprendimiento (2016)
29. Samaniego, A., Blanco-Oliver, A., Palacin-Sánchez, M.: Gender differences in loan portfolio management: evidence for the loan officer position in a development environment. Working paper (2022)

30. Samaniego, A., Espinoza, V., Salas, E., Armas, R.: Acceso al crédito en el Ecuador: caso de estudio provincia de Zamora Chinchipe. Opción: Revista de Ciencias Humanas y Sociales (18), 1479–1500 (2018)
31. Vizhñay, A., Samaniego, A.: Determinantes del acceso al crédito en el Ecuador. Revista Espacios **40**(13), 25–36 (2019)

Technical Innovation and Management

Empowering Collaborative Learning Processes in Educational Institutes

Rita Aguilar González[1], Janio Jadán-Guerro[2](✉), and Marcos Chacón-Castro[3,4]

[1] Doctora en Ciencias de la Educación, Docente de Posgrado en Educación y Derecho, Directora Académica del Centro Capacitador IMETAL y Evaluador de CONOCER, Jalisco, Mexico

[2] Centro de Investigación en Mecatrónica y Sistemas Interactivos (MIST), Universidad Tecnológica Indoamérica, Machala and Sabanilla, Quito, Ecuador
janiojandan@uti.edu.co

[3] Facultad Ciencias de la Educación - FACED, Maestría en Entornos Digitales, Universidad Indoamérica, Bolívar y Quito, Ambato, Ecuador
mchacon10@indoamerica.edu.ec

[4] Fundación Universitaria Internacional de La Rioja, Grupo Investigación GIECO Bogotá, Bogotá, Colombia
marcos.chacon@unir.net

Abstract. This research article proposes to illustrate the advantages of the collaborative learning process that motivates the creation of harmonious environments in the educational environment. In this analysis the educational institute is considered as a company. For this reason, the traditional educational model known as "Master Class" where the professor gives an explanation, and the student takes notes or simply listens the class to acquire knowledge is obsolete. Nowadays the education is more participative where the students have a main role in the learning process. The students collaborate in the learning process sharing knowledge, ideas and information as they are part of a team work. Moreover, leadership types such as the methodology of accompanying leadership which helps professors in their daily labors in aspects such as. Situational leadership, Empowerment, role games, shared vision and work tools. Finally, the results illustrate the advantages of the collaborative learning process which motivates the creation of harmonic environments in the educational environment.

Keywords: Leadership · collaborative work · situational leadership · empowerment · shared vision

1 Introduction

Leadership has a main role in any company or institution and it requires the use of different strategies to achieve the company or institutional expected goals. The fact that we live and work together in society makes that results are barely impossible to achieve individually. It is important to obtain group strategies to accomplish desirable outcomes.

This paper only focus to empower the leadership in the educational institutions. Normally, the goal of any educational institution is the formation and development of

C. Montenegro et al. (Eds.): ICMTT 2023, LNNS 774, pp. 231–238, 2024.
https://doi.org/10.1007/978-3-031-43733-5_21

the student. This happens in an integral way, and during the process the student will acquire knowledge, skills, aptitudes and abilities to be competent in the society.

Regarding the goal of any educational institute, one of the strategies more convenients to empower the leader labor inside the institution is the collaborative work. This strategy makes that all the members have an active role focusing in its job with the aid of everyone meanwhile the leader coordinates the process succeeding in an excellent team work. This investigation is divided in several parts. Section 2 of this investigation continues with the precedent work of art about this topic. Section 3 describes the investigation method used, the participants, tools and the procedure. Section 4 describes the leadership types in an educational institute. Section 5 explains the results and discuss the limitations of the investigation.

2 Work of Art

This work is based and it correlates with several referents which have been monitored and that represent a tool for its development. The identification of these investigations it has allowed us the possibility to organize what topic are necessary to investigate deeply; and to ensure the applicability, results and effectivity of the project.

Ferreiro et al. [1] quote "The active schools of the early 20th century, pioneered by John Dewey (1859–1952), also emphasized the need for student interaction, mutual support, and cooperation within their framework".

This generates better results step by step in the learning process in the system of competences. The student commits to develop its own duties inside the class framework individually and in group.

To empower leadership in the educational institute it is necessary the labor of situational leaders and to modernize the empowerment with a shared vision. These concepts are defined in the next paragraph for a better understanding.

What does collaborative work mean? It is the union of knowledge, skills, attitudes, aptitudes and abilities of a group of people focused to achieve a common goal and to generate new knowledge.

W. Johnson, P. Johnson et al. [2] "Collaborative learning is the use of small teams where students work together to maximize learning for themselves and others".

What does leadership mean? It is composed by the verb lead which means the initiative in an action; an example for others to follow. Leadership more than a word is a concept. It is the function to guide, motivate, to take assertive decisions to take a group of people to fulfill objectives.

Chacón et al. [3] describes an interesting educational perspective. The efficacy of the problem-solving has been an issue in every level of study. Not only in a regional level, but a national and international level. Metrics as results of references, the deficiency of this competence points that is necessary make a review of the student performance. For this reason, it is important to identify the creative work with rural mathematics professors. Skill to check and solve problems with decisions.

Garcia et al. [4] realized a study to determine if the learning strategies are based in projects which use interactive technological tools helping the students to develop advance skills in collaborative work. This study used a multi-case method. The analytic unit is integrated by 10 students from 18 to 25 years and 2 professors from the careers of engineer and business from an organization. This was realized from February to April 2013. One group has to show empathy and respect of the ideas of the rest. The richness of working in collaboration with people gives to the students the opportunity to prepare for real working and be more versatile in the real world.

Sierra and Robles [5] have developed a study that shows the collaborative environmental formation impact in the incidental thought development in a public school grade 10 in Córdoba, Colombia. This program is based in the inconvenient that generates the process of teaching and random thought. This reflects the performance of the apprentices in internal and external tests. This is available in SPSS software 25 version. Furthermore, there were not significant differences in skill development terms in the group. Interpretation, representation and reasoning. Nevertheless, the increase of the GPA ranges is well known and shows the efficacy of the correct use of collaborative learning strategies in order to develop random thought within the students in research institutes.

Pérez, Farfán, Delgado and Baylon-Chavagari analyzed et al. [6]. The cooperative learning impact within high school students in educational institutes. They take into consideration the importance of the learning process and the development of basic skills. Furthermore, cooperative learning makes possible the implementation of pedagogic strategies to empower the learning of the students in the context of personal interaction, active interdependences, shared responsibilities and variable environment. Therefore, this is one of the most popular and accepted strategy and it is applied in several high-schools and courses. The collaborative learning gives the necessary support for the holistic development of the students in the class environment where requires mutual support to achieve the expected goals. In this perspective, collaborative learning increases the emotional support of the students in a learning environment and critical learning; creating educative well-being between the students in the educational institutes.

Mendoza J. and Mendoza B et al. [7] had as objective the determination to create cooperation strategies for secondary students in Cajamarca. The study used a qualitative perspective with a representative and transversal design. They realized a questionnaire used as a not listed tool. This tool was approved by 90% by an expert report. Reliability coefficient by Kuder-Richardson were equal to 0.83 and 0.81, therefore the dispositive reliability is very high. Methodologic strategy of the collaborative work is divided in four phases: programming, action, reflection, and puzzle or group investigation method, which creates favorable conditions for the collaboration of students. They confirmed the hypothesis that the appropriate methodologic strategies lead to adequate learning, which benefits the methodology strategies of the students.

3 Methodology

This investigation has an exploratory focus that wants to design and to validate a strategy to empower digital skills in students during its first university cycle. The strategy is based in the ABC methodology that follows the line investigation-action: planning, observation, action and reflection. There are three stages in order to apply this methodology. First, the planning phase where it is necessary to realize the design of the investigation both theoretical and methodological. Second, the phase of action with the development of the investigation. Third, the reflection phase where the results of the investigation are considered.

3.1 Participants

To validate the ABC methodology, four experts in the field of pedagogy participated where they followed a strict process of evaluation. A population of 4 official educational institutions in Mexico was established, composed of 172 teachers, 4786 students, and a convenience sample of 120 teachers and 357 participants was selected.

3.2 Materials/Resources

The tools used for this investigation were interviews and observation panels.

3.3 Procedure

Each tool makes questions about the connectivity theory to ease the use of ICT as a tool to empower the participant cooperation, influencing the learning and the attitudes in the class.

- Planning: this phase identifies the learning goal to achieve with the method. Besides, it plans the size of the group who will collaborate. Moreover, it develops the evaluation process which will be used through individual evaluation tools, pair evaluation and so on. In the end, we develop sources for interdependence.
- Action: this phase shows the presentation of the tasks and common goals to the students. Also we describe the evaluation criteria and desirable performance. We advise and help with the stipulated tasks with interventions to guide the development of collaborative skills. We did closing activities and conclusions.
- Observation: this phase realizes a follow up of the level of the students who participated in the investigation. We evaluate the obtained results through self-assessment, pair self-assessment and observation protocols. We end with an evaluation of the performance of the team giving feedback about the whole process

  - Reflection: The learning obtained by the participants will be evaluated through collaborative learning as a methodological strategy that directly involves the teacher, students and the didactic environment. From the point of view of the socio-constructivist theory, it allows students, through collaborative work, to propose collective solutions to everyday life situations.

4 Types of Leadership

Leadership classifies in several types such as charismatic leadership, managerial grid, but in effects of this article it only focuses in the situational leadership, the empowerment and the shared vision.

4.1 Situational Leadership

Situational leadership consists that the leader has to consider several aspects and contexts of the company, institution, staff, time, situation, etc. to act and manage to achieve desirable goals.

This leadership is one of the more common ones, due to its variety and flexibility depending on the circumstances. There are several points to develop this leadership [8]:

- It is not a matter of more work, it is a matter of effectivity
- There is nothing more unfair that treat unequal persons as equals
- A leader does not go in front, the leader is behind each person and team work to offer support and guide them in the required moment.

The situational leader has, normally, four styles. Each style may be a process, and it depends on the maturity of the followers, workers or personal in charge. It is important to know what style in front of the workers is; and to know how to reach the style number four, as it is showed in the Fig. 1

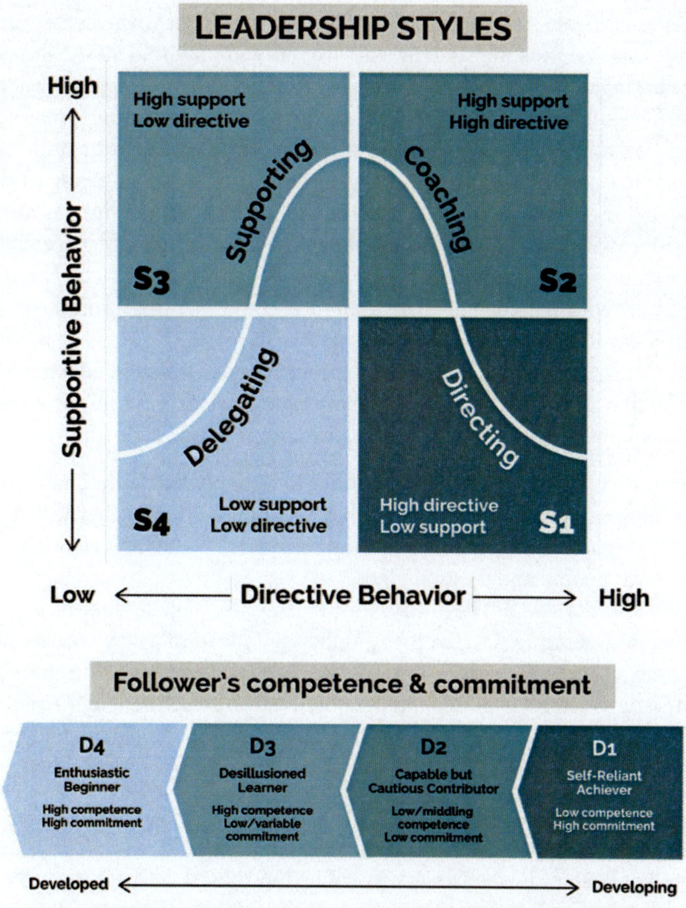

Fig. 1. Situational leadership styles. Authors: Hersey & Blanchard

Situational leadership focuses completely towards the collaborator. Each person is different, therefore this leadership is flexible; and the leader will choose styles depending the collaborator and the circumstance. This makes the leader a flexible one. In the classroom happens the same; the professor must analyze and take into consideration the personality, knowledge, and all the individual and group context to choose and determine the leadership style to use with each student to achieve goals and reach the style number four.

4.2 Empowerment

The English term is used in the last years to encourage the workers and members of a Company or institution to actively take part in the decision-making process as a group with the director or leader generating 'power' between the members. In the crucial decision moments of the company or institution the worker or member has all the right of opinion, which makes mandatory to leader to hear and take that opinion into consideration.

This type of leadership is behind in the traditional hierarchy pyramid, which gives pass to the net or circle structure. In this structure the client is the center of the net, in the educative institution is the student; and around you may find all the members of the institution sharing responsibilities and assisting the students in growing its educative knowledge. The student shapes the center of the net as the main integrant; and turning around him you may found directives, professors, auxiliary, coordinators, parents and classmates. Without a specific order, everyone works for the well-being of the student, thus he may achieve its objectives in concordance with the educative institution.

This model Works better with the collaborative work strategy, a space where there is leadership without authoritarianism. A sharing leadership and team work where everyone is compromised with the educative institution and the learning process of the student working for a common aim.

Se debe de especificar el rol de cada integrante de la institución educativa para poder lograr el objetivo dentro del trabajo colaborativo, existen en términos específicos los colaboradores y los líderes, siendo muy importante el rol de cada uno.

- Collaborators (students): This role is about to actively participate in the continuous learning process and the aid to the other classmates. It has the following characteristics. a) Responsible with the learning process, b) Motivated with learning, c) Collaborative people and d) strategists
- Leader: The leader will not be always the professor. The leader may be the student/s as individual or as a group. Its role is to emphatically lead in a way where everyone, the team, group and even the institution achieve the expected goals. The characteristics of the professor as leader is a) explain the cooperation structure and the required social skills b) monitor and intervene according the situation c) evaluate and approach the issues.

4.3 Shared Vision

For a strong leadership, it is necessary to define the shared vision. Shared vision allows to an educative institution that everyone may change its actions and interests towards the same institution. This makes both the person and the institution feel connected with the duties, perceptions and ideals of the organization and create a common link. There could only exist a shared vision a prior where everyone stablishes a personal vision; the shared vision is the sum of the individual view.

The leadership inside the collaborative work plays an important role in the shared vision, because it must make that everyone, even the collaborators achieve the expected shared institutional vision.

The following tools are required to achieve collaborative work with the correct direction and shared vision of the situational leader and empowerment, as well as the members of the educational institution:

- Fluid and appropriate communication
- The motivation on the part of the leader mainly

To complete the management process, there is an evaluation that allows you to understand the level of performance and find the best evaluation tool, and it is easy to use. [8, 9].

5 Discussion and Conclusions

To sum up, there are several advantages to empower the leadership through the collaborative work strategy (collaborative learning) in the educative practice, due to the several considerations. This happens mainly because the achieved learning is unmeasurable. Everyone learn from everyone and they sum efforts to achieve a common goal.

Knowing and including the situational leadership, the empowerment and the shared vision as part of the collaborative work, the leader transforms as a guide who goes together with the students. It motivates itself and the student work. It makes the environment harmonious, emphatic, pleasant obtaining better results for the institution and to the students.

The students will participate sooner or later in a course as leaders when they do collaborative work, maximizing its skills as leaders and collaborators. This will make to learn the true meaning and sense of leadership. Leadership is not about to command, it is about to guide and work in collaboration. Ferreiro, Margarita, Ramón & Espino et al. [1] quote "Distributed Leadership Principle: All students should understand, learn, and develop leadership tasks that perform a variety of roles in teamwork and promote others as leaders, reporters, and especially cheerleader roles".

The students maximize the leadership with the collaborative strategy work, reaffirming its moral values, respect, gratitude, love, friendship, dignity, honesty, humility, solidarity, prudence, perseverance, loyalty and so on. Values that are not only practiced inside the classroom, but in the society. "When the leaders go forward, the others will follow later" [10].

The classes are part of the real challenge for students. They are the main actors of the construction of their own knowledge. They learn from other students and not only from the professor [11, 12].

References

1. Flores, M.E., Alemán-Saravia, A.C., Deroncele-Acosta, A.: Liderazgo científico investigativo del docente para la transformación del contexto universitario. Revista Venezolana de Gerencia **27**(99), 1151–1168 (2022). https://doi.org/10.52080/rvgluz.27.99.19
2. Johnson, D.W., Johnson, F.P.: Joining Together: Group Theory and Group Skills. Allyn & Bacon, Needham Heights, MA (1997)
3. Chacón., M.: Estrategia didáctica para fortalecer la competencia en la resolución de problemas en estudiantes de ingeniería de un curso de ecuaciones diferenciales de una universidad privada. [online]. Disponible en: http://hdl.handle.net/20.500.12749/13921 (2021)
4. García Disla, I.Y.: Aprendizaje por proyecto: incidencia de la tecnología de la información para desarrollar la competencia de trabajo colaborativo. Ciencia y Sociedad **38**(4), 691–717 (2013) ISSN: 0378- 7680. Disponible en: https://www.redalyc.org/articulo.oa?id=87029731004 (2022)
5. Sierra Llorente, L.J., Robles González, J.R.: Aprendizaje colaborativo mediado por un ambiente de liderazgo, en el desarrollo del Pensamiento Aleatorio, ASN, vol. 6 (10), pp. 9–29, Jun (2021)
6. Pérez Salgado, L.N., Farfán Pimentel, J.F., Delgado Arenas, R., Baylon Chavagari, R.G.: El aprendizaje cooperativo en la educación básica: una revisión teórica. Revista Metropolitana de Ciencias Aplicadas **5**(1), 6–11 (2021)
7. Mendoza Sánchez, J. R., Mendoza Sánchez, B.F.: Cajamaíca. Desafios **12**(1), e254 (2021)
8. Sánchez, E., Rodríguez, A.: 40 años de la teoría del liderazgo situacional: una revisión. Revista Latinoamericana de Psicología **42**(1), 29–39 (2009)
9. Rodríguez Estrada, M.: Liderazgo (Desarrollo de habilidades directivas), Serie Capacitación integral, El Manual Moderno (1995)
10. Denny, R.A.: Motivar para ganar, Editorial Selector. Las estrategias y técnicas didácticas en el rediseño: Aprendizaje Colaborativo, http://www.sistema.itesm.mx/va/dide/inf-doc/estrategias/colaborativo.html (2001)
11. Rivera Medina, Y.D.C., Aparicio Molina., C.A.: Características de las prácticas de liderazgo pedagógico en programas de integración escolar que favorecen el trabajo colaborativo entre docentes. Perspectiva Educacional **59**(2), 27–44 (2020). https://doi.org/10.4151/07189729-vol.59-iss.2-art.1055
12. Senge, P.: La quinta disciplina. El arte y la práctica de la organización abierta al aprendizaje. Ediciones Juan Granica, España (1996)
13. Medina Bustamante, S.M.: El aprendizaje cooperativo y sus implicancias en el proceso educativo del siglo XXI. INNOVA Res. J. **6**(2), 62–76 (2021). https://doi.org/10.33890/innova.v6.n2.2021.1663

Efficiency of Permanent Crops: Bananas, Cocoa, Coffee in Ecuador. An Analysis During the COVID 19 Pandemic

Tania Morales Molina[1](✉), Juan Pablo Martínez Mesías[2], Darwin Aldas Salazar[2], and Ximena Morales-Urrutia[2]

[1] Universidad Tecnológica Indoamérica, Ambato, EC 180103, Ecuador
taniamorales@uti.edu.ec

[2] Universidad Técnica de Ambato, Ambato, EC 180104, Ecuador

Abstract. Ecuador is in a privileged climatological zone. It admits that in its territories permanent crops of high demand are generated both locally and internationally, as is the case of bananas, coffee, and cocoa. For this reason, it is important to control the resources involved in its production and verify the level of efficiency in order to maintain competitive standards worldwide. Thus, in the present study the technical efficiency of the aforementioned crops is calculated and analyzed in a period of great importance such as 2020–2021, since it significantly marked the world economies due to the COVID 19 pandemic. To calculate the efficiency, the Data Envelopment Analysis (DEA) was used, that works with inputs and outputs, which were identified in the database of the Continuous Surface and Agricultural Production Surveys (ESPAC) report, generated by the Directorate of Economic Statistics and the Agricultural Statistics Unit of the National Institute of Statistics and Censuses (INEC). Determining production as the output variable and the main input variables: planted area, workers, fertilizers, and insecticides. The main results indicate that banana has not undergone significant changes in its production efficiency, while cocoa and coffee maintained lower levels of efficiency in 2020 compared to 2021 in most of the provinces of Ecuador where it is grown. Evidencing that the management of resources during the pandemic were limited and, in many cases, scarce.

Keywords: Technical efficiency · Permanent crops · Agricultural production

1 Introduction

In recent decades, the agricultural sector has achieved an important position in the global economy, being considered as one of the activities that plays an important role in the growth and economic development of rural areas of the countries [1].

Agricultural production has registered significant growth in recent years. Thus, by the year 2050 the demand for food is expected to exceed 9,100 million, twice as much as is currently the case [2]. Therefore, it is necessary to cover the internal and external demand with a sustainable productivity of the sector.

C. Montenegro et al. (Eds.): ICMTT 2023, LNNS 774, pp. 239–247, 2024.
https://doi.org/10.1007/978-3-031-43733-5_22

In this context, agriculture is projected as a clearly competitive market. For this, a mandatory production is required with the use of resources and inputs. Such as: fertilizers, labor, land among others, which, in turn, must present a uniform level of competitiveness [3]. Some studies have shown that the increase in agricultural productivity is considered a condition for the development and growth of demand [4].

Agricultural efficiency has some implications, if the existing inputs and resources are not used properly, it would not be profitable and even less sustainable over time [5]. Therefore, the level of efficiency reached by farmers makes it possible to determine what is the possibility of increasing benefits to the economies based on the use of existing inputs [1].

The importance that the agricultural sector has reached worldwide has led to a series of techniques being proposed to enhance crop yields [6]. Likewise, this activity maintains a central role in the economy, being one of the main sources of inputs for the poorest sectors of rural areas [7]. In this sense, the efforts to optimize agricultural production are especially relevant when making efficient use of available resources, in such a way that, greater productivity, efficiency and sustainability can be generated [7].

In this regard, the Food and Agriculture Organization of the United Nations shows that in developed countries, the use of agricultural yield analysis techniques through the implementation of new technologies is generating increases and above all high productivity rates for the agricultural section [8]. While, in the case of developing countries, they are still in the process of implementing technology for their agricultural sectors. Therefore, the use of yield analysis methods is considered a key tool to improve the efficiency of this sector [9].

The case of Latin America is not isolated from that of many countries and continents, since they have considered agriculture as an engine that boosts the economy of some sectors, through the generation of economic resources and job creation in rural areas [10]. Therefore, the region has promoted policies that emphasize increasing the yields of the agricultural sector [11].

In Ecuador, the situation is not so encouraging, since there are certain drawbacks that have prevented the sector from taking off and reaching its maximum performance [12]. In this regard, the inequitable distribution of land stands out, which means that productive land is concentrated mainly in large estates, resulting in extremely unproductive [13] and, the main problem of the sector is the lack of capital to invest in cutting edge technology.

However, agriculture is characterized by its low productivity, due to the misuse of different factors, whether these are technical or socioeconomic. In a large part of the world, farmers are forced to work depending on the available resources, producing different yields per hectare, due to inefficiency and ignorance of new practices and use of techniques and technology appropriate to each product [5].

In Ecuador, the cultivation of bananas, coffee and cocoa represents 2.1% of the Gross Domestic Product (GDP), these are considered as traditional products of high demand in the international market [14]. However, the performance of agricultural capacity in production has been a permanent difficulty in the sector. Given the scarcity of capital and technological resources, which is evident in the productive infrastructure such as irrigation canals and modern water reservoirs that provide basic water resources for agricultural areas [15].

Therefore, the purpose of this article is to estimate the technical efficiency of the production of banana, cocoa and coffee permanent crops during the COVID-19 pandemic.

2 Methodology

This article develops the calculation and analysis of technical efficiency in the production of each of the permanent crops: banana, coffee and cocoa for the different provinces of Ecuador. The study was developed under a quantitative approach and the data envelopment analysis (DEA) technique focused on CCR outputs was used to measure efficiency. This technique allows evaluating producers' yields, without considering random factors such as: market value, quality of products and inputs, climate, among others. We worked with input and output variables for each of the products, which were obtained from the report of the Continuous Agricultural Surface and Production Surveys (ESPAC), generated by the Economic Statistics Directorate and the Agricultural Statistics Unit of the National Institute of Statistics and Census, with respect to permanent crops [12, 15].

Variables were coded as indicated in Table 1.

Table 1. Variable Coding.

Variable description	Coding
Tons of production	TP
Hired workers	HW
Family workers	FW
Planted surface	PS
Insecticides	I
Fertilizers	F(PKN)

The data that was filtered for each variable of the analyzed products, allowing to build the database for this research, being defined the input and output variables as shown in Table 2.

Table 2. Variables

Tipo de variable	Variable
Dependent variable (Ouputs)	TP
Independent variables (Inputs)	PS HW/FW F I(PKN)

Once the input and output variables of the model have been defined. The calculation of productivity was generated for each province that cultivates the mentioned products, this by establishing the weighted sum of each indicator, where Eq. 1 was used [16].

$$Productivity = \frac{\sum_{r=1}^{s} U_{rj} * Y_{rj}}{\sum_{i=1}^{m} V_{ij} * X_{ij}} \tag{1}$$

where:

X: represents the inputs
Y: Represents the outputs
U: Percentage weight of each of the outputs
V: Percentage weight of each of the entries

After obtaining the productivity of each province. The technical efficiency was calculated through Eq. 2, which is equal to the division between the productivity of each of the provinces, for the province that has obtained the highest productivity [17]. With this, the technical efficiency of each of the provinces will be obtained compared with the province that is most efficient for the production of each of the permanent crops analyzed [18].

$$Efficiency = \frac{\sum_{r=1}^{m} U_{rj} * Y_{rj} / \sum_{i=1}^{m} V_{ij} * X_{ij}}{\sum_{r=1}^{s} U_{rj} * Y_{rj} / \sum_{i=1}^{m} V_{ij} * X_{ij}} \tag{2}$$

3 Results

To have an overview of the current state of production of permanent crops under study, as well as the use of resources such as insecticides and fertilizers, a descriptive study was carried out through the averages of the same of all the provinces of Ecuador in the years 2020 and 2021. For a better description of the resources, the natural logarithm was determined to standardize the data and the graph, since they are in different units of measure. The results are shown in Fig. 1, in which, it is shown that in the case of banana production decreased by approximately 72,000 Ha on average between 2020 and 2021. On the contrary, there is evidence of a notable increase in the use of resources such as fertilizers, while the use of insecticides remained constant.

In the case of cocoa, there was also a significant decrease in production for the years studied in approximately 130 tons on average at the national level. However, the use of fertilizers is increasing, and the use of insecticides is decreasing. Finally, in the production of coffee, unlike a growth from 2020 to 2021, the use of fertilizers decreases, and the use of insecticides increases.

These results directly influence the technical efficiency of production, since it depends on the resources, its availability and use to maintain, increase, or decrease, both planted and harvested areas.

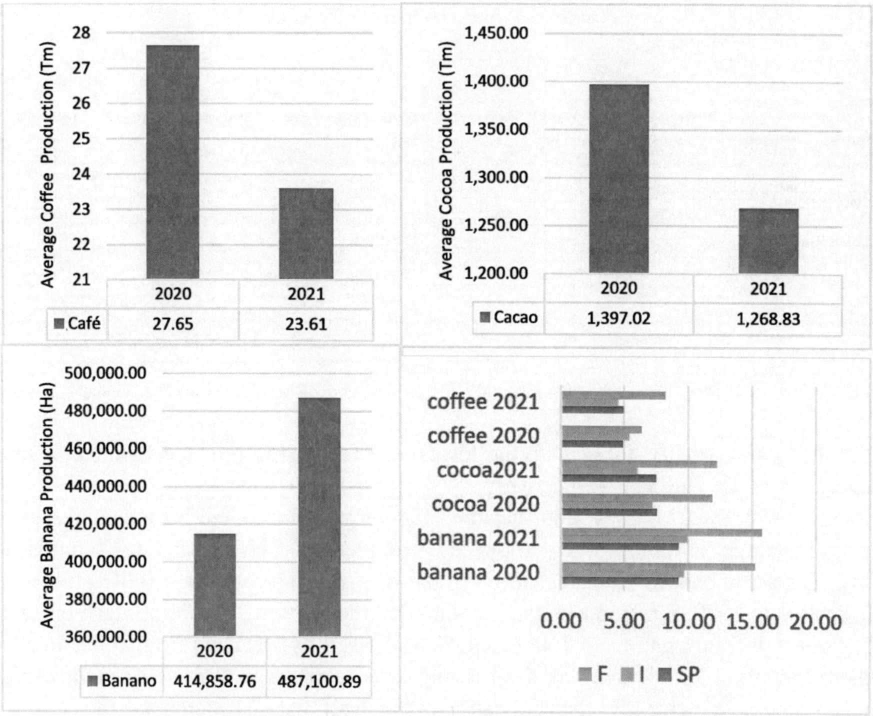

Fig. 1. Production and resources in Ecuador years 2020–2021.

3.1 Technical Efficiency of Banana Cultivation

According to the results, both in 2020 and 2021, a technical efficiency of 100% is reached in banana production in all study provinces, where it is assumed that during the time of the COVID 19 pandemic, they used adequately the factors of production (Table 3). This result also indicates that efficiency has been maintained since bananas are a countercyclical crop in the Ecuadorian economy and a large part of the product is destined for export. For this reason, the economic crisis caused by the pandemic was not deeply reflected in banana cultivation activities, due in large part to exports to world markets.

Therefore, this sector was able to continue carrying out its activities in a normal manner and with high demand during COVID-19, which allowed efficiency not to drop.

3.2 Technical Efficiency of Cocoa Cultivation

The technical efficiency calculated in cocoa crops, for the year 2020 (Fig. 2), shows that the average of all the provinces analyzed is 82%. While, by 2021, 87% making it clear that, in this last year, resources were used more efficiently. During 2020 it is observed that in the provinces of Bolívar, Chimborazo, Guayas, Imbabura, Napo and Orellana they present the highest efficiency. That is, 100%, which defines that there is the correct use of the factors. For its part, the province of Los Ríos indicates the second-best efficiency

Table 3. Banana technical efficiency

2020			2021		
Province	Range	Theta	Province	Range	Theta
Cañar	1	1	Cañar	1	1
Cotopaxi	1	1	Cotopaxi	1	1
El Oro	1	1	El Oro	1	1
Esmeraldas	1	1	Guayas	1	1
Guayas	1	1	Los Ríos	1	1
Los Ríos	1	1			

with 90%; and, finally, those with the lowest efficiency this year are Esmeraldas and Azuay with 46% and 39% respectively.

For the year 2021, it is seen that the efficiency in cocoa cultivation increases in general average, it should be noted that the provinces of Bolívar, Cañar, Cotopaxi, El Oro, Guayas, Imbabura, Manabí, Morona Santiago, and Napo maintain 100% efficiency. Compared to 2020, it is important to mention that the province of Chimborazo presents the largest decrease gap from 100% to 43%. Followed by Orellana from 100% to 52% where the study shows that this drop is due to a decrease in the use of planted area and an excessive increase in the application of fertilizer and insecticide. On the contrary, Esmeraldas increases its efficiency in these two years from 46% to 87%, and Manabí from 60% to 100%, where growth is denoted, especially in the area planted and improvement in the use of its resources.

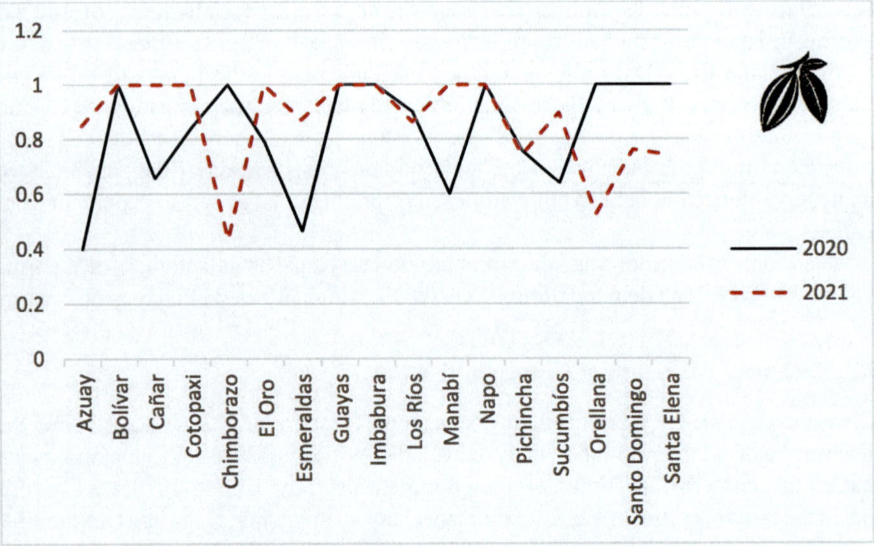

Fig. 2. Technical efficiency of cocoa in Ecuador years 2020–2021.

3.3 Technical Efficiency of Coffee Cultivation

After calculating the technical efficiency of coffee in the provinces of Ecuador, (Fig. 3), it is observed that, for 2020, the average efficiency of coffee cultivation was 77% and for 2021 it was 97%, which reveals improvement in production efficiency during the study period.

It is also evident that in all provinces there is growth from 2020 to 2021, except for Loja, which decreases from 100% to 71%. In this context, it stands out that the greatest growth in efficiency presents Azuay and Bolívar from 40% to 100%, Los Ríos from 45% to 100% and Sucumbíos from 58% to 100%, this is mainly due to the decrease in the use of resources such as fertilizers and better management of the planted area.

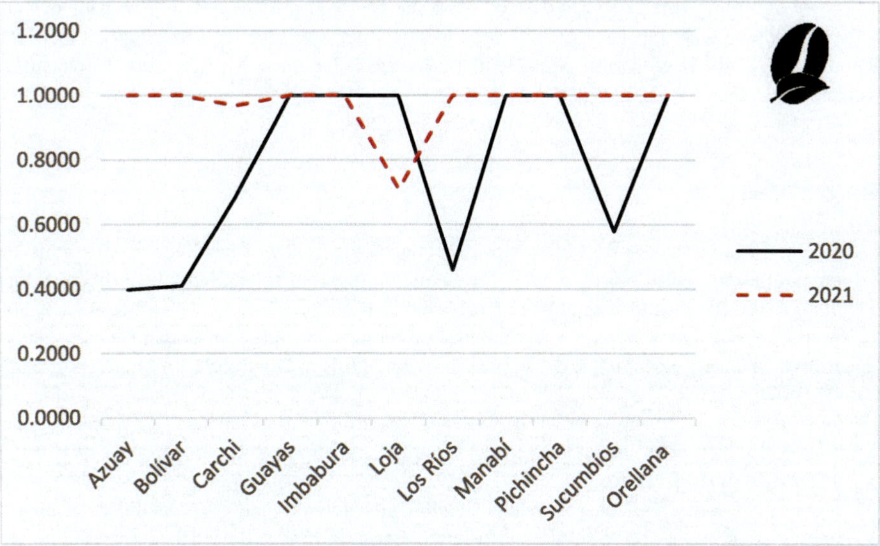

Fig. 3. Technical efficiency of coffee in Ecuador years 2020–2021.

4 Conclusions

The Covid 19 pandemic marked a significant milestone in the world economy, leaving important results that must continue to be analyzed in this case with respect to agricultural production. Technical efficiency allows us to know how optimal the use of resources for production is and where this use is being generated.

Based on the results obtained by the DEA, in banana cultivation the efficiency reached 100% in the two study periods. On the other hand, for cocoa in 2020 it was 82% and in 2021 87% and finally coffee with an efficiency in 2020 of 77% and in 2021 of 97% in general average of all the provinces, showing an evident growth in these two years, this is explained since 2020, being a year of recession due to the pandemic, all the economies were affected, however a recovery is noted for 2021.

Even though the economy had a strong recessionary impact from COVID -19 in 2020 and 2021, did not profoundly affect this type of crops, since these are in great demand in world markets. Therefore, these sectors had to continue carrying out their activities in such a way that they did not allow efficiency to decline.

The main limitation of the study was taking data from a database that did not have information on many parameters and even though there is a great variety of indicators in permanent crops, they did not present quality information so that they could obtain a satisfactory analysis of the yields of each product.

The research traces a baseline to carry out future efficiency studies for other types of crops such as transitory, pastures and flowers, the same information that can be obtained from the base worked in the present investigation.

Acknowledgment. This article is derived from the research project called "Behavior of agricultural production during the COVID-19 pandemic", approved through Resolution No. UTA-CONIN-2022-0017-R with code SFFCAUD 02 by the DIDE from the Technical University of Ambato, Ecuador.

References

1. Abate, T.M., Dessie, A.B., Mekie, T.M.: Technical efficiency of smallholder farmers in red pepper production in North Gondar zone Amhara regional state, Ethiopia. J. Econ. Struct. **8**(1), 1–18 (2019). https://doi.org/10.1186/s40008-019-0150-6
2. Manogna, R.L., Mishra, A.K.: Agricultural production efficiency of Indian states: evidence from data envelopment analysis. Int. J. Fin. Econ. **27**(4), 4244–4255 (2022). https://doi.org/10.1002/ijfe.2369
3. Manogna, R.L.: Innovation and firm growth in agricultural inputs industry: empirical evidence from India. J. Agribusiness Dev. Emerg. Econ. **11**, 506–519 (2021). https://doi.org/10.1108/JADEE-07-2020-0156
4. Ngango, J., Kim, S.G.: Assessment of technical efficiency and its potential determinants among small-scale coffee farmers in rwanda. Agriculture **9**, 161 (2019). https://doi.org/10.3390/agriculture9070161
5. FAO/WFP: special report FAO/WFP crop and food security assessment mission to ethiopia (2012)
6. Cepal, Fao, Iica: Perspectivas de la agricultura y del desarrollo rural en las américas (2021)
7. FAO: El estado mundial de la agricultura y la alimentación 2020 (2020). https://doi.org/10.4060/cb1447es
8. CEPAL: Efectos económicos y financieros en América Latina y el Caribe del conflicto entre la Federación de Rusia y Ucrania. Rev. Com. Económica para América y el Caribe, pp. 1–10 (2022)
9. Khati, K.S., Bhandari, A.K., Mukherjee, D.: Productive efficiency of agricultural sector in Uttar Pradesh. In: Chakraborty, T., Mukherjee, D., Saha, S. (eds.) Contemporary Issues in Sustainable Development: The Case of India, pp. 50–71. Routledge India, Abingdon, Oxon; New York, NY: Routledge (2020). https://doi.org/10.4324/9781003141020-5
10. Castellanos, R.M., Morales-Pérez, M.: Análisis crítico sobre la conceptualización de la agricultura de precisión. Cienc. en su PC 23–33 (2016)
11. Ho, T.Q., Hoang, V.N., Wilson, C.: Sustainability certification and water efficiency in coffee farming: the role of irrigation technologies. Resour. Conserv. Recycl. **180**, 106175 (2022). https://doi.org/10.1016/j.resconrec.2022.106175

12. INEC: Encuesta de Superficie y Producción Agropecuaria Continua (ESPAC). ESPAC, pp. 1–49 (2020)
13. Banco Central del Ecuador: Reporte de Coyuntura II. Rep. Coyunt. Sect. Agropecu, p. 50 (2020)
14. CAMAE: Café y cacao: aliados tradicionales de la reactivación económica – Camae. http://www.camae.org/cacao/cafe-y-cacao-aliados-tradicionales-de-la-reactivacion-economica/?fbclid=IwAR15onGw80k0leijYoLY_lVV4u5YByBYjCGMK_7ZUwc-Ma6aemzTlsc7W48. Last accessed 8 Feb 2023
15. Instituto Nacional de Estadística y Censos (INEC): Ecuestas de Superficie y Producción Agropecuaria y Continua (ESPAC)
16. Jaime, J.A.: Formulaciones en el Análisis Envolvente de Datos (DEA). Resolución de casos prácticos. Depósito Investig. – Univ. Sevilla 93 (2016)
17. Charnes, A., Cooper, W.W., Rhodes, E.: Measuring the efficiency of decision making units. Eur. J. Oper. Res. **2**, 429–444 (1978). https://doi.org/10.1016/0377-2217(78)90138-8
18. Sperat, R.R., Brugiafreddo, M.P., Raña, E.: Eficiencia técnica en la agricultura familiar: Análisis envolvente de datos (DEA) versus aproximación de fronteras estocásticas (SFA). Nova Scientia **9**, 342–370 (2017). https://doi.org/10.21640/ns.v9i18.697

Technology in Tourism and Tourist Experience

Culinary Trends in the Ambateña Gastronomic Tourism Offer

Francisco Torres-Oñate[1](✉), Sonia López-Pérez[2], Mario Romo-Rojas[1], and Carmen Vaca-Vaca[1]

[1] Facultad de Ciencias Humanas y de la Educación, Universidad Técnica de Ambato, Ambato, Ecuador
{cf.torres,mg.romo,carmenivacav}@uta.edu.ec
[2] Centro de Idiomas, Universidad Técnica de Ambato, Ambato, Ecuador
sda.lopez@uta.edu.ec

Abstract. The research studies the gastronomic tourist offer of establishments registered in the Municipality of the city of Ambato with the distinctive Q for quality and the application of culinary trends used in restaurants. The work will allow gastronomy professionals and fans to visualize valuable characteristics for the creation of new culinary experiences. The methodology responds to a non-experimental design with a quantitative approach, descriptive-correlational scope to determine the influence between the variables. Two instruments were applied, a register of establishments to determine culinary trends (type of food) and setting (style of decoration); There is also a questionnaire based on two models: Fishbein's Weighted Attitude Model and Norma Castrezana's Multidimensional Scales Model that establishes preferences in food establishments. A validation of the instrument was carried out using Cronbach's Alpha with a "very good" reliability (0.825), for the analysis of the information the statistical program IBM SPSS Statistics 21 was used. The correlation of Spearman's Rho is being applied, obtaining a "moderate correlation" (0.471), which establishes one in the study variables, accepting the correlation of culinary trends and the gastronomic tourist offer in the city of Ambato.

Keywords: Traditional Cuisine · Motivation · Q for quality · Gastronomic Tourism · Experience

1 Introduction

Gastronomy and its relationship with tourism is much more than satisfying a biological need in humans, in many cases it even becomes the main motivation and central axis of the trip, this type of knowledgeable consumer is called "Gourmet". This has given way to the creation of culinary trends worldwide, standing out in this segment the techno-emotional, molecular, and avant-garde cuisine whose principles are similar in the use of chemicals and high-tech equipment to obtain dishes with particular characteristics by modifying their organoleptic characteristics. Local cuisine has a special mention, it is being one of the trends that inspires tourists to make cooking the main part of their experience in the destination they visit.

C. Montenegro et al. (Eds.): ICMTT 2023, LNNS 774, pp. 251–260, 2024.
https://doi.org/10.1007/978-3-031-43733-5_23

The study was carried out in the city of Ambato and is located in the central part of the Republic of Ecuador. It is also characterized as an important commercial area due to its strategic location. It has an average climate of 17 °C and stands out mainly for the production of fruits and bakery, which are two nationally recognized products.

Certainly Ecuador, to achieve a solid gastronomic offer focused on tourism, this must be consolidated through establishments that offer their consumers not only food, but also experiences where accessories such as the setting (music, decoration) and excellent service make the stay of the customer a memorable moment. Studies carried out in Mexico on gastronomic tourism reveal that tourists allocate 20% of their budget to the purchase of food [1], distributed in restaurants and snacks in general. It is focusing on the current consumer, it is worth mentioning the Millennials whose habits set trends, one of them has to do with sustainability in the establishments they visit, so that they value the social responsibility of the business they consume.

The gastronomic tourist offer of the city of Ambato is constantly growing, with international gastronomy having a preponderant place, led by Mexican, Japanese and Italian fusion cuisines. In the background is the local gastronomy, whose presence in high-level establishments is developing [2].

2 State of the Art

2.1 Culinary Trend

Trends are ideas applied to daily life by liking or attraction to an object, a way of behaving and gastronomically speaking, it is the way of preparing and eating food. It is a change or evolution adapted to the life of society with individual tastes coming together to make collective preferences [3].

Just as there is a predisposition for fashion or new cultures, this also indirectly affects eating practices, since it has occurred more quickly in urban areas, but rural locations have not been immune to this process, mainly those established in the peripheral areas. When it is close to cities, trends obviously make a difference, which is why some communities choose to revamp. That is, in order to modernize some dishes while others keep their recipes as they are due to their tradition [4].

The gastronomic trend is created from a series of decisions made by a certain group of people, which in a certain way can predict future behavior. These are not entirely accurate because they depend on a series of factors that determine habits. Gastronomy is a cyclical orientation, currently evidencing a course towards the cuisine of origin as a travel motivation [5].

Molecular or avant-garde cuisine is one of the recent trends that is based on changing the appearance of preparations (textures, color, and appearance), using chemicals and high-tech equipment to achieve it without drastically modifying the original flavor [6]. Local gastronomy is the cultural identification of a certain place, channeled correctly and can be established as a tourist product of its own. In fact, the regional cuisine represents traditional products that combine the geographical and cultural environment, expressing the uniqueness of its food production, the surprising flavors and the knowledge of its people [7]. Authentic, traditional and/or innovative culinary experiences, gastronomy tourism can also include other types of activities, such as visiting local producers, participating in gastronomic festivals, or attending cooking classes [8].

In the tourist activity, culinary orientations are identified that seek to create experiences where creative proposals are required that feed the intrinsic need of the human being [9]. Thus, food constitutes a tourist attraction that motivates to visit certain places, reflecting the gastronomy, culture, and traditions of the areas where they are located [10]. In fact, food is reaffirmed as a key factor in tourism, local traditions, the habits of the inhabitants and lifestyles. The typical gastronomy is identified as the main factor of destination diversification, capable of impressing the traveler and influencing their purchasing preferences [11].

Changes in tourist behavior have been constant, for example, in the 1970s and 1980s gastronomic tourism focused on the offer of consumption centers that had a gourmet product and a chef of recognized prestige. In the early 1990s this changed, as tourists were offered a variety of experiences that took various forms, such as gastronomic celebrations and festivals, direct participation in the preparation and tasting of food products, even planting and harvesting with agro tourism and tours of gastronomic routes [12]. On the other hand, Gastronomic tourism can be understood as a modality of cultural tourism that allows to know the habits and way of life of the visited community through its gastronomic representations. This has as objective that locals and strangers adopt new ways of living, considering that the trip influences positive ideas for their daily life [13].

Gastronomy has set a trend over time, it is positioning itself more and more when choosing the destination, at this point, the growth of a gastronomic offer based on quality local products and the consolidation of its own market for the gastronomic tourism [13]. Therefore, it also constitutes by itself a product of increasing relevance that is recognized as an important factor in the improvement of the tourist offer and as an element of economic development of a territory [9], which means that tourists today prefer places with a greater gastronomic propose.

Since pre-Hispanic times, food has played a fundamental role in forging ethnic groups, populations, cities, countries, and cultures. It has been an inheritance transmitted turned into an art of cooking, this knowledge has been jealously guarded and shared only with people linked to said society, who have developed techniques, knowledge and flavors that have delighted palates of all kinds [14]. In Europe during the 20th century, the first routes or "gastronomic guides" were carried out, which consisted of detailing the set of restaurants and inns with the greatest notoriety of those times, giving rise to the famous "gourmets" or "gourmets", who were people with great gastronomic knowledge and a privileged palate, they were in charge of classifying and categorizing the restaurants [15].

In Latin America, food is full of subjectivities. It is classified in an economic and social status according to its components, it is worth noting the cultural richness and the plurality of its components and productive processes that they have [6]. The valuation of gastronomy within the tourist experience is not a recent phenomenon since local gastronomy are considered expressions of people's identity, they acquire relevance and capture the interest of the traveler from the 60s of the 20th century. Thus, the rejection of the standardization of food pushes individuals to the fervent search for what is authentic and natural, in fact, to the roots of the cuisine called terroir [16].

Aspects such as: the creativity of chefs, new cooking techniques, globalization, sustainable development and French gastronomy that thrives on large scale global consumption. In addition to other examples such as molecular cuisine, the "food truck" (fast food) and the "slow food" (long cuisine) which are based on a fusion of international flavors that are new alternatives that chefs implement to satisfy the demand for diverse palates [17].

2.2 Gastronomic Offer

The offer refers to the correlative phenomenon of the demand, it can be constituted in a product, service or experience. These can vary according to various factors (social, cultural and economic) that intervene in the applicant's perception, for this reason, it is constantly changing [18]. The gastronomic proposal refers to establishments in the commercial sector such as restaurants, cafeterias, bars and inns. It is determined as the set of food and beverages offered to the consumer. Thus, the constant improvement of this situation promotes local gastronomy, promoting and positioning its consumption and even internationalizing it [19].

A good gastronomic proposal cannot be limited solely to food, it must be surrounded by an appropriate environment, decoration and setting. The success of any destination as a recipient of gastronomic tourism is based on close collaboration between institutions and all the necessary links to develop tourism products such as agricultural and livestock producers, fishermen, suppliers, markets, restaurants, cooks, hotels and sectoral associations [20]. In addition, traditional gastronomy is made up of knowledge transmitted from generation to generation, it is also an important part of the local culture and reflects the use of raw materials from the area, which directly influences the culinary offer of the destination, since the result of typical dishes and destination fusion food [21].

Ambateña gastronomy is determined by the customs of the parishes that make up the city. First of all, in Atocha, empanadas de viento (simple dough made of refined wheat flour, stuffed with cheese and fried) and colada morada (thick and sweet purple corn flour drink with regional fruits) are offered daily. Then, in the central market you can taste the iconic "Llapingacho" (potato omelette stuffed with cheese, accompanied by sausage and fried egg), which is becoming a benchmark in local gastronomy. Finally, "El Pan de Pinllo" (it is a dough made of white and integral flour with lard) and Gallinas de Pinllo" (chickens cooked and roasted in local seasonings accompanied by boiled potatoes and peanut sauce) [22].

Therefore, the Ambateño culinary heritage has great potential to become a tourist attraction and be a main motivation for visits for nationals and foreigners [23]. Thus, the importance of the local culinary offer is characterized by its unique products, of high gastronomic and nutritional value; its techniques and methods of handling products that differentiate it from other culinary cultures in the world [24].

Gastronomic motivation and satisfaction studies contribute to a better approach to the visitor's profile, which makes it possible to develop an adequate offer of the destination that satisfies the consumer. The strategic sectors incorporate local cultural and heritage aspects that position it as a true tourist product [25]. Gastronomy is a strategic element in the positioning of a tourist destination, it is becoming a growing motivation for visitors and even without being the main reason it can become a travel experience or simply part of a routine [26].

Elite tourist destinations such as France, Italy and Spain have stood out due to the great demand for their gastronomy, which is turning this aspect into a priority activity. In Latin America, Peru is at the forefront achieved by its varied and delicious gastronomy. On the other hand, Ecuador is projected as a destination that focuses its tourist attraction on gastronomy [2].

The 21st century tourist not only wants to satisfy his/her need to eat, he/she goes further in the search for sensory experiences through the consumption of native and exotic foods in an environment different from his/her reality. They are creating a connection between the tourist and the experience with the local culture, environment and food [15].

3 Methodology

The research had a non-experimental-transectional design, it was carried out in a single period of time with the quantitative approach and the descriptive-correlational scope responds to the instruments used to collect the information. The study population focuses on the consumers of the Ambateña culinary offer through a formula for an indeterminate population with a sample of 203 people. The statistical program IBM SPSS Statistics 21 was used for the application of the non-parametric statistical test of Spearman's Rho.

The developed instrument was built from two models: Fishbein's Weighted Attitude Model, where satisfaction with the service is measured, and Norma Castrezana's Multidimensional Scales Model, which is used to establish satisfaction with a destination. Cronbach's alpha was applied to the instrument and it showed a value of 0.825, it is determining a *good* reliability, as shown in Table 1.

Table 1. Cronbach's Alpha Reliability statistics

Cronbach's Alpha	No. of elements
0,825	21

An evaluation was implemented using a Likert-type scale where parameter preferences are identified in the gastronomic offer such as: price, environment, menu variety, flavor, presentation, quality, prestige, speed, hygiene, attention – service and brand.

4 Results

The results presented below are a summary of the 21 questions asked through the instrument described in the previous chapter and explained the preferences and criteria of those surveyed in percentages, according to the items indicated below:

According to scale of preferences in the gastronomic proposal in establishments with quality Q mark in the city of Ambato (Fig. 1).

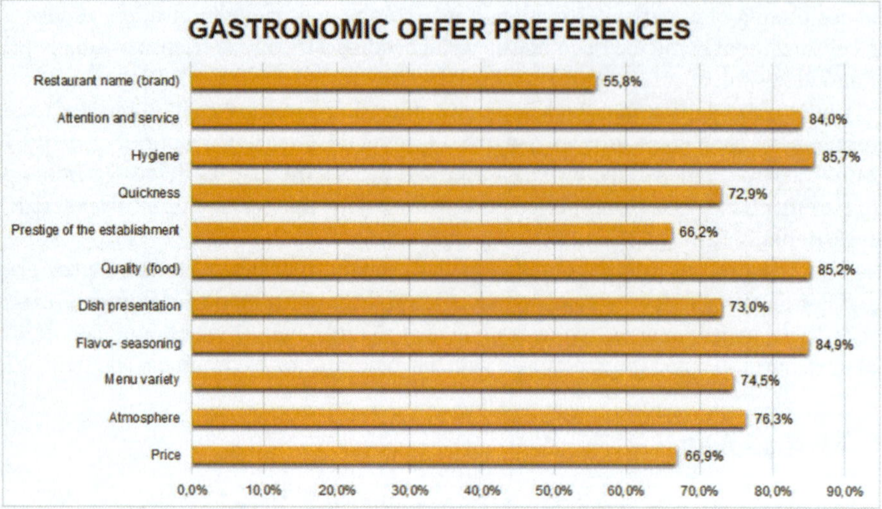

Fig. 1. Preferences of the gastronomic offer, Source: Ambato quality Q restaurant customers

The importance is determined in aspects such as hygiene (85.7%), food quality (85.2%), seasoning-flavor (84.9%), attention and service (84%). All the aforementioned aspects were considered of great importance within the offer of the establishments. It should be noted that the dimensions are directly related to food preparation and commercial service.

The establishments detail their gastronomic offer as follows (Fig. 2).

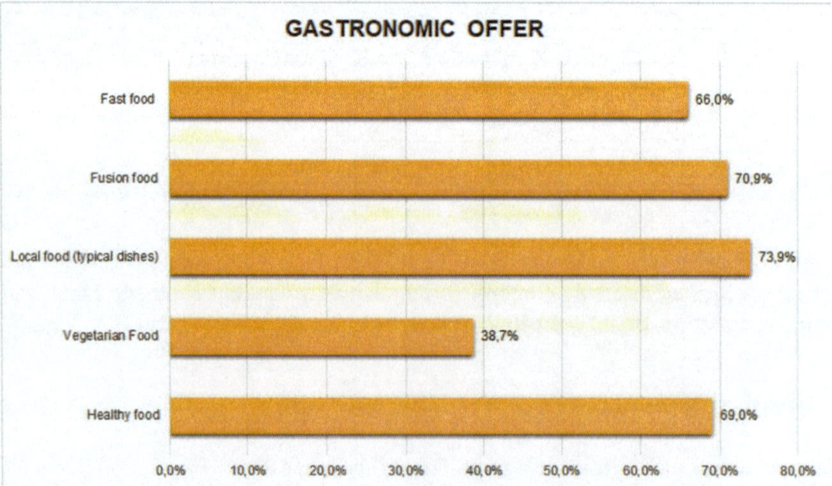

Fig. 2. Culinary proposal preferences, Source: Clients restaurant Q quality Ambato

The preference for local food (typical dishes) is 73.9%, fusion cuisine (international gastronomy) is 70.9%, predominant trends in the consumption of national and international gastronomy over other styles are identified. It is also observed that for the currently established demand there is a supply of establishments such as pizzerias, seafood restaurants and cafeterias that supply the demand.

The style preferences regarding the setting of the premises (Fig. 3).

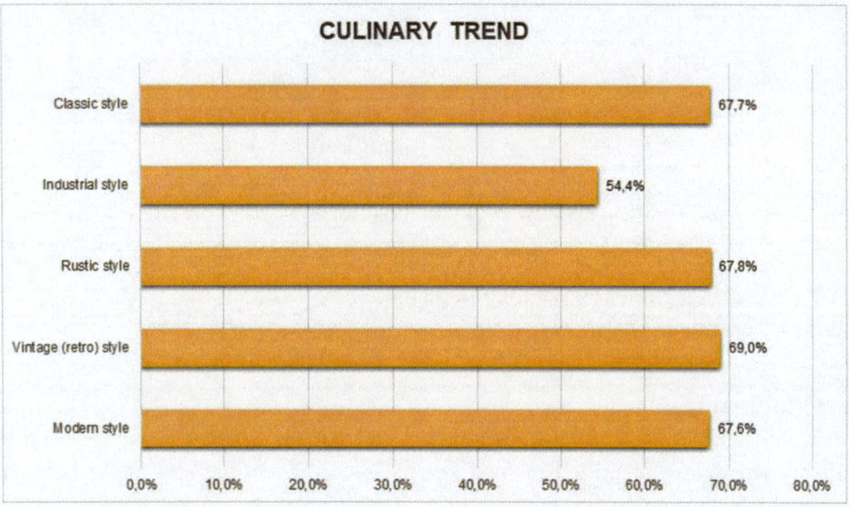

Fig. 3. Scenery preferences, Source: Ambato quality Q restaurant customers

The preference of vintage style (retro – old) 69% and rustic style 67.8% is determined. They are identified as the environments that the respondents like the most. It is also giving priority to retro and informal trends that are understood as the style in vogue in the city and with a growing trend among the new proposals.

The correlation analysis is carried out, in (Table 2) the calculation of Spearman's Rho correlation is shown.

The correlation coefficient in the gastronomic offer and culinary Trend variables is 0.471, which establishes a moderate correlation of these variables in the study carried out. In fact, culinary trends *affect* the gastronomic tourism proposal in food establishments with a quality Q distinctive in the city of Ambato.

Table 2. Spearman's Rho non-parametric correlation

Correlations				
			GASTRONOMIC PROPOSAL	CULINARY TREND
Spearman's Rho	GASTRONOMIC_PROPOSAL	Correlation coefficient	1,000	,471**
		Sig. (bilateral)		,000
		N	202	202
	CULINARY_TREND	Correlation coefficient	,471**	1,000
		Sig. (bilateral)	,000	
		N	202	203

** The correlation is significant at the 0.01 level (bilateral).
Source: SPSS 21 statistical program

5 Conclusions

Culinary trends are identified in food service providers in the city of Ambato. It is observed that the preference in the visit of establishments is in restaurants (68.50%) over other types of food businesses, the preference for places of consumption of local gastronomy (73.9%) and fusion cuisine (70.9%). According to its setting, aged (69.0%) and rustic (67.8%), defining with these characteristics the preferences in the culinary trends of the city. Following the Likert scale instrument, the consumers of the establishments indicate the importance of aspects such as Hygiene (85.7%), Food Quality (85.2%) and Taste (84.9%). Being the most relevant and directly related to supply, they are determining factors when deciding consumption. Finally, the correlation of culinary trends and the offer through Spearman's Rho statistical test (0.471) determines a moderate correlation, that is, there is dynamics between the variables studied.

Acknowledgement. The authors would like to thank the Universidad Técnica de Ambato (UTA) and the Dirección de Investigación y Desarrollo (DIDE) for their support for this work through the PROMOTION OF QUALITY-OF-LIFE research group. We would also like to thank the research project: GASTRONOMIC AND INNOVATION TOURISM BASED ON THE USE OF METAVERSES AND VIRTUAL REALITY FOR THE CENTRAL ZONE OF ECUADOR, approved by Resolution No. UTA-CONIN-2023-0050-R.

References

1. González Figueroa, J.: La identidad de la cocina mexicana. CULINARIA Revista virtual especializada en Gastronomía, pp. 03–15 (2018)

2. Huertas, T., Suárez, E., Cuétara, L.: Perfil del cliente gastronómico del cantón Mocha. UNIANDES Episteme, pp. 497–506 (2016)
3. Mejía López, L.S., Mejía Castillo, S.E., Bravo Rodríguez, M.: Tendencias gastronómicas: La encrucijada entre lo tradicional y lo innovador. CULINARIA Revista virtual especializada en Gastronomía (8), 27–40 (2014)
4. Meléndez Torres, J.M., Cañez de la Fuente, G.M.: La cocina tradicional regional como un elemento de identidad y desarrollo local. Centro de Investigación en alimentación y Desarrollo, pp. 182–204 (2010)
5. Hernández Heredia, C., Tamayo Salcedo, A., Castro Ricalde, D., Muñoz, I.: Tendencias gastronómicas predominantes en la producción de revistas científicas de Iberoamérica. Ciencia Ergo Sum **23**, 76–84 (2016)
6. Camacho, M., Mejía, M.: Turismo Gastronómico del cacao y el chocolate, tendencias latinoamericanas. Gran Tour: Revista de Investigaciones Turísticas 58–76 (2018)
7. Hernández, R., Dancausa, M.: TURISMO GASTRONÓMICO La gastronomía tradicional de Córdoba (España). Estudios y Perspectivas en Turismo **27**, 413–430 (2018)
8. Oliveira, S.: Guía para el desarrollo del turismo gastronómico. Estudios y Perspectivas en Turismo **20**(3), 738–752 (2011)
9. Aguirregoitia Martínez, A., Fernández Poyatos, M.: La simplicidad como proceso creativo culinario: una tendencia en el turismo gastronómico. PASOS. Revista de Turismo y Patrimonio Cultural **17**(5), 875–888 (2019)
10. Vázquez, G., Hernández, R., Dancausa, G.: Análisis del potencial en materia de turismo industrial gastronómico dentro de la provincia de Córdoba según localidad. Turismo, Desarrollo Y Buen Vivir. Revista De Inventigación De La Ciencia Turística 4–27 (2016)
11. Hernández Mogollón, J.M., Di Clemente, E., López Guzmán, T.: El turismo gastronómico como experiencia cultural. El caso práctico de la ciudad de Cáceres (España). Dianlnet (68), 407–427 (2015)
12. Reyes, A., Guerra, E., Quintero, J.: Educación en gastronomía: su vínculo con la identidad cultural y el turismo. El periplo sustentable (32) (2017)
13. Falcón, J.P.: Tendencias globales de desarrollo del turismo gastronómico aplicadas al caso de Argentina. Redmarka Rev. Mark. Apl. **1**, 35–67 (1970)
14. Cuevas, M., Pérez, M., Pecero, E.: Preservación de la herencia cultural a través del turismo gastronómico. Hitos de Ciencias Económico Administrativas **24**(68), 177–189 (2018)
15. Fernández-Sánchez, L., Rodríguez, Z., Pozo, J., Espinosa, J.: Estrategias para el Fortalecimiento del Turismo Gastronómico en el Cantón Pastaza, Ecuador. Rev. Amaz. Cienc. Tecnol. **5**, 118–136 (2016)
16. Jeambey, Z.: Rutas Gastronómicas y Desarrollo local: un ensayo de conceptualización en Cataluña. Revista de Turismo y Patrimonio Cultural **14**, 1187–1198 (2016)
17. Casas Mateus, J.A., Albarracín Tunjo, I.L., Cortés González, C.E.: Gastronomía molecular. Una oportunidad para el aprendizaje de la química experimental en contexto. TED: Tecné, Episteme y Didaxis **2**, 125–142 (2017)
18. Salazar, D., Burbano, C.: Análisis de la oferta gastronómica, una perspectiva comercial a través del marketing de servicios: Caso sector La Rumipamba, Pichincha, Ecuador. Revista Interam. de Ambient. Tur. **13**(1), 2–14 (2017)
19. Carabias, L.: Ofertas gastronómicas. Paraninfo, Madrid (2017)
20. Lemoine, F., Otero, B., Carvajal, G., Almeida, E.: El servicio de gastronomía y su impacto para el desarrollo turístico en Bahía de Caráquez. TRANSITARE **4**, 53–68 (2018)
21. Hernández-Ramírez, J.: Cuando la alimentación se convierte en gastronomía. Procesos de activación patrimonial de tradiciones alimentarias. Cultura-hombre-sociedad **28**(1), 154–176 (2018)
22. Gobierno Provincial de Tungurahua, H.: Ambato Ruta Agroturística Artesanal. H. Gobierno Provincial de Tungurahua, Ambato (2015)

23. Torres-Oñate, F., Viteri, M.F., Infante-Paredes, R., Donato-Moreira, S.: Heritage cooking as tourist motivation: ambato case study. In: Stankov, U., Boemi, S.-N., Attia, S., Kostopoulou, S., Mohareb, N. (eds.) Cultural Sustainable Tourism: A Selection of Research Papers from IEREK Conference on Cultural Sustainable Tourism (CST), Greece 2017, pp. 109–114. Springer International Publishing, Cham (2019). https://doi.org/10.1007/978-3-030-10804-5_11
24. Páez-Quinde, C., Torres-Oñate, F., Viteri, M.-F., Porras, M.-E.: M-learning as support tool in the diffusion of the traditional food: case study ambato. In: Auer, M.E., Hortsch, H., Sethakul, P. (eds.) ICL 2019. AISC, vol. 1134, pp. 327–335. Springer, Cham (2020). https://doi.org/10.1007/978-3-030-40274-7_33
25. Poveda, T., Jijón, G.: Nuevas tendencias alimentarias enfocadas en la comida típica ecuatoriana. Caso: huecas tradicionales de la ciudad de Ambato, Provincia de Tungurahua. EPISTEME **5**, 1400–1413 (2018)
26. Carvache, M., Carvache, W., Torres-Naranjo, M.: Análisis de Satisfacción La gastronomía de Samborondón – Ecuador. Estudios y Perspectivas en Turismo **26**, 731–745 (2017)

Integration of 4.0 Technology for Customer Service at the Huella-Verde Tourist Center in Pastaza-Ecuador

Fernández Villacrés Gustavo Eduardo[1](✉), Minga Gómez Juan Carlos[2], Trujillo Quinto Wilfrido Amilcar[2], and Arcos Naranjo Gissela Alexandra[1]

[1] Instituto Superior Tecnológico España, Ambato, Ecuador
{gustavo.fernandez,gissela.arcos}@iste.edu.ec
[2] Instituto Superior Mayor Pedro Traversari, Chillogallo Sector, Ecuador
{juan.minga,wilfrido.trujillo}@institutotraversari.edu.ec

Abstract. The purpose of this scientific article is to summarize the implementation of a process of adoption of various technologies in the hotel "Huella Verde" for the improvement of customer service. The starting point was the problem related to the scarce use of technologies as an element of operational and touristic support, based on this, the following research question was formulated: How to improve customer service in the hotel? In the methodological part, customers were investigated for their preferences and employees, both agreeing that the hotel needed more technician processes. It is considered that the internet should be implemented and technologies such as augmented reality, QR codes, and web pages should be adopted to put information related to the flora and fauna near the hotel. In the end, it is concluded that customer service can be improved based on the adoption of new 4.0 technologies in the hotel.

Keywords: Tourism · 4.0 technology · augmented reality · customer service · hotel

1 Introduction

The technological progress of the last 10 years has been exuberant, there has been the emergence or rather the maturity of certain technologies that were being developed little by little, among these technologies we have: artificial intelligence, robotics, the internet of things, and more. To these have been added other technologies such as big data, business intelligence, augmented reality, virtual reality, 3D printers, drones, cloud computing, blockchain, and more.

All this has given rise to the 4.0 technologies that are what have begun to support the new phenomenon of Industry 4.0. The fourth industrial revolution, also known as Industry 4.0, was launched in 2013 in Germany, under the endorsement of business and industry leaders, politicians, and academics who envisioned it as a means to raise the competitiveness of the German manufacturing industry through the gradual integration of cyber-physical systems into the different phases of d production [1].

C. Montenegro et al. (Eds.): ICMTT 2023, LNNS 774, pp. 261–271, 2024.
https://doi.org/10.1007/978-3-031-43733-5_24

On the other hand, Amezquita [2] points out that: in the fourth industrial revolution, robotics and artificial intelligence are used in industrial production, which affects the existence of human labor, whether repetitive or creative. Since there is a need for new forms of work, distribution, and exchange, this has a great impact on how companies are organized, and this is directly or indirectly related to the industrial revolution [2].

All 4.0 technology is having a direct impact on different areas of human endeavor, for example, augmented reality and virtual reality are greatly influencing education and marketing, on the other hand, drones and robots are becoming part of modern business logistics, cloud computing is used by small and medium-sized companies, robots are frequently used in production and so on in all areas.

Within the 4. o technologies, augmented reality is one of the most important, as it can be used to build mobile applications that can generate new experiences in tourism. Most mobile applications have focused on cultural aspects and urban environments, but unfortunately, they have had a little deployment in rural tourism [3].

The incorporation of augmented reality allows a lot of information of interest to the tourist destination to be added. On the other hand, augmented reality contributes to increasing attraction and increasing the visitor market. It can also be pointed out that virtual reality is in full development, especially the one oriented to cultural heritage and virtual visits to museums [4].

In recent years, tourism has grown considerably, this is essential because the means of transportation have been modernized, today it is very easy to take a plane and travel, also the internet has become the main means of communication and tourism promotion, there are also platforms or mobile applications to buy airline tickets, rent hotels, rent cars, buy bus tickets and more, this means that undoubtedly the use of information and communication technologies are increasingly affecting all areas of the hotel businessl [5].

All this can be summarized in the sense that the Internet boom is transforming the hotel business. This statement is evident in the complexity of the tourism infrastructure and the change in tourist behavior, which has brought new challenges induced by demand, changes in organizations, and an advance in the technological development of the tourism sector. The perspective of deepening knowledge in scenarios where the tourist is an active agent due to his interaction with digital technology is necessary to understand the role of the tourist as a co-creator of this technology [6].

Tourism activity is a global service, the tourism sector is a fundamental pillar in the economy of many countries because it means a very important percentage of the Gross Domestic Product. According to the secretary of the World Tourism Organization in his report in 2018, tourism represents 10.4% of the world's Gross Domestic Product. It is also stated that these percentages have been increasing due to globalization and that they have transformed the productive model of several economies to the point of turning them into true "tourist industries". However, having a high percentage of GDP does not mean that they generate higher revenues. As the World Economic Forum indicates in its annual report on tourism competitiveness, the leading tourism industries according to turnover are the United States, China, Germany, Japan, England, and France [7].

The growth of tourism has not only improved the overall economy but has also contributed to job creation and a healthy balance of payments. These last two advantages have helped modernize countries with backward economies and have created new opportunities for travel and new experiences thanks to some economic advantages [8].

According to new data from the World Tourism Organization [9] in the past year 2022, about 900 million people made international trips, which means that the number of travelers doubled compared to 2021. All parts of the world recorded an increase in international travel. The Middle East had the highest growth and can already be compared to pre-pandemic travel figures. The countries of North, Central, and South America recovered 80% of their pre-pandemic tourist numbers. Europe also recovered 75% of its former travelers.

The hotel Huella-Verde is considered a "Rainforest Lodge" which is nothing more than a lodge in the rainforest, the hotel is in the parish of "Canelos" which is an hour by bus from Puyo, has 4 Bungalow with a capacity for 16 people in total. It also has 4 cabins for 2 people. Among its services, we have a private bathroom, hot water shower, extra-large bed, and veranda with hammocks. The main activities to be carried out at the hotel include a visit to a river beach, a tour of a hidden waterfall, a tour of the jungle, a visit to an indigenous community, and a cacao tour. Despite the sustainable tourism philosophy, it is considered necessary to use technology, especially the Internet, and new elements such as virtual and augmented reality.

Many tourists, even though they are warned about the lack of internet, request it or often cancel their visits because of it. In addition, the presence of a guide is required to explain aspects of plants and animals found in the vicinity of the hotel. It should be remembered that most of the processes are manual and that tourists make contact through their web page (https://www.huella-verde.org/); in addition, the services have not been personalized, which makes the guest feel insufficiently welcomed. Although the number of clients remains stable, the owners feel that service needs to be improved. [10].

The following research question arises from the above: How to improve customer services at the Hotel "Huella Verde" in the province of Pastaza in the Republic of Ecuador?

To solve the problem, a project was proposed with the objective of: "Implementing innovative technological strategies to improve customer service at the jungle lodge "Huella Verde". Essentially a proposal is made in the sense of technological improvement to generate new user experiences and facilitate processes aimed at raising the level of customer satisfaction.

According to the hotel owners, it is possible to think that a digital transformation process is beginning, which implies that it is necessary to review some fundamental concepts of digital transformation.

According to Vacas [11], digital transformation implies business disruption, which means that there must be a profound change in products and services, and the emergence of value-added services that will take advantage of new technologies to have a direct impact on customer experience and income statements. It is considered that channels and processes should be optimized; new business models and revenue streams should be sought based on customer demands; and, naturally, the internal culture of the company should be changed to put digital at the center of everything it does [12].

Digital transformation should mean increased productivity and efficiency, as well as a new source of innovation and competition. Globalization and recent social policies have forced companies to embrace digitalization, demonstrating the benefits of doing so for both consumers and individual companies, while embracing the challenges that still need to be overcome to successfully adopt and implement these technologies. [13].

According to Moreno Gutiérrez (2021), this evolutionary process slows down or is valued at superfluous levels due to the lack of definition of the concept and the absence of identifiable objectives. It is therefore essential to establish a clear strategy with progressive tactics and to identify the key factors of digital transformation. [14].

In general terms, it can be pointed out that digital transformation is supported by four fundamental aspects, which are:

- Systematization of operational processes
- New digital business models
- Innovative user experiences
- Digital competencies of human talent [15]

Digital transformation processes in the hotel sector have been mostly assumed in the previous year, because of overcoming the pandemic process Covid-19. The following are some conceptualizations related to these transformation processes in tourism and hotel companies.

2 Methods

A qualitative-quantitative cross-sectional study was carried out (according to Gómez, et al. [16] the qualitative aspects are based on interviews and observation while the quantitative aspects are based on surveys), the research involved observation and a direct interview with the hotel manager, Mr. Marco López, while the quantitative aspects were based on a customer survey. Bibliographic research was carried out as a state-of-the-art to develop the introduction according to the subject matter. The research methodology applied was the survey and the interview. The instruments associated with the methodology are the electronic questionnaire and the interview guide. For Hernández S [17], The population is made up of all those involved in the problem, either directly or indirectly, for this research we worked with a population made up of 43 clients from the last 12 months, 6 employees and 1 manager: 43 customers from the last 12 months, 6 employees and 1 manager. In the data collection plan, a period of 30 days was stipulated for them to answer the electronic questionnaires sent to the different clients; after this period, we worked with the people who answered the questionnaires. An exclusion was made for those clients who were friends or relatives of the manager.
The hypotheses proposed are:

Ho = With the integration of 4.0 technology, customer service will be improved, in the tourist center Huella-Verde in Pastaza-Ecuador.
H1 = With the integration of 4.0 technology, customer service will not be improved at the Huella-Verde tourist center in Pastaza-Ecuador.

3 Results

After the development of the client surveys, the following results have been obtained:

Question No 1. How much did you miss the internet during your stay? (Fig. 1)

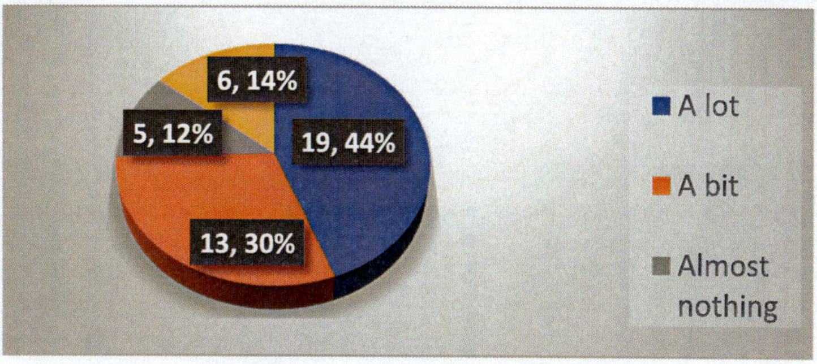

Fig. 1. Results of Question 1

Question No. 2. Was the amount due for your lodging and consumption issued promptly? (Fig. 2)

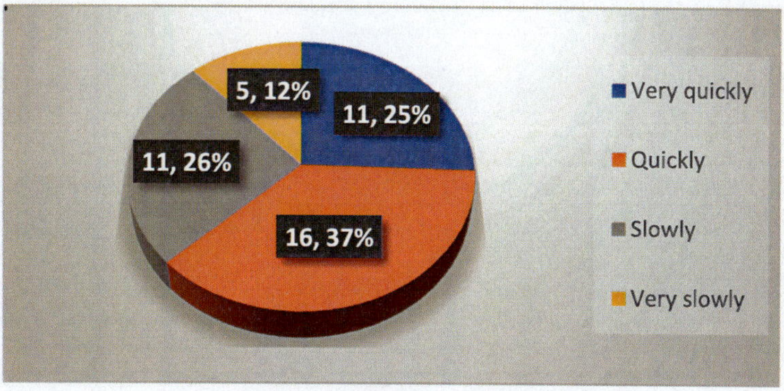

Fig. 2. Results of Question 2

Question No. 3. In your nearby tours, did you have all the technical information about plants, flowers, animals, the history of the region, and more? (Fig. 3)

The main results of the employee survey are published below.

Question No. 1. Has the lack of internet access in the hotel been noticed by customers? (Table 1)

Question 2. Do you think the hotel should rely more on technology? (Table 2)

The following is the interview with the manager.

Question No. 1. Mr. Manager, what is the conception of the hotel?

Answer: The philosophy of the hotel is to provide a great service to the client and to isolate him/her totally from civilization to provide comfortable lodging but isolated

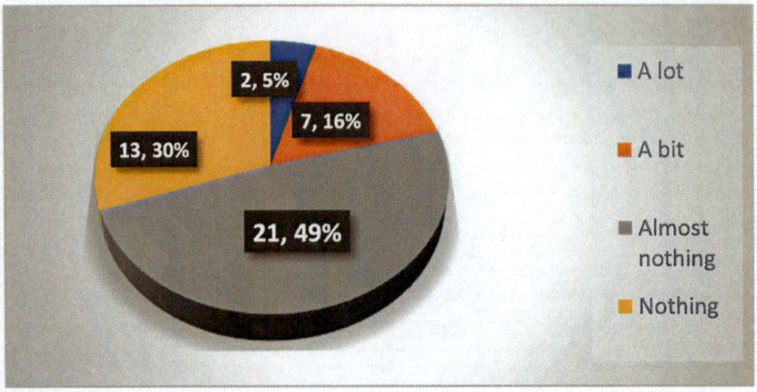

Fig. 3. Results of Question 3

Table 1. Results of question 1 to employees

Response	Frequency	Percentage
very noticeable	3	50%
notorious	2	33%
little noticeable	1	17%
Nothing noticeable	0	0%
Total	6	100%

Table 2. Results of question 2 to employees

Response	Frequency	Percentage
Yes	5	83%
No	1	17%
Total	6	100%

from the environment in which each person lives, and also to rediscover the jungle and to make sustainable community tourism.

Question No. 2. Why think about technological support then?

Answer: Personally, I am aware that many people have requested the possibility of internet access, there have been cases in which families have stayed and the parents do not want internet, but the children do, so there has been a family conflict. On the other hand, in the nearby jungle walks we need to provide information about certain plants, about history, about fables that are generated at the site, and I believe that with technological aspects we could provide this information. On the other hand, since the invoicing processes are manual, on several occasions' mistakes have been made and we have been undercharged. In addition, I think clients must have a minicomputer center, where they can consult and print what they need.

4 Discussion

First, a summary of the findings will be made:

Regarding the clients that have required the internet during the stay, 44% indicate that it is a lot and 14% that a little, this means that 58% of the clients are requiring the internet.

As for the speed of invoicing, 25% of the customers indicated that it was slow, while 11% indicated that it was very slow, which means that 36% of the customers are dissatisfied with the speed of issuance of their respective invoices.

Regarding the information obtained during the tourist tours in the vicinity of the hotel, 49% indicated that they obtained very little information and 30% indicated that they received no information at all, which means that 79% of the clients did not receive information about plants, animals, biodiversity, history, fables and more during their visits to the sites near the hotel.

As for the employees, 83% of them indicate that the request for internet by customers has become noticeable, and 84% personally consider that the hotel needs to rely more on technology to improve customer service at the hotel.

The manager also notes in his interview that: even though the hotel is a sustainable environmental conservation enterprise, the main requirement of the clients is the provision of Internet access for different activities.

The following is a comparative process with several studies related to technology in hotel management.

For example, the one developed by Alberto Martínez and Jhoana Moreno [18] with their topic "Using information and communication technologies (ICT) as a competitive factor in the management of the hotel sector in Villavicencio, Colombia" which is a descriptive study conducted in Villavicencio, Colombia, in which 34 hotels are evaluated in 7 aspects related to technology, these aspects are characterization, technological infrastructure, connectivity, frequency of internet access, ICT usability in hotel management, appropriation of ICT for hotel management and ICT contribution to hotel management: Characterization, technological infrastructure, connectivity, frequency of internet access, the usability of ICTs in hotel management, appropriation of ICTs for hotel management and contribution of ICTs to hotel management. The instrument had 31 questions and the main results are: 100% of the hotels use the internet, 75% of the customers logged on to the internet, about the services currently offered by companies worldwide on topics such as SaaS (software as a service), PaaS (platform as a service), IaaS (infrastructure as a service), very little use is made of data analytics and/or big data, with 2.9%, or IT security also with 2.9%. As for backup as backup and the Internet of Things, there is a higher use, but it is still low, with 14.7% for both cases. The highest percentage of use is in terms of cloud storage with 20.6%. All these data perfectly validate the present research since the Internet must exist in every hotel. The study shows that 75% of hotel users access the Internet, while in this research 80% of customers required the use of the Internet, and 85% of employees indicate that there should be greater support from technologies as an element of support in hotel customer service. From all this, it can be deduced that the competitiveness of a hotel depends on the level of technology it uses [18].

We have also reviewed the work of Carmen Martínez [19] in her paper "Customer Satisfaction with the Arrival of new technologies in the hotel sector" in which she defines the use of technologies such as robots for check-in, access to services or systems with the use of the digital fingerprint in the aspects of biosecurity [20]. Making the simile with the present research, we are aware that the idea is the same because we want to put more technological elements for the improvement of the quality of customer service.
The hypothesis is then demonstrated.
The hypotheses defined are:

Null hypothesis. Ho = With the integration of 4.0 technology, customer service will be improved at the Huella-Verde tourist center in Pastaza-Ecuador.
Research hypothesis. H1 = With the integration of 4.0 technology, customer service will be improved at the Huella-Verde tourist center in Pastaza-Ecuador.

$$\mathrm{Zo} = (\mathrm{X} - \mathrm{nPo})/(\mathrm{nPo}(1 - \mathrm{Po}))^{1/2}$$

where:

n = Sample size.
X = Elements that meet the condition.
Po = Proportion to test.

A significance level of 5% is assumed and it is considered that if more than 90% of the respondents affirm that: with the integration of 4.0 technology, customer service will be improved.
The criteria for rejection of H0 are as follows:
If H1: $p > p0$ H0 is rejected if $Z0 > z\alpha$
Which in other words would be:

If H1: $p < p0$, H0 is NOT rejected if
$Z0 < z\alpha$
Replacing values in the formula we have

$$\mathrm{Zo} = (43 - 50 * 0,9)/(50 * 0,9(1 - 0,9))^{1/2}$$

$$\mathrm{Zo} = 2/14,46$$

$$\mathrm{Zo} = 0,138$$

The value of Za deduced from the tables is 1.68. The conditions for acceptance and rejection of the hypothesis are illustrated below (Fig. 4).

Being Zo < Za the null hypothesis Ho formulated above is accepted, therefore, it is accepted that: With the integration of technology 4.0 IF customer service in the hotel will be improved.

Technological Integration Proposal
As the main element of technological integration is the Internet, it is proposed:

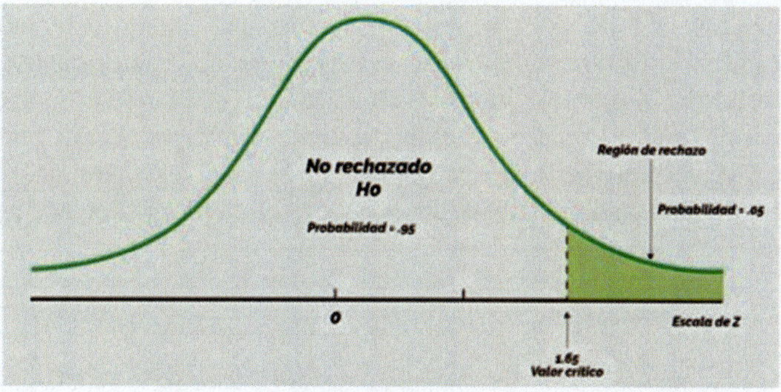

Fig. 4. Gaussian bell

1. Acquisition of a dedicated internet communication channel through antennas from Canelos to the hotel: Canelos to the hotel, a bandwidth of at least 100 Mbs is required. If this is not technically possible, a satellite channel can be considered.
2. Creation of a public Wi-Fi zone, preferably outside the hotel. Enhance this wi-fi network with network expanders such as Ubiquiti nano station.
3. Implement a small computer center with 3 networked computers and a network printer that can be accessed by customers via Wi-Fi.
4. Install a web portal with information about animals of the region, plants, people, climate, fables, and others so that customers can access them through a QR code. Then in the areas near the hotel QR codes will be implemented and accessed through phones connected via Wi-Fi.
5. Implementation of an augmented reality mobile application associated with QR codes means that if the QR code is read, the plant appears in a 3D video or on a descriptive web page.
6. Install a customer registration system with billing for services, the billing will be electronic, and the system will also generate a database of customers for subsequent business intelligence with them.
7. The business intelligence process will be done with big data and from there data will be obtained such as date of birth, food preferences, and requirements in terms of types of medicine, e.g., insulin, and blood pressure pills.
8. The proposal is complemented by a slight remodeling of the entrances to make the hotel more accessible for people with disabilities.

The technological implementation will have a maximum budget of US$12,000 and the investment should be amortized in a maximum of 2 years.

5 Conclusions

The following conclusions were drawn from the research conducted:

There is a need for technological implementation in some respects related to customer service, aspects such as: in the tourist tours they lack information and this can be complemented with augmented reality, high-speed internet is required in the hotel, operational processes such as billing should also be automated.

The integration of technology will not change the hotel's philosophy of developing community and conservation tourism. This is very important because the hotel also tries to generate a culture of conservation in its clients based on the love of nature.

It is very important to point out that the technology integration will generate better customer service and a better image before potential customers, therefore, the proposal is in line with the expectation that it will generate more income to become a sustainable development enterprise.

References

1. Joyanes, L.: Industria 4.0. Alfaomega, Ciudad de México (2017)
2. Amézquita, P.: La Cuarta Revolución Industrial y Algunas Implicaciones en las Escuelas de Negocios. Palermo Bus. Rev. 185–200 (2018)
3. López, E., Sevilla, J.: Realidad aumentada en destinos turísticos rurales: oportunidades y barreras. Int. J. Infor. Syst. Tourism (IJIST) **4**(2), 56–70 (2021)
4. Bernard, M.: Nuevas tecnologías y difusión del turismo cultural: descubriendo a Goya con realidad aumentada. Repositorio digital Universidad de la Coruña (2020)
5. Salas, B., Juarez, G., Carmona, T.: Índice de Madurez Digital para Empresas Hoteleras Familiares de Veracruz, México. HAMUT'AY **7**(3), 11–29 (2021)
6. «La neutralidad del turismo electrónico: su huella en el turista digital y en la transformación digital de la industria turística. Latin Am. J. Comput. 1–20 (2021)
7. Sanchez, A.: Transición digital y reajuste del sector turístico en la Unión Europea. Revista Internacional de Turismo, Empresa y Territorio **6**(2), 1–24 (2022)
8. Vasi, S., Escudero, C.: La transformación digital en el sector turístico, vol. 1, nº 1, pp. 1–30. Las redes sociales. Universidad de Comillas (2018)
9. OIT: Organización Mundial del Turismo, El turismo va camino de recuperar los niveles prepandémicos en algunas regiones en 2023, 17 Enero 2023. https://www.unwto.org/es/news/el-turismo-va-camino-de-recuperar-los-niveles-prepandemicos-en-algunas-regiones-en-2023. Último acceso: 28 Febrero 2023
10. Esprilla, M., Valderrama, M.: El rol de la transformación digital en las prácticas de comercialización del sector hotelero colombiano . 12 Octubre 2022
11. Vacas Aguilar, F.: Transformación digital: del lifting ala reconversion. TCyE 135–143 (2018)
12. Cabra, C., et al.: Plan de transformación digital para la promoción turística en el viceministerio de turismo. Caso de estudio: Departamento de Boyaca. Turismo y Sociedad **28**, 151–185 (2021)
13. Moreno Gutiérrez, N.: Transformación digital en procesos productivos. Transformación Digital en las PYMES Manufactureras 117–144 (2021)
14. García, J.: Las 6 claves de la Transformación Digital Definición y desarrollo. IEBS, Madrid (2020)
15. Castro, A., Delgado, T., Ash G.: Diagnóstico. De transformación digital sobre hotelería. Caso de estudio en el Hotel Nacional de Cuba, pp. 542–552. Universidad y Sociedad (2022)
16. Gómez, et al.: La investigación científica y sus formas de titulación. Aspectos conceptuales y prácticos. Editorial Jurídica del Ecuador. Uniandes, Ambato (2017)
17. Hernández, S.: Metodología de la investigación. McGrawHill, México (2017)
18. Caro, M., et al.: Calidad de las tecnologías de la información y competitividad en los hoteles de la península de Yucatán. Contaduría y Administración **4**(1), 11–19 (2018)

19. Martínez, C.: La satisfacción del cliente con la llegada de las nuevas tecnologías en el sector hotelero. 1 Julio 2021. https://repositorio.comillas.edu/xmlui/handle/11531/64647. Último acceso: 6 Enero 2023
20. Joyanes, L.: Industria 4.0. La cuarta revolución industrial. AlfaOmega, Madrid (2017)

After the Hiking Tour: Walkers' Behaviour

Sara Silva(✉), Tiago Marques, Jorge Novais, Saulo Folharini, and António Vieira

University of Minho, Av. da Universidade, 4800-058 Guimarães, Portugal
sara.silva@eaad.uminho.pt, vieira@geografia.uminho.pt

Abstract. Pedestrianism is a recreational, leisure, sporting, and tourist activity that requires the creation of strategies to ensure the sustainability and profitability of resources. It is essential to understand, beyond the motivations for seeking a particular region for nature hiking, whether there is concern about signalling problems detected along the route after completing the course. As part of the CLICTOUR project – Climate change resilient tourism in protected areas of Northern Portugal, we seek to understand the reasons why hikers do not repeat their choice of pedestrian route and whether there is a concern to signal detected problems to local authorities along the pedestrian route. The research we conducted allowed us to understand that long-distance hikers tend to report situations of absence and vandalism of signalling, as well as rubbish dumping sites. As for the reasons for not repeating the pedestrian route, participants who prefer to do local routes stop doing them due to the lack of maintenance. Those who prefer local routes give reasons such as the lack of maintenance, degree of difficulty, lack of support infrastructure, slippery, muddy and/or rocky paths, and paths with poor vegetation.

The offer of pedestrian routes is fundamental to organize the use of forest spaces, especially reserve and nature protection spaces. For the sustainability of natural areas, it is essential to have communication strategies capable of receiving feedback from hikers in order to improve the experience of others.

Keywords: Pedestrianism · Sustainability · Feedback

1 Introduction

After the pandemic period, people started to search for natural and nature spaces with greater interest, giving them new value and importance. The contact with and appreciation of nature and open spaces has gained a growing demand from the general population, especially those originating from urban areas. Nevertheless, hiking is not a recent trend. In fact the practice of organized hiking tours dates back the 18th century, when consisting mainly in closely associated with England. The need to travel between small population centres in natural areas led to the creation of footpaths and connecting trails. The absence of motorized vehicles led to the construction of "natural" paths that emerged from the constant trampling by people and animals [1]. The advent of modernity brought new forms of transportation and created new, less winding and more easily accessible routes. However, the itineraries that have been created have remained in the natural culture of

C. Montenegro et al. (Eds.): ICMTT 2023, LNNS 774, pp. 272–286, 2024.
https://doi.org/10.1007/978-3-031-43733-5_25

hiking between the villages that inhabit the natural areas. According to Daniel Svensson, Sverker Sörlin, and Katarina Saltzman [1], pedestrian routes are a sociocultural process that arises naturally thanks to the historical burden of their purpose and function and should be considered heritage for this reason. The trails that wind through the natural landscapes are not just the result of human trampling. Herbivorous mammals are the main generators of trampling vegetation, leaving multiple paths along the routes they trace through nature [2].

In Portugal, it was only in the 1980s that the first pedestrian routes began to be established [3, 4]. There are numerous studies that address pedestrian routes and the impact of nature on people, the impact of people on territories, and management procedures to modify unwanted impacts [5–13].

2 Infrastructure for Hiking Practice

As a basis for the construction of knowledge on hiking, the reading of various documents from the Portuguese Camping and Mountaineering Federation (FCMP), which is responsible for the National Register of Hiking Trails, was used. According to information provided by the FCMP [14], hiking is a sports practice that has been on the rise in recent years. As it is a sport, it has its own regulations and is subject to locations prepared for its practice, which can be compared to a sports facility. 'Hiking is generally practiced on traditional and ancient paths that deserve to be preserved as they are a privileged means of contact with nature and interpretation of the environment, promoting sustainable development and nature conservation' [14]. According to the National Nature Tourism Program, applicable in the National Network of Protected Areas, created through Resolution of the Council of Ministers No. 112/98, of August 25, hiking is an 'activity of hiking distances in nature, in which tourist, cultural, and environmental aspects intervene, usually developing along well-defined paths, marked with internationally accepted codes and signs. Great Route hiking trails are routes with a length of more than 30 km and requiring more than one day of travel, designated by the letters GR followed by the registration number and the white and red marking colour. When they are trans-European (starting or ending in Portugal and running through more than three countries), the numbering is completed with the letter E (Europe) and the respective European numbering. Small Route hiking trails are routes with a length of less than 30 km, not exceeding a day's journey, designated by the letters PR, sometimes followed by the registration number and letters designating the municipality, with yellow and red colours. Finally, Local Route do not have a fixed length, but are identified on the ground with the letters PL and green and white colours. In their genesis, it is a route whose path is entirely or more than half in an urban environment [14]. The homologation of hiking trails is based on the philosophy that 'properly marked hiking trails in areas of great interest, promoted and monitored for maintenance, become attractive as active tourism products' [14].

One of the main gaps we observe is the lack of official information on the number of approved trails, with dates of approval, inspection dates, and information on evaluations. The homologation of hiking trails should be synonymous of the quality of facilities for hiking practice, as well as the safety of practitioners and the protection of the environment where the sport takes place. These rules are particularly important to be followed when implementing trails in nature protection areas. According to research by MacLeod [15], having numerous hiking trails does not necessarily mean quality of experience. There are requirements that must be taken into consideration 'the scale of the trail will determine the mode of transport used, but it is possible to promote more environmentally responsible forms of transport on long distance trails as well as on shorter hiking tours' [15].

In this article, we aim to present the research conducted with hikers in nature, crossing their characteristics with post-trail perceptions and the reasons why they do not repeat the choice of the hiking trail.

3 Materials and Methods

The present research is part of the CLICTOUR project – Climate change resilient tourism in protected areas of Northern Portugal [NORTE-01-0145-FEDER-000079]. The focus of this study was to understand the behaviour of nature walkers after completing marked routes (e.g., GR, PR, PL). The aim of this study was to verify with participants their behaviour after completing the route in situations such as the absence of signage, acts of vandalism, littering areas or the evaluation of the experience of hiking the pedestrian route. Still within the context of post-route completion, we sought to understand the reasons why participants do not repeat their choice of route.

Between June and September 2022, a public opinion survey was conducted using a semi-structured qualitative questionnaire implemented with the google forms platform. This is an approach that requires low levels of financial and human resources, and is increasingly used by researchers [16–18]. The questionnaire was disseminated through the project's social media channels, various Facebook pages of nature walkers, as well as through promotional posters with a QR Code that linked to the online survey.

The questions were designed according to the general objectives for building knowledge about the characteristics of pedestrians and their post-route completion behaviours'. The questions were constructed based on a literature review on the preferences and motivations of pedestrians [19]. Two questions were constructed with a list of items to be selected by participants. The analysis was carried out considering three groups: participants who prefer GR; participants who prefer PR; participants who prefer PL.

Post-route completion feedback is of significant relevance, thus the following hypotheses were developed:

H_1: the preferences of the survey respondents for pedestrian routes and the post-route completion signalling of detected problems are independent, that is, not related.
H_a: the preferences of the survey respondents for pedestrian routes and the post-route completion signalling of detected problems are not independent, that is, they are related.

The state of nature conservation, the characteristics of the pedestrian routes and the state of the pedestrian route infrastructure may be motivations for taking the route, but post-route completion may be the reasons for not repeating it. Assessing the reasons for not repeating a route is fundamental to understanding the potential of the equipment and for managing future resources to be built. Based on the responses obtained from the participants, we aim to understand the reasons why they do not repeat their choice of pedestrian route, and develop the following hypotheses:

H2: the preferences of the survey respondents for pedestrian routes and the reasons why participants do not repeat their choice of route are independent, that is, not related.
H_b: the preferences of the survey respondents for pedestrian routes and the reasons why participants do not repeat their choice of route are not independent, that is, they are related.

The results of the public opinion survey were analysed and processed in the IBM SPSS Statistics 27 software. As the variables have qualitative characteristics, the non-parametric statistical method of the Chi-square independence test [20, 21] was used for comparison. Nominal qualitative variables were tested based on the contingency table analysis that relates the categories. All conditions for the application of the test were met in most variables (e.g., there are no cells with an expected frequency of less than five observations; all cells have an expected frequency of more than one).

4 Results

The sample consists of 337 participants, with (GR = 66; PR = 153; PL = 118). The participants' profile is quite homogeneous regarding their preference for pedestrian routes, with the PR being the preferred pedestrian route (see Table 1). As for the sociodemographic characteristics of the participants, we have a group mostly composed of women (n = 210, % = 62.3), whose preference for pedestrian routes mostly varies between PL (41%) and PR (42.4%). In the case of men (n = 127), the majority prefers PR (50.4%). The participants' average age is 43.12, and most of them are in the age group between 25 and 64 years old (84%). The age group between 25 and 44 mostly prefers PR (44.4%), as well as the age group between 45 and 64 (49.6%). The participants mostly have higher education (n = 227), and this group mostly prefers PR (48.9%). These are participants who are employed (n = 278, % = 82.5), and most of them are married (n = 173, % = 51.3). The monthly net total income of the household is mostly in the range of €1001 to €2500 (56.4%). Considering the profile of hikers outlined by several studies (22–25), we can conclude that the participants in our study have a very similar profile.

Table 1. Characteristics of the participants.

	Preferred signposted hiking routes					
	GR		PR		PL	
	n	%	n	%	n	%
Gender						
Female	35	16,7%	89	42,4%	86	41,0%
Male	31	24,4%	64	50,4%	32	25,2%

(continued)

Table 1. (*continued*)

	Preferred signposted hiking routes					
	GR		PR		PL	
	n	%	n	%	n	%
Age (years)						
15–24	9	28,1%	11	34,4%	12	37,5%
25–44	29	19,6%	66	44,6%	53	35,8%
45–64	25	18,5%	67	49,6%	43	31,9%
≥65	3	13,6%	9	40,9%	10	45,5%
Educational level						
Primary education	0	0,0%	1	20,0%	4	80,0%
First cycle (High school)	0	0,0%	1	50,0%	1	50,0%
Second cycle (High school)	4	18,2%	7	31,8%	11	50,0%
Third cycle (High school)	23	28,4%	33	40,7%	25	30,9%
Higher education	39	17,2%	111	48,9%	77	33,9%
Employment situation						
Domestic worker	2	66,7%	0	0,0%	1	33,3%
Unemployed	4	33,3%	3	25,0%	5	41,7%
Employed	52	18,7%	134	48,2%	92	33,1%
Reired	3	13,6%	10	45,5%	9	40,9%
Student	5	22,7%	6	27,3%	11	50,0%
Marital status						
Married	27	15,6%	80	46,2%	66	38,2%
Single	30	24,2%	53	42,7%	41	33,1%
Divorced	8	22,9%	17	48,6%	10	28,6%
Wid owed	1	20,0%	3	60,0%	1	20,0%
Household income						
until 500€	4	36,4%	3	27,3%	4	36,4%
500€–1000€	15	21,7%	28	40,6%	26	37,7%
1001€–2500€	34	17,9%	91	47,9%	65	34,2%
2501–3000€	4	10,8%	20	54,1%	13	35,1%
≥3000€	9	30,0%	11	36,7%	10	33,3%

GR – Great Route. PR – Small Route. PL – Local Route.

4.1 After the Hiking Tour

Considering verifying whether the preference of pedestrian routes indicated by respondents influences their signalling behaviour after completing the route, the Chi-square test of independence was performed for each response option. At a significance level of <0,05, there is statistical evidence to assert that the preferences of pedestrian routes of the respondents and the signalling behaviour after detecting problems during the route are not independent, i.e., they are related [accepted H_a]. The relationship is present with variables such as the absence of signage [$X^2_{(2)} = 10{,}229$; p-value = 0,006], situations of vandalism [$X^2_{(2)} = 6{,}673$; p-value = 0,037], and locations for waste disposal [$X^2_{(2)} = 10{,}743$; p-value = 0,004] (Table 2).

Table 2. Contingency table of the items listed to the local authorities in relation to the hiking routes of preference of the participants.

Variables	GR	PR	PL			
	Yes*	Yes*	Yes*	Person x^2	df	Exact Sig
Signal the local authorities						
Absence of signage				10,229	2	0,006
Count (%)**	49_a (74,2%)	124_a (81,0%)	108_b (91,5%)			
Expected Count	55,0	127,6	98,4			
Residual	−6,0	−3,6	9,6			
Standardized Residual	−0,8	−0,3	1,0			
Adjusted Residual	−2,2	−1,1	2,9			
Situations of vandalism				6,673	2	0,037
Count (%)**	42_a (63,6%)	$117_{a,b}$ (76,5%)	95_b (80,5%)			
Expected Count	49,7	115,3	88,9			
Residual	−7,7	1,7	6,1			
Standardized Residual	−1,1	0,2	0,6			
Adjusted Residual	−2,5	0,4	1,6			

(continued)

Table 2. (*continued*)

Variables	GR	PR	PL			
	Yes*	Yes*	Yes*	Person x^2	df	Exact Sig
Locations for waste disposal				10,743	2	0,004
Count (%)**	42_a (63,6%)	$113_{a,b}$ (73,9%)	100_b (84,7%)			
Expected Count	49,9	115,8	89,3			
Residual	−7,9	−2,8	10,7			
Standardized Residual	−1,1	−0,3	1,1			
Adjusted Residual	−2,5	−0,7	2,9			
Evaluation the experience				1,017	2	0,652
Count (%)**	62_a (93,9%)	148_a (96,7%)	112_a (94,9%)			
Expected Count	63,1	146,2	112,7			
Residual	−1,1	1,8	−0,7			
Standardized Residual	−0,1	0,1	−0,1			
Adjusted Residual	−0,7	1,0	−0,4			
Do not signal				1,595	2	0,466
Count (%)**	45_a (68,2%)	91_a (59,5)	71_a (60,2%)			
Expected Count	40,5	94,0	72,5			
Residual	4,5	−3,0	−1,5			
Standardized Residual	0,7	−0,3	−0,2			
Adjusted Residual	1,3	−0,7	−0,3			
Never find problems				5,088	2	0,080
Count (%)**	51_a (77,3%)	117_a (76,5)	77_a (65,3%)			

(*continued*)

Table 2. (*continued*)

Variables	GR	PR	PL			
	Yes*	Yes*	Yes*	Person x^2	df	Exact Sig
Expected Count	48,0	111,2	85,8			
Residual	3,0	5,8	−8,8			
Standardized Residual	0,4	0,5	−0,9			
Adjusted Residual	0,9	1,4	−2,3			

GR-Great Route. PR-Small Route. PL-Local Route. * Positive responses analysis. **% routes of preference.

As independence of variables was not rejected for the present investigation, an analysis of adjusted residuals was conducted. For a significance level of 0,05 an adjusted residual with a value less than −1,96 or greater than 1,96 indicates a relationship between the variables: absence of signage (GR Z = −2,2; PL Z = 2,9), vandalism of signage (GR Z = −2,5), and locations for waste disposal (GR Z = −2,5; PL Z = 2,9). However, the Z value needs to be adjusted, as it is a 2x3 comparative grid. Following MacDonald & Gardner's (2000) post hoc analysis, a Bonferroni correction was applied, resulting in a required significance level (α) of 0,0083 instead of 0,05 [22].

After Bonferroni adjustment, the adjusted critical Z-score (Zcrit = 2,63) (see Fig. 1) with standardized residual value resulted in only statistically significant difference between the observed and expected values for the absence of signage (Zcrit = 2,9) and garbage dump sites (Zcrit = 2,9).

$$\text{Zcrit} = \frac{adjustedalpha}{(n.^{\underline{o}}Lines \times n.^{\underline{o}}Columns)} \tag{1}$$

Another possible observation is pairwise post hoc analysis using Bonferroni correction [22]. The alpha level will maintain adequate control of the error rate for the contingency table size. In Table 2, a letter is associated with each count value. If all the letters are the same in a row, it means that for each participant's preference-marked route (column value), there is no significant difference among them at the 0,05 level.

The comparison of proportions indicates no differences in the evaluation of the experience between participants who did not mark problems and those who never encountered problems based on pedestrian route preference. However, the remaining variables have proportions that differ in at least one pair. This is the case for the variable of absence of signage, where the frequency of preferences for GR and PR are different from the frequency of LR.

4.2 Post-Route Completion Feedback Reported to Local Authorities

Considering the preference of pedestrian routes indicated by the respondents, in order to verify whether it influences the reasons why respondents do not choose to repeat the pedestrian route. The independence test for each response option revealed a relationship with: Lack of maintenance of the route [$X^2_{(2)} = 8{,}774$; p-value = 0,012]; High degree of difficulty [$X^2_{(2)} = 28{,}501$; p-value = 0,000]; Absence of a panoramic view [$X^2_{(2)} = 8{,}774$; p-value = 0,012]; Routes with little vegetation and no character [$X^2_{(2)} = 10{,}798$; p-value = 0,005]; Slippery, muddy and/or rocky routes [$X^2_{(2)} = 8{,}883$; p-value = 0,012].

As explained for the previous hypothesis, since the independence of the variables was not rejected, an analysis of adjusted residuals was performed for each variable (Tables 3 and 4). There are adjusted residuals with statistically significant differences between observed and expected values at a significance level of 0.05: Lack of trail maintenance (PR Z = 2,9; PL Z = −2,5); Degree of difficulty (GR Z = −4,4; PL Z = 4,4); Lack of support infrastructure (GR Z = −2,4; PL Z = 3,1); Trails with poor vegetation and character (PL Z = −3,2); Slippery, muddy, and/or rocky trails (PL Z = 2,7).

Table 3. Contingency table of the items listed to reasons for not repeating the route in relation to the hiking routes of preference of the participants (part I).

	GR	PR	PL			
Variables	Yes*	Yes*	Yes*	Person x^2	df	Exact Sig
Reasons not to repeat the route						
Overcrowding of users				0,400	2	0,817
Count (%)	24_a (36,4%)	49_a (32,0%)	40_a (33,9%)			
Expected Count	22,1	51,3	39,6			
Residual	1,9	−2,3	0,4			
Standardized Residual	0,4	−0,3	0,1			
Adjusted Residual	0,5	−0,5	0,1			
Lack of maintenance of the trail	8,774	2	0,012			
Count (%)	$25_{a,b}$ (37,9%)	76_b (49,7%)	38_a (32,2%)			
Expected Count	27,2	63,1	48,7			
Residual	−2,2	12,9	−10,7			

(continued)

Table 3. (*continued*)

	GR	PR	PL			
Variables	Yes*	Yes*	Yes*	Person x^2	df	Exact Sig
Standardized Residual	−0,4	1,6	−1,5			
Adjusted Residual	−0,6	2,9	−2,5			
High level of difficulty	28,501	2	0,000			
Count (%)						
Expected Count	7_a (10,6%)	49_b (32,0%)	58_c (49,2%)			
Residual	22,3	51,8	39,9			
Standardized Residual	−15,3	−2,8	18,1			
Adjusted Residual	−4,4	−0,6	4,4			
Absence of a panoramic view	8,774	2	0,012			
Count (%)	9_a (13,6%)	21_a (13,7%)	16_a (13,6%)			
Expected Count	9,0	20,9	16,1			
Residual	0,0	0,1	−0,1			
Standardized Residual	0,0	0,0	0,0			
Adjusted Residual	0,0	0,0	0,0			
Absence of a trails objective	0,002	2	1,000			
Count (%)	9_a (13,6%)	12_a (7,8%)	14_a (11,9%)			
Expected Count	6,9	15,9	12,3			
Residual	2,1	−3,9	1,7			
Standardized Residual	0,8	−1,0	0,5			
Adjusted Residual	1,0	−1,4	0,7			

(*continued*)

Table 3. (*continued*)

	GR	PR	PL			
Variables	Yes*	Yes*	Yes*	Person x^2	df	Exact Sig
Absence of support infrastructure	2,089	2	0,349			
Count (%)	2_a (3,0%)	$14_{a,b}$ (79,2%)	22_b (18,6%)			
Expected Count	7,4	17,3	13,3			
Residual	−5,4	−3,3	8,7			
Standardized Residual	−2,0	−0,8	2,4			
Adjusted Residual	−2,4	−1,1	3,1			

GR-Great Route. PR-Small Route. PL-Local Route. * Positive responses analysis. **% routes of preference.

As explained earlier, it is necessary to adjust the alpha ($\alpha = 0.0083$) and the critical Z (Zcrit = 2,63). The statistically significant difference between observed and expected values is observed through the adjusted residual with a value below −2,63 or above 2,63. Thus, we have the variables: Absence of trail maintenance (PR Z = 2,9); Degree of difficulty (GR Zcrit = −4,4; PL Zcrit = 4,4); Lack of support infrastructure (PL Zcrit = 3,1); Trails with poor vegetation and character (PL Zcrit = −3,2); Slippery, muddy and/or rocky trail (PL Zcrit = 2,7) as statistically significant reasons.

The post hoc analysis (Pairwise Analysis) [22]. The comparison of proportions indicates no differences between preferences for pedestrian routes and reasons: pedestrian routes are overcrowded, do not have a panoramic view, do not have an associated objective, lack signs indicating the direction of the route, are dominated by burnt areas, are shared with motorized vehicles, or lack safety on the routes. All listed motivations are proportional among the groups with no statistically significant differences. However, in the categories of: lack of maintenance of the route (PR n = 76_b; GR n = $25_{a,b}$; PL n = 38_a); degree of difficulty (GR n = 7_a; PR n = 49_b; PL n = 58_c); routes are disfigured and have poor vegetation (GR n = 25_a; PR n = 50_a; PL n = 21_b); slippery and/or rocky routes (GR n = 13a; PR n = 43a,b; PL n = 47b) show different proportions in group comparison.“

In short, we can conclude that according to the residual analysis, people who prefer to take local hiking routes indicate that the degree of difficulty, a route that has lost its character and has poor vegetation, and slippery, muddy and/or rocky paths are the main reasons for not choosing to repeat the route. People who prefer to take shorter routes also cite the lack of maintenance of the route as a reason not to repeat it. With these conclusions, we can accept hypothesis H_b.

Table 4. Contingency table of the items listed to reasons for not repeating the route in relation to the hiking routes of preference of the participants (part II).

Variables	GR	PR	PL			
	Yes*	Yes*	Yes*	Person x^2	df	Exact Sig
Reasons not to repeat the route						
Absence of signs indicating the direction of the trails				1,851	2	0,386
Count (%)	13_a (19,7%)	39_a (25,5%)	30_a (28,8%)			
Expected Count	16,8	39,0	30,1			
Residual	−3,8	0,0	3,9			
Standardized Residual	−0,9	0,0	0,7			
Adjusted Residual	−1,2	0,0	1,0			
Trails depersonalize with little vegetation				10,798	2	0,005
Count (%)	25_a (37,9%)	50_a (32,7%)	21_a (17,8%)			
Expected Count	18,8	43,6	33,6			
Residual	6,2	6,4	−12,6			
Standardized Residual	1,4	1,0	−2,2			
Adjusted Residual	1,9	1,6	−3,2			
Trails where the burnt landscape predominates				5,051	2	0,082
Count (%)	24_a (36,4%)	54_a (35,3%)	37_a (23,7%)			
Expected Count	20,8	48,1	37,1			
Residual	3,2	5,9	−9,1			
Standardized Residual	0,7	0,8	−1,5			
Adjusted Residual	1,0	1,4	−2,2			
Slippery, muddy and/or stony ground path				8,883	2	0,012
Count (%)	13_a (19,7%)	$43_{a,b}$ (28,1%)	47_b (39,8%)			
Expected Count	20,2	46,8	36,1			

(continued)

Table 4. (*continued*)

Variables	GR	PR	PL			
	Yes*	Yes*	Yes*	Person x^2	df	Exact Sig
Residual	−7,2	−3,8	10,9			
Standardized Residual	−1,6	−0,6	1,8			
Adjusted Residual	−2,1	−0,9	2,7			
Shared trails with motorised vehicles				3,213	2	0,201
Count (%)	33_a (50,0%)	80_a (52,3%)	49_a (41,5%)			
Expected Count	31,7	73,5	56,7			
Residual	1,3	6,5	−7,7			
Standardized Residual	0,2	0,8	−1,0			
Adjusted Residual	0,3	1,4	−1,8			
Lack of safety on the trails				0,746	2	0,696
Count (%)	36_a (54,5%)	93_a (60,8%)	69_a (58,5%)			
Expected Count	38,8	89,9	69,3			
Residual	−2,8	3,1	−0,3			
Standardized Residual	−0,4	0,3	0,0			
Adjusted Residual	−0,8	0,7	−0,1			

GR-Great Route. PR-Small Route. PL-Local Route. *Positive responses analysis. **% routes of preference.

5 Conclusions

The offer of nature tourism, especially through pedestrian routes, implies the constant commitment and effort of evaluating the infrastructure and monitoring it. The different types of pedestrian profiles present motivational and particular preference peculiarities. Thus, it is essential to have a recurrent evaluation of the experience of various types of pedestrians. It is imperative to have a high-quality offer where there is respect for natural heritage, landscapes, threatened wildlife and floral species. The design of distinct routes, directed at different market segments, contributes to increasing the experience and satisfaction of the walker by adapting the physical characteristics of the route. It is also fundamental to develop routes where heritage resources are effectively integrated, and facilities and services are increasingly specific in their offering. The Eco-trail may be a solution where there is concern for an offer of pedestrian routes that respect the demands of pedestrians, the expectations of local communities and the tourism sector grows sustainably [23]. Environmental education arises as a fundamental need with advertising and promotion strategies focused on the target audience. Walking needs to be understood as the main activity, only then can the maturity of pedestrianism be achieved with the benefits already achieved by other countries in the construction of the walking tourism product [7].

Acknowledgments. This research was conducted within the framework of project CLICTOUR, code NORTE-01-45-FEDER-000079 funded by FEDER Norte2020, Portugal2020.

References

1. Svensson, D., Sörlin, S., Saltzman, K.: Pathways to the trail – landscape, walking and heritage in a scandinavian border region. Nor. Geogr. Tidsskr. – Nor. J. Geogr. **75**(5), 243–255 (2021). https://doi.org/10.1080/00291951.2021.1998216
2. Croft, D.B.: Walking in each other's footsteps: do animal trail makers confer resilience against trampling tourists? Environments **6**(7), 83 (2019). https://doi.org/10.3390/environments6070083
3. Braga, T.: Pedestrianismo e Percursos Pedestres; Amigos dos Açores – Associação Ecológica, Ed.; Nova Gráfica, Lda. (2007). ISBN 978-972-8144-27-2
4. Rodrigues, Á.L. de, O. Trilhos Pedestres, Turismo: Análise Exploratória Ao Mercado Dos Trilhos Pedestres Em Portugal, Universidade de Aveiro (2004)
5. Buckley, R.: Nature tourism and mental health: parks, happiness, and causation. J. Sustain. Tour. **28**, 1409–1424 (2020). https://doi.org/10.1080/09669582.2020.1742725
6. Gao, Y., Zhang, T., Sasaki, K., Uehara, M., Jin, Y., Qin, L.: The spatial cognition of a forest landscape and its relationship with tourist viewing intention in different walking passage stages. Urban For. Urban Green. **58**, 126975 (2021). https://doi.org/10.1016/j.ufug.2020.126975
7. Gómez-Martín, M.B.: Hiking tourism in Spain: origins, issues and transformations. Sustainability **11**(13), 3619 (2019). https://doi.org/10.3390/su11133619
8. Kelley, H., van Rensburg, T.M., Jeserich, N.: Determinants of demand for recreational walking trails in Ireland. Tour. Manag. **52**, 173–186 (2026). https://doi.org/10.1016/j.tourman.2015.05.028
9. Lukoseviciute, G., Pereira, L.N., Panagopoulos, T.: Assessing the income multiplier of trail-related tourism in a coastal area of Portugal. Int. J. Tour. Res. **24**, 107–121 (2022). https://doi.org/10.1002/jtr.2487
10. Monz, C.A., Gutzwiller, K.J., Hausner, V.H., Brunson, M.W., Buckley, R., Pickering, C.M.: Understanding and managing the interactions of impacts from nature-based recreation and climate change. Ambio **50**, 631–643 (2021). https://doi.org/10.1007/s13280-020-01403-y
11. Owen, N., Humpel, N., Leslie, E., Bauman, A., Sallis, J.F.: Understanding environmental influences on walking: review and research agenda. Am. J. Prev. Med. **27**, 67–76 (2004). https://doi.org/10.1016/j.amepre.2004.03.006
12. Ritchie, B.W., Tkaczynski, A., Faulks, P.: Understanding the motivation and travel behavior of cycle tourists using involvement profiles. J. Travel Tour. Mark. **27**, 409–425 (2010). https://doi.org/10.1080/10548408.2010.481582

13. Russell, B., McDaid, A., Toscano, W., Hume, P.: Moving the lab into the mountains: a pilot study of human activity recognition in unstructured environments. Sensors (Switzerland) **21**, 1–14 (2021). https://doi.org/10.3390/s21020654
14. Federação de Campismo e Montanhismo de Portugal RNPP – Registo Nacional de Percursos Pedestres (2016)
15. MacLeod, N.: Self-guided trails – a route to more responsible tourism? Tour. Recreat. Res. **41**, 134–144 (2016). https://doi.org/10.1080/02508281.2016.1147222
16. Mayer, K., Lukács, A.: Motivation and mental well-being of long-distance hikers: a quantitative and qualitative approach. Heliyon **7**, e06960 (2021). https://doi.org/10.1016/j.heliyon.2021.e06960
17. Molokáč, M., Hlaváčová, J., Tometzová, D., Liptáková, E.: The preference analysis for hikers' choice of hiking trail. Sustainability **14**(11), 6795 (2022). https://doi.org/10.3390/su14116795
18. de Melo, R.J.E.S., Gomes, R.A.M.: Understanding nature sports organizations in Portugal. Open Sports Sci. J. **9**, 13–25 (2016). https://doi.org/10.2174/1875399x01609010013
19. Cairo, H., Lois, M.: Re-signifying smuggling. In: Michael Hall, C., Ram, Y., Shoval, N. (eds.) The Routledge International Handbook of Walking, pp. 215–222. Routledge, Abingdon, Oxon; New York, NY: Routledge, 2017. (2017). https://doi.org/10.4324/9781315638461-21
20. Laureano, R.: Teste de Hipóteses Com o SPSS Statistics – O Meu Manual de Consulta Rápida; 3.ª Edição.; Edições Sílabo (2022)
21. Reis, E., Andrade, R., Calapez, T.m Melo, P.: Estatística Aplicada – Vol. 2; Edições Sílabo, Ed.; 6ª Edição (2019)
22. MacDonald, P.L., Gardner, R.C.: Type I error rate comparisons of post hoc procedures for I j chi-square tables. Educ. Psychol. Meas. **60**, 735–754 (2000). https://doi.org/10.1177/00131640021970871
23. Hugo, M.L.: A comprehensive approach towards the planning, grading and auditing of hiking trails as ecotourism products. Curr. Issues Tour **2**, 138–173 (1999). https://doi.org/10.1080/13683509908667849

Tourism in the Different Scientific Areas

Analysis of Citizen Satisfaction in Municipal Services

Sónia P. Nogueira[1,2](✉), Nuno Ribeiro[1,3], and Sílvia Remondes[1]

[1] Instituto Politécnico de Bragança, Campus de Santa Apolónia, 5300-253, Bragança, Portugal
sonia@ipb.pt

[2] Research Center in Political Science, Bragança, Portugal

[3] Management Applied Research Unit (UNIAG), Bragança, Portugal

Abstract. Local authorities play a vital role in community development and are one of the sources closest to the population. Today's citizens have become more attentive and demanding and do not accept undifferentiated treatment from local elected officials. In fact, as users of the public services provided by local authorities, citizens are continuous evaluators of both their services and the public policies adopted by the elected representatives. As voters and, therefore, decision-makers, citizens certainly cast their votes for whomever they expect to meet their needs.

This study seeks to evaluate the satisfaction and define a Satisfaction Index (SI) of the citizens towards their municipality of residence, namely the municipality of *Alfândega da Fé*, Portugal.

With a quantitative approach, this case study applied a questionnaire survey to a sample of residents in the selected municipality. In total, 350 questionnaires were collected, corresponding to approximately to 0.076% of the population, assuming a sampling error of 5.03%.

The results obtained regarding the SI show that the citizens of *Alfândega da Fé* are globally satisfied with their municipality of residence, scoring 73.29 points out of 100 for overall satisfaction. The results also show that the dimensions 'Quality of life in terms of leisure, access to culture, environmental quality and safety', 'Image of the municipality' and 'Quality of basic infrastructures and services' have the most significant weight in the overall satisfaction of the citizens with their municipality of residence.

Keywords: Citizen satisfaction · Municipal services · Satisfaction Index · Perceived quality

1 Introduction

Satisfaction seems to be related to the subjective perception of citizens and the adequate performance of local government. Although quality and satisfaction are two distinct concepts, there is a closely linked and actual construction between them to understand citizens' satisfaction levels.

Satisfaction can be seen as the ramification of a governing partnership. As a knowledgeable citizen, the local elected official represents the majority's will for the community's welfare. Knowledge of the territory's characteristics and proximity means that

C. Montenegro et al. (Eds.): ICMTT 2023, LNNS 774, pp. 289–299, 2024.
https://doi.org/10.1007/978-3-031-43733-5_26

local elected representatives are closer to them. Understanding the specific weaknesses of citizens and the potential of the territory helps make decisions that are swift, timely, and capable of developing the community more quickly and efficiently.

Living in sparsely populated territories can be an advantage for living a better quality of life, both environmentally and socially. The high degree of proximity in these territories between local elected officials and citizens can be an asset since, in places with larger populations, it is more difficult to understand the individual needs, especially those of the minority [1]. These territories have a greater tendency towards cooperation and social, cultural, and recreational well-being.

Knowing the citizen's perception of the local authority is highly relevant to all actors involved [2, 3]. Citizens are users of services and therefore have an opinion about them. In turn, if mayors become aware of these evaluations, this may be of great benefit not only for current public policies but also for the redefinition and definition of new public policies that meet the citizen's well-being [4–11].

Local governments should seek to adopt and adapt public services according to citizens' needs and preferences; they are inevitably continuous evaluators of decision-makers efforts and results. A previous study showed that the better the citizens' general living conditions, the more they support the local government [12]. Cultural and leisure facilities, essential collective services, and others significantly impact the citizens' evaluation. The study of the mentioned authors shows that the more financial resources municipalities have at their disposal and the more investments they make, namely in physical structures, the satisfaction also tends to increase. Citizens have a relative capacity to influence municipal decisions and activities; however, perhaps the most important thing is not the expenditure on services per se but the citizens' perception of their expenditure.

In turn, making these satisfaction assessments known to the municipalities can also be valuable because by involving them in political and social life, they can better evaluate local public policies and the performance of those whom they elected to represent them, and this may bring about a more realistic scenario between the electoral promises and reality.

The main objective of this study was to evaluate the resident citizens' satisfaction in different areas, including the municipality's image, the basic infrastructures and services, the quality of life in economic and leisure terms, the access to culture, the environmental quality and safety in the *Alfândega da Fé* municipality. Specifically, we sought to define an index to assess citizen satisfaction with their municipality of residence. A questionnaire was applied to the *Alfândega da Fé* municipality citizens to achieve the defined objective.

It is expected that the obtained results will enable the signaling of alert indicators that constitute management instruments for the redefinition of public policies taken by local elected representatives to contribute to improving the performance of public management. Knowing the citizens' satisfaction level may provide beneficial information for elected representatives helping them make decisions, solve existing problems and guide them toward results.

Given that citizens can be potential agents in the joint dynamization of the territory, growth, the development of organizations, the land, and people and, as a result, the country, it is also hoped that this study will be of great importance to increased civic participation. Citizens need to realize that this role is not unilateral but rather bilateral; there must be a reciprocal give and take. To this end, the citizen should try to do their part by finding out about local elected representatives' decisions and participating in them whenever possible. Without effective participation, there can be no conscious and coherent degree of satisfaction with public policies. Therefore, this study's results can be a demonstration framework for citizens and local elected representatives to learn about the citizens' evaluation of the operating services and policies and encourage them to become true partners in local government.

2 Customer Satisfaction Index

There are several converging definitions for one where it encompasses something in conformity with a sure thing and according to expectations.

The Consumer Satisfaction Index (CSI) measures the interaction between customer/citizen, product, service, and result/quality/performance. These indices can demonstrate the quality achieved and consumer satisfaction through the disconfirmation theory, characterized previously [13].

Disconfirmation theory [13] is often used to study customer/citizen satisfaction and dissatisfaction, in which disconfirmation is the assessment that the customer/citizen does regarding the performance of a product/service/policy considering their expectations [14].

CSIs were built either at a national scale [(first gained relevance in Sweden and subsequently applied in several countries (USA and Europe)] or for different types of products and services [15]. Based on economic behavior theories (disconfirmation of expectations theory by [13]), the CSI tries to identify the determinants of quality, how consumers perceive a certain quality and the outcomes. One problem of this index may arise from the fact that past experiences and expectations are directly related, i.e., good past experiences produce higher expectations, and bad experiences produce lower expectations.

For public administration, expectations are even more challenging issues. While bureaucracy bashing tends to lower expectations, political preferences may distort the perceived quality of services.

The most influential CSI is the American Customer Satisfaction Index (ACSI) [16]. The ACSI model [16 p.9] is presented as a cause-and-effect model with indices for drivers of satisfaction on the left side (customer expectations, perceived quality, and perceived value), satisfaction in the center and outcomes on the right (for businesses, these are customer complaints and customer loyalty, including customer retention and price tolerance, for governments there is no indication).

The ACSI is an econometric model that integrates perceptions of quality and satisfaction by analyzing specific behaviors such as complaints, i.e., based on the expectation disconfirmation theory [13]. It assumes that customers have expectations about service quality and that these are formed from previous experience or the reputation of the service. The index provides elected officials and citizens with a unique perspective on how citizens evaluate what is being done.

In another study [17], the ACSI reveals which services influence people's general quality evaluations and satisfaction with local government services and potential behavioral consequences. This index gives policymakers with a different perspective on public services and their effects on citizens. It can help elected officials identify areas where they can make improvements or additional efforts based on citizen satisfaction to improve citizens' quality of life.

The Norwegian Customer Satisfaction Barometer (NCSB) [18] went further by eliminating expectations from the model and focusing on image and loyalty. The NCSB introduced a new way of measuring quality through the SERVQUAL model [19]. The core of the SERVEQUAL consists of five dimensions of service quality (assurance, reliability, responsiveness, empathy, and tangibles) that allow indirect comparisons between expectations and perceived performance to assess quality.

3 Data Collection

Each location should be studied individually to understand satisfaction since each has a particular characteristic context. Applying a case study will obtain more factual data regarding a particular population's satisfaction levels [20].

Data collection was carried out through a questionnaire to a random sample of 350 residents in the *Alfândega da Fé* municipality (approximately 0.076% of the population) aged 18 years or older. The sampling error was 5.03%, assuming a significance level of 95%, making the sample representative of the study population.

Alfândega da Fé is a low population density territory with an aged population and reduced opportunity of maintaining or increasing the municipality's young population. For example, in 2001, the resident population by age group was: 0–14 (745); 15–24 (719); 25–64 (2,845); $\geq$65 (1,654), while in the year 2011, the distribution by age group was: 0–14 (496); 15–24 (497); 25–64 (2,451); $\geq$65 (1,660). Besides the aging of the population, in the *Alfandega da Fé* municipality, one can also notice scarce labor opportunities for young people to settle in, since according to the same report (INE, 2011), the little fixation of young people in the municipality is due to the lack of employment opportunities in the region, as the unemployment rate is the highest in the 15–24 age group. According to the PORDATA database (www.pordata.pt), in 2018, the *Alfândega da Fé* municipality had an aging index of 393 points.

The questionnaire seems to be the most appropriate data collection instrument since the citizens, as users and continuous evaluators of the public policies adopted by the decision-makers are the main "clients" of the local administration.

A previous study noted that customer satisfaction should not be measured in isolation but together with service quality [21].

The data collection method considered the evaluation of different dimensions of satisfaction: 'Image of the Municipality', the 'Quality of Basic Infrastructures and Services in the Municipality', the 'Quality of Life in Economic Terms', and the 'Quality of Life in Terms of Leisure, Access to Culture, Environmental Quality and Safety'.

4 Results

4.1 CSI and Analysis of the Impact of its Determinants

We analyzed structural equations using the PLS estimation method to analyze the results obtained in each of the scales of assessment of the dimensions under analysis and the relationship between them. The mean scores of each item of the dimensions analyzed were calculated. Their convergent and divergent validity were determined using the mean-variance extracted (MVE) and the heterotrait–monotrait ratio of correlations (HTMT), respectively.

The predictive capacity of the model explaining Global Satisfaction was also calculated using the Blindfolding analysis.

As shown in Fig. 1, the proposed model defined 'Image of the Municipality', 'Quality of Infrastructures and Basic Services', 'Quality of Life in Economic Terms' and 'Quality of Life in Terms of Leisure, Access to Culture, Environmental Quality and Safety', to explain the effect of the variables related to the dimensions.

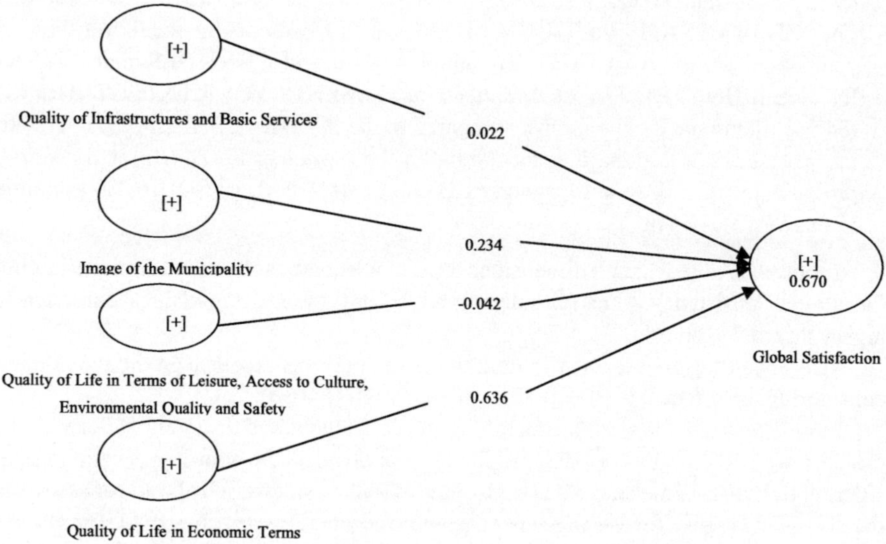

Fig. 1. Structural equation model and cause-effect relationships.

These results allow us to evaluate the determinants of Global Satisfaction of residents with the *Alfandega da Fé* municipality regarding their convergent and divergent validity and reliability.

According to Table 1, it can be stated:

- Concerning the 'Image of the Municipality' dimension, the item that contributes most to it is 'A Municipality where one LIVES WELL *(i.e., where citizens/residents have a quality of life)*' (22.5%). In contrast, the 'The municipality has NATIONAL RELEVANCE *(i.e., contributes to the country GDP, is important for the national economy...)*' item was the least relevant.

- The item that best defines the 'Quality of Infrastructures and Basic Services' dimension is the 'MUNICIPAL SERVICES: Quality of MUNICIPAL SERVICES in serving the citizens, good service, professionalism … of the Parish Council, Municipality,…' (26.6%), while 'SANITATION AND BASIC PUBLIC SERVICES: Quality of PUBLIC SERVICES (water, sanitation, energy, waste collection treatment…)' is the most critical item (23.9%).
- Concerning the 'Quality of Life in Economic terms' dimension, it is observed that the 'EMPLOYABILITY *(labor supply – employment…)*' item is the largest contributor to this dimension (27.3%). The 'How do you rate the relation QUALITY/COST OF LIFE in the Municipality of *Alfândega da Fé*?' item is the least relevant (22.3%).
- The 'Quality of Life in Terms of Leisure, Access to Culture, Environmental, Quality and Safety' dimension, the item 'Environmental Quality and Safety' stands out as the most significant contributor to the 'Quality of PUBLIC SECURITY services *(has good police and security services, security, surveillance e prevention,…)*' (31.4%) item, the least relevant being the 'Good "AIR" and ENVIRONMENTAL QUALITY *(non-polluted,… Green spaces, gardens, Prado amphitheater, biological ornamental lake, Castle sightseeing…)*' (5.5%).
- The 'GLOBAL SATISFACTION' (34.3%) and 'It represents the ideal, fulfilling all your expectations, what is in your opinion when you CONSIDER that LIVING IN THE MUNICIPALITY of *Alfândega da Fé* MEETS YOUR EXPECTATIONS' (34.3%) items stand out the most compared to the 'GLOBAL SATISFACTION with the work that local authorities have done in the *Alfândega da Fé (its local council, parish council,… in general)* given PUBLIC FINANCIAL RESOURCES available to them?' (31.4%) item.

It should be noted that all dimensions present adequate factor reliability considering the internal consistency values (Cronbach's Alpha and composite reliability) consistently higher than 0.7 (Table 1).

Also, according to the data in Table 2, the model presents excellent predictive validity, considering the effect size ($F = 0.58$), which is high (>0.35).

Table 2 presents the results concerning the convergent and divergent validity of the dimensions studied. Thus, all dimensions present an adequate convergent validity considering the MVE value above 0.50. Also, the HTMT result, which allows for assessing the divergent validity, reveals that it is sufficient since it is always below 0.90. The HTMT ratios range from 0.48 to 0.88, which means no correlation between the constructs. A correlation exists between constructs, but this does not affect the results, as evidenced by the statistic $t >$ p-value $= 0.000 < 0.05$. It should be noted that the construct 'Global Citizen Satisfaction' was built from the variables 'Global satisfaction that the municipality presents regarding the *Alfândega da Fé* municipality', 'Satisfaction with resources taking into consideration the public financial resources available in the municipality' and 'Fulfilment of expectations'.

After analyzing each of the determinants likely to influence the satisfaction of the municipalities and the Global Satisfaction of the Citizens with the municipality of residence (i.e., *Alfândega da Fé*), we analyzed the explanatory model of satisfaction results according to the different dimensions.

Table 1. Measures for evaluating the quality of the proposed model.

	Index (0–100 points)	% Importance	Cronbach' Alpha	Composite reliability	*F*
Visibility	68.57	17.4%			
Development	75.83	22.1%			
Living well	84.29	22.5%			
Modernization	74.22	22.2%			
National relevance	64.35	15.6%			
Image of municipality	**73.46**	**100.0%**	**0.89**	**0.91**	-
Teaching	82.23	24.0%			
Health	77.60	25.4%			
Sanitation and basic public services	76.23	23.9%			
Municipal services	74.40	26.6%			
Quality of infrastructures and basic services	**77.61**	**100.0%**	**0.87**	**0.91**	-
Quality/Cost of life	64.74	22.3%			
Employability	61.77	27.3%			
Attracting companies	54.86	26.9%			
Housing	74.11	23.4%			
Quality of life in economic terms	**63.87**	**100.0%**	**0.89**	**0.93**	-
Cultural events	92.45	28.5%			
Leisure services	91.45	26.9%			
Public safety	79.25	31.4%			
'Air' and Environmental Quality	97.83	5.5%			
Natural resources	98.00	7.5%			
Quality of life in terms of leisure, access to culture, environmental, quality and safety	**91.82**	**100.0%**	**0.80**	**0.84**	-

(*continued*)

Table 1. (*continued*)

	Index (0–100 points)	% Importance	Cronbach' Alpha	Composite reliability	*F*
Global satisfaction that the citizen has with the municipality of *Alfândega da Fé*	76.86	34.3%			
Satisfaction with resources given the public financial resources available to the municipality	70.91	31.4%			
Fulfillment of expectations	72.11	34.3%			
Total satisfaction	**73.29**	**100.0%**	**0.94**	**0.96**	**0.58**

Table 2. MVE[a] and HTMT[b] test results.

Dimension	1	2	3	4	5
1-Image of municipality	0.68[a]	0.71[a]	0.76[a]	0.53[a]	0.89[a]
2-Quality of infrastructures and basic services	0.83[b]	-	-	-	-
3-Quality of life in economic terms	0.86[b]	0.79[b]	-	-	-
4-Quality of life in terms of leisure, access to culture, environmental quality, and safety	0.65[b]	0.62[b]	0.65[b]	-	-
5-Global Citizen Satisfaction	0.78[b]	0.67[b]	0.88[b]	0.48[b]	-

Note: [a] MVE; [b] HTMT

4.2 Multiple Linear Regression Analysis of the Effect of Dimensions on Global Citizen Satisfaction

According to Table 3, the set of determinants explains 67.0% of the citizens' Global Satisfaction with their municipality of residence.

More specifically, it can be stated that concerning the global satisfaction of the citizen:

– The dimension 'Image of the Municipality' contributes to the highest variance of satisfaction with the municipality, explaining 51.1% of the effect. Specifically, the variance of the overall satisfaction with the municipality is expected to increase by 0.23 points (HTMT = 0.23).

- The second-largest contribution comes from the dimension 'Quality of Life in Economic Terms', which explains 13.9% of the satisfaction, with an increase in its variation of 0.64 points (HTMT = 0.64).
- The third-largest contribution concerns the dimension 'Quality of Infrastructures and Basic Services', which explains 1.8% of satisfaction and contributes to its variance by 0.02 points (HTMT = 0.02).
- Finally, the most negligible contribution, which is almost null, is due to the dimension 'Quality of Life in Terms of Leisure, Access to Culture, Environmental Quality and Safety', which only explains 0.2% of the citizens' global satisfaction. Its increase is expected to contribute to a 0.04 points decrease (HTMT = -0.04) in the citizens' global satisfaction with their municipality of residence.

Table 3. Analysis of the impact of its determinants.

Dimensions	Index (0–100 points)	HTMT	p-value	R^2 (% importance)	Importance 100%
Image of municipality	73.46	0.23	0.000	0.511 (51.1%)	76.3%
Quality of infrastructure and basic services	77.61	0.02	0.662	0.018 (1.8%)	2.7%
Quality of life in economic terms	63.87	0.64	0.000	0.139 (13.9%)	20.7%
Quality of life in terms of leisure, access to culture, environmental quality, and safety	91.82	−0.04	0.400	0.002 (0.2%)	2.9%
Global Citizen Satisfaction	73.29	-	-	0.670 (67.0%)	100.0%

5 Conclusions, Limitations, and Future Research

The present research aimed to propose an index of citizens' satisfaction with their municipality of residence – *Alfândega da Fé*. The study is based on the citizens' importance that the local political decision-makers should recognize since they make value judgments regarding satisfaction with the services provided and the local public policies adopted for their residence territories. The competitiveness between municipalities is increasingly evident when discussing the attractiveness of municipalities to citizens.

The SI provides local elected representatives with results concerning citizens'/people's satisfaction since it shows them their opinions of their political actions. The SI also defines or redefines the local public policies adopted to meet the citizens' expectations. Letting citizens know about these evaluations (if they are primarily positive) can also be valuable because it can appeal to potential citizens looking to settle and/or install new companies.

Notably, the municipality of *Alfândega da Fé*, as the municipality of residence, obtained an overall citizen satisfaction score of 73.29 points. The score obtained allows us to state that the citizens of *Alfândega da Fé* seem to be entirely satisfied with their municipality of residence.

Finally, even though this research may constitute an analysis on a micro-scale, given that it is a small municipality in the interior of Portugal, overall, citizens are satisfied with their municipality of residence. It is therefore hoped that these results will provide support and incentives for the continuous improvement of local public policies in the municipality and, consequently, a continuous social and political well-being for resident citizens/people and that it may also constitute a potential attraction for the continuity of residents, new residents, and the installment of new companies.

Even though the previously defined objective – to know the satisfaction of the citizens of *Alfandega da Fé* – was considered to have been achieved, naturally, there are limitations in this research, starting with the application of questionnaires that could have been carried out online to cover a larger sample. Another limitation may be the use of the questionnaire since the results are based on the respondents' perceptions, not allowing the veracity of the respondents' answers to be ascertained.

As a suggestion for future research, we suggest exploring new variables, namely complaints, trust, and intention to stay or change the municipality of residence. The study's extension to other municipalities may also contribute to further investigating the theme, broadening it, allowing a more comprehensive discussion, and even presenting a global satisfaction ranking of the citizens of different municipalities.

Acknowledgments. The authors are grateful to the Foundation for Science and Technology (FCT, Portugal) for financial support by national funds FCT/MCTES to UNIAG, under Project no. UIDB/04752/2020.

References

1. Van de Walle, S., Van Ryzin, G.: The order of questions in a survey on citizen satisfaction with public services: lessons from a split-ballot experiment. Public Adm. **89**(4), 1436–1450 (2011)
2. James, O.: Evaluating the expectations disconfirmation and expectations anchoring approaches to citizen satisfaction with local public services. J. Public Adm. Res. Theory **19**(1), 107–123 (2009)
3. Kopalle, P., Lehmann, D.: Strategic management of expectations: the role of disconfirmation sensitivity and perfectionism. J. Mark. Res. **38**(August), 386–394 (2001)
4. Beeri, I., Uster, A., Vigoda-Gadot, E.: Does performance management relate to good governance? A study of its relationship with citizens' satisfaction with and trust in Israeli Local Government. Public Perform. Manag. Rev. **42**(2), 241–279 (2019)
5. Brans, M., Hondeghem, A.: Competency frameworks in the Belgian governments: causes. Construc. Contents. Public Admin. **83**(4), 823–837 (2005)

6. Hibbing, J.R., Theiss-Morse, E.: Stealth Democracy: Americans' Beliefs About How Government Should Work. Cambridge University Press (2002)
7. Jilke, S.: Citizen satisfaction under changing political leadership: the role of partisan motivated reasoning. Governance **31**(3), 515–533 (2018)
8. Roch, C., Poister, T.: Citizens, accountability and service satisfaction: the influence of expectations. Urban Aff. Rev. **41**(3), 292–308 (2006)
9. Rose, L., Pettersen, P.: The legitimacy of Local Government – what makes a difference? Evidence from Norway. In: Hoggart, K., Clark, T. (eds.) Citizen Responsive Government. Research in Urban Policy, vol. 8. JAI Press, Amsterdam (2000)
10. Van Ryzin, G.: Expectations, performance, and citizen satisfaction with urban services. J. Policy Anal. Manage. **23**(3), 433–448 (2004)
11. Van Ryzin, G.: The measurement of overall citizen satisfaction. Public Perform. Manag. Rev. **27**(3), 9–28 (2004)
12. Roos, K., Lidström, A.: Local policies and Local Government legitimacy. The Swedish case. Urban Res. Pract. **7**(2), 137–152 (2014)
13. Oliver, R.: A cognitive model of the antecedents and consequences of satisfaction decisions. J. Consum. Res. **42**, 460–469 (1980)
14. Hossain, M.: Redefining expectancy disconfirmation theory using LIS SERVQUAL: an integrated framework for evaluating library service quality and user satisfaction. Int. Inform. Libr. Rev. **51**(3), 203–216 (2019)
15. Fornell, C.: A national customer satisfaction barometer: the Swedish experience. J. Mark. **56**, 6–21 (1992)
16. Fornell, C., Johnson, E., Anderson, J., Cha, J., Bryant, E.: The American customer satisfaction index: nature', 1996, purpose, and findings. J. Mark. **60**(4), 7–18 (1996)
17. Van Ryzin, G., Muzzio, D., Immerwahr, S.: Drivers and consequences of citizen satisfaction: an application of the american customer satisfaction index model to New York City. Public Adm. Rev. **64**(3), 331–341 (2004)
18. Parasuraman, A., Zeithaml, V., Berry, L.: SERVQUAL: a multiple-item scale for measuring consumer perceptions of service quality. J. Retail. **64**(1), 12–40 (1988)
19. Şandor, S.: Determinants and outcomes of citizens' satisfaction with public services in Cluj-Napoca. Transylvanian Review of Administrative Sciences, 21 E/2007, pp. 103–112 (2007)
20. DeHoog, R., Lowery, D., Lyons, W.: Citizen satisfaction with Local Governance: a test of individual, jurisdictional, and city-specific explanations. J. Politics **2**, 807–837 (1990)
21. McCall-Kennedy, J., Schneider, U.: Measuring customer satisfaction: why, what and how. Total Qual. Manag. **11**(7), 883–896 (2003)

Tourism Industry and Ecology

The Ecotourism Landscape of Qatar

Alan S. Weber(✉)

Weill Cornell Medical College in Qatar, Doha 24144, State of Qatar
alw2010@qatar-med.cornell.edu

Summary. The small oil and gas-rich State of Qatar has been propelled into the international tourism spotlight with its successful hosting of the 2022 FIFA World Cup football championship. In order to showcase its additional ecotourism offerings, Discover Qatar and Qatar Tourism, who promote, manage, and regulate the tourism sector in Qatar, have embarked on a strategic plan to promote Qatar's natural spaces and unique flora and fauna as tourist resources. For example, Qatar boasts a large whale shark population and one of the largest populations in the world of Dugongs (sea cows, a marine mammal related to the Manatee). The Arabian Oryx has also been successfully re-introduced into the wild and tourists may explore Qatar's mangrove forests by boat. However, Qatar's limited geographical size, and the increasing urbanization and development of this small peninsula, as well as the harsh, arid desert climate which can reach temperatures of 50-52 degrees Celsius, place hard limits on the development of ecotourism in Qatar. Based on the limited amount of tourism research on Qatar, and several international symposia on the ecology of Qatar, in addition to the author's own original research on desert tourism in the region, this contribution reviews the current state of both Ecotourism and Sustainable Tourism in Qatar and offers practical recommendations for their successful implementation without long-term harm to Qatar's extremely fragile ecosystems, which are under threat from building construction, marine development (artificial islands and channels), hydrocarbon pollution, and natural warming and increasing salination of the Persian Gulf waters.

Keywords: Ecotourism · Qatar · Sustainable tourism

1 What is Ecotourism?

Although the terms 'Ecotourism' and 'Sustainable Tourism' are sometimes used synonymously, there are several important distinctions between these two concepts. Sustainable tourism is one form of the economic and socially sustainable practices known as 'sustainable development' which have been widely adopted internationally to secure both inter- and intra-generational equity, as first delineated in the Brundtland Report (*Our Common Future*) in 1987. The United Nations World Tourism Organization (UNWTO), argues that "sustainable tourism should:

1. Make optimal use of environmental resources that constitute a key element in tourism development, maintaining essential ecological processes and helping to conserve natural heritage and biodiversity.

C. Montenegro et al. (Eds.): ICMTT 2023, LNNS 774, pp. 303–309, 2024.
https://doi.org/10.1007/978-3-031-43733-5_27

2. Respect the socio-cultural authenticity of host communities, conserve their built and living cultural heritage and traditional values, and contribute to inter-cultural understanding and tolerance.
3. Ensure viable, long-term economic operations, providing socio-economic benefits to all stakeholders that are fairly distributed, including stable employment and income-earning opportunities and social services to host communities, and contributing to poverty alleviation" [1].

Fennell's definition of ecotourism has been widely cited in the literature as a "sustainable form of natural resource-based tourism that focuses primarily on experiencing and learning about nature, and which is ethically managed to be low-impact, non-consumptive and locally oriented. It typically occurs in natural areas and should contribute to the conservation or preservation of such areas" [2]. Thus ecotourism can be simply defined as a type of sustainable tourism practice, with the natural environment (including flora, fauna, geology, and natural ecosystems, as well as the built heritage of indigenous peoples) as the focus of activity. What is often overlooked in definitions of sustainable tourism and ecotourism is that humans now are participants in most natural ecosystems on earth, and the economic and cultural needs of indigenous peoples as resource consumers must be factored into ecotourism strategies, and balanced against preservation and conservation of natural resources. Severely limiting human activity in conservation zones is not practical in many parts of world in which local peoples rely on natural resources for their livelihoods.

2 The State of Qatar

2.1 History, Sociology, Economy

A brief history of the development of the State of Qatar is important for understanding its current national tourist strategy and the natural and environmental limits placed upon that developmental plan. Qatar emerged as a modern, high-income country (ranked number 6 in GDP at purchasing power parity per capita in 2022 according to the IMF) in the 1990s due to the exploitation of its North Field natural gas deposits, which comprise 14-15% of the world's known gas reserves [3]. The country is a constitutional monarchy governed by Shariah law, and the majority of citizens are Sunni Muslims with a small community of Shias originating from the Persian empires. The hereditary rulers of Qatar, the Al Thani family (the current Emir is H.H. Tamim bin Hamad Al Thani), also consult with the Majlis al-Shura, an elected body of advisors from the local tribes. Al-shura is a respected custom in Qatar and the Gulf region, and insures that local stakeholders will be partners in ecotourism development.

Arabic is the official language of the country, although English is widely spoken in education and business settings along with Hindi, Nepali, Tagalog, and Urdu due to large expatriate populations. For political reasons, no accurate census of the Qatari population has been carried out, but a rigorous 2010 study estimated the number of Qatari citizens to be 245,770 out of a population of 2.6 million [4]. This odd demographic mixture means that only 9% of the population of Qatar are passport carrying citizens, and 91% are expatriates working in all of Qatar's major industries at all skill levels. Further research needs to be conducted to determine the precise environmental attitudes of this diverse mix of cultures, and their views on ecotourism.

The form of Islam practiced in Qatar is Salafism or Wahhabism, one of the most conservative branches of Muslim practice related to the Hanbali school of Shariah law; thus, long-standing cultural and religious patterns in Qatar are largely intact which will be of great interest to ecotourists, who are also very often interested in indigenous local cultures in addition to the environment. Despite the westernization and modernization brought by the massive influx of expatriate workers from western countries and Asia and South Asia during the past two decades, Qataris have maintained their traditional clothing, foods (although American fast food has recently become popular), distinctive Arabic dialect (Khaleeji, or Gulf Arabic), and their religious customs such as strict observation of Ramadan, regular prayer five times per day, and following Zakat (charity) and Hajj (pilgrimage to Mecca).

2.2 The Tourism Sector in Qatar

Due to Qatar's heavy reliance on oil and gas exports as sources of wealth (approximately 60% of the country's GDP derives directly from hydrocarbon exports and closely related industries), the country experiences periodic price shocks from fluctuating oil and gas prices, which directly impacts government revenues and the state's ability to fund education, government, and Qatar's extensive social programs. These 'boom and bust' cycles, known among economists as the 'Dutch Disease,' or 'Resource Curse', have led economic planners in Qatar to formulate a series of ambitious national development strategies to shift Qatar's economy away from natural resource extraction towards a knowledge-based economy which produces knowledge products including education, patents, training, processes, software, and biotechnological innovation.

The most recent general developmental plan is entitled *Qatar National Vision 2030* [5]. The most recent detailed 5-year plan, the *National Development Strategy 2018-2022*, called for a more diversified economic approach to Qatar's economy, including tourism development: "a more competitive, productive and diversified economy and a more dynamic private sector with greater contribution to the national economy," and to "develop and coordinate priority sectors strategies (manufacturing, professional and scientific activities, logistics, financial services, tourism and ICT)" [6].

The COVID-19 epidemic had a profound effect on tourist arrivals to Qatar, with 1,053,000 land, air and sea arrivals in 2019 which declined to 100,00 arrivals in 2021 due to quarantine and strict entrance policies. Qatar successfully hosted the FIFA World Cup football matches in December, 2022 after the lifting of COVID-19 restrictions, and the 2022 number of arrivals will undoubtedly surpass the 2019 figures when the full 2022 statistics are issued by the Public Statistics Authority (there were 725,000 arrivals for the half-year 2022, January to June) [7].

3 Ecotourism in Qatar

3.1 The Desert Environment and Challenges

Qatar's climate is classified in the Köppen-Geiger system as a 'Hot desert climate' (*BWh*). Daytime mean temperatures can vary from 46-52 °C during the summer months. Some of the driest parts of the country such as Al Wakrah report on average 0 mm (0.0 inches) of rain in May, and 15 mm (0.6 inches) in December, generally the wettest month in Qatar [8]. Thus Qatar's biomes are extremely fragile and water-scarce, with sparse ephemeral vegetation appearing after limited and sporadic winter rains, and substantial parts of the country are covered by barchan sand dunes, an extension of the extremely arid Rub al-Khali (Empty Quarter). The rest of the country consists of limestone gravel desert, with very limited soils and clays for agriculture, which requires artificial irrigation from wells.

Qatar is also experiencing rapid urbanization and the loss of natural habitats. Doha has developed into a major metropolis of over 2 million people where a majority of the population now lives. The population grew from 700,000 in the early 2000s to over 2.6 million in 2023 from immigration of foreign workers in the construction, services and oil and gas sectors. A large-scale artificial island project called the Pearl Island, developed from 2004-2023, was constructed from reclaimed shallow water ocean beds surrounding the northeast coast of Qatar, destroying reefs and local mangrove forests. However, the project was built with environmentally sound principles in mind, such as the circulation of seawater to prevent harmful bacteria and algae blooms, and to protect marine life. The area has now seen recovery of corals and sea grasses after the initial negative environmental impact of dredging, and the relocation of millions of tons of sand and rock to the seafloor.

The main challenge to the desert ecotourism of Qatar thus is the harsh desert climate, and the vulnerability of the limited number of animal and plant species who have adapted to its water and food scarce ecosystems. In fact, several species have faced extinction or near extinction in recent times, including the Arabian Oryx, which has been successfully reintroduced through government programs. Many other endemic species of interest to tourists in Qatar, such as the hedgehog, Sand Cat (*Felis margarita*), Arabian Horned Viper, etc. are either nocturnal or so scarce as to be rarely encountered in the wild. Domestic camels, however, are extremely abundant and tourists may interact with them at Camel Racing sites, camel riding companies, or in the wild. Qatar's marine environment also contains whale sharks [9], and a large population of Dugongs, or sea cows, a slow-moving marine mammal of the Dungongidae family, of which only 4 species remain world-wide, including Manatees. The shores of Qatar are also the breeding grounds for Hawksbill Turtles, and the Qatar Natural History Club sponsors trips to turtle nesting sites carried out in an environmentally-responsible manner with biologists present.

In 2007, recognizing the threat of urban sprawl from Doha, the government of Qatar and UNESCO designated 120,000 hectares of protected land as the Al Reem Biosphere in the Zekreet region of northwest Qatar. This area is open to tourists and contains free-roaming populations of gazelles and Oryxes, as well as wind-sculpted limestone mesas [10]. This park comprises almost 10% of the entire land area of Qatar, making Qatar one of the world's top countries for protected areas.

3.2 Strategies for Development

Hot desert environments have been underestimated and underutilized as tourist destinations undoubtedly due to the fact that they were previously very difficult to access, and due to negative attitudes towards deserts as hostile and barren places of little interest to environmentally-inclined ecotourists. Some of the most successful ecotourism destinations, such as Costa Rica, Ecuador and areas of the South Pacific occur in lush, tropical destinations; thus the concept of ecotourism is often associated with temperate and humid tropical biomes.

However, Qatar's deserts, and desert regions in general, boast several unique features to draw tourists in the ecotourism market segment. As Weber notes, "certain desert features such as barchan sand dunes, salt flats and ventifacts only exist in drylands. Ventifacts or yardangs are statue-like rocks that have been sculpted into aesthetically pleasing forms by wind erosion. A striking example is 'Lot's Wife', a pillar of almost pure salt on Mount Sodom, Israel....and [since Qatar's] deserts border areas of salt or fresh water, water sports and water activities including swimming, boating, jet skiing, bird watching, water animal safaris, and fishing can be combined with desert recreation" [11]. In addition, Qatar has 'singing sand dunes' in the south of the country: "by walking on the surface, sand dunes under certain dry conditions produce a range of wailing, thumping, screeching and singing noises that Bedouins attributed to djinn (genies). The experience of hearing a singing sand dune is highly unique and unforgettable, a key tourist asset" [12].

Adding to the flora and fauna resources mentioned in the previous section, Qatar maintains many of its traditional *beddu* (Bedouin) customs, such as camel, goat, and sheep production, camel racing, use of traditional goat-hair tents (for camping), traditional foods, and unroasted coffee-making rituals, etc. New buildings are now being designed according to traditional architectural principles which originated in the Sassanian period, which include energy efficient thick walls (originally mud, limestone, native gypsum, and coral rock), small windows, and wind towers. Wind towers, tall structures with channels to capture cooler upper air flows and draw out heat from buildings, have been used in the Gulf for centuries. Although they cannot completely cool a modern building, wind towers can substantially reduce air conditioning costs.

Falconry (breeding and hunting with falcons) is also a highly regarded sport in Qatar, with some prize falcons being sold in excess of 1 million USD. The Marmi Festival featuring falconers and other bird breeders, such as homing pigeons, is held annually at Sabkha Sealine, near the Inland Sea, which is a protected area for camping and recreation [13]. In Souq Waqif, a traditional Gulf market, can be found one of the world's few falcon hospitals, as well as traditional Arab trade goods including musical instruments (ouds and rebabs), honey, saffron, spices, coffee, rice, etc. Although often visited by foreigners as a tourist attraction, Souq Waqif is an authentic and traditional shopping center, where locals and expatriates meet for meals and sheesha smoking and to purchase normal household items and hardware.

4 Conclusions and Recommendations

The arid desert environment of Qatar, and growing urbanization and reduction of natural spaces dictates that ecotourism should only play a modest role in Qatar's overall tourism strategy to promote economic diversification. Other approaches–such as MICE tourism (the DECC exhibition center in downtown Doha is one of the largest of its kind in the Middle East), niche medical tourism (Aspetar Sports Medicine specialty hospital), and 72-hour layover tourism through Qatar Airways–should form the focus of tourism development in Qatar. All tourism strategies in Qatar due to its desert environment should follow the recommended sustainable guidelines issued by the United Nations World Tourism Organization (UNWTO) in 2006 in the document *Sustainable Development of Tourism in Deserts: Guidelines for Decision Makers* [14]. Unregulated mass tourism in Qatar's arid deserts could cause substantial and potentially irreversible damage (for example, species extinction) lasting for centuries.

References

1. United Nations World Tourism Organization (UNWTO): Sustainable Development, https://www.unwto.org/sustainable-development, last accessed 12 March 2023
2. Fennell, D.A.: Ecotourism: An introduction. Routledge, London (2004)
3. International Monetary Fund (IMF): World Economic Outlook Database, 2022. (11 October 2022), https://www.imf.org/en/Publications/WEO/weo-database/2022/October/weo-report, last accessed 14 March 2023
4. Winckler, O.: How Many Qatari Nationals Are There? Middle East Quarterly, pp. 1–16. Spring (2015)
5. State of Qatar: Qatar National Vision 2030, Government Communications Office, https://www.gco.gov.qa/en/about-qatar/national-vision2030/, last accessed 14 March 2023
6. State of Qatar: Public Statistics Authority, Qatar Second National Development Strategy, 2018-2022, https://www.psa.gov.qa/en/knowledge/Documents/NDS2Final.pdf, last accessed 14 March 2023; see also, Weber, A.S.: The World Cup 2022 and Qatar's Mass Tourism Challenge: A Sustainable Future?, Center for International and Regional Studies, https://cirs.qatar.georgetown.edu/the-world-cup-2022-and-qatars-mass-tourism-challenge-a-sustainable-future/, last accessed 14 March 2023
7. Qatar Tourism: Highlights of Tourism in Qatar: Half Year (H1) 2022 Tourism Report, https://www.qatartourism.com/content/dam/qatar-tourism/qatar-tourism-reports/Qatar-Tourism_H1_vf.pdf, last accessed 14 March 2023
8. Climatedata.org Homepage: https://en.climate-data.org/asia/qatar/al-wakrah/alwakrah55833/#climate-graph, last accessed 14 March 2023
9. Discover Qatar: Whale Sharks of Qatar, https://www.discoverqatar.qa/whale-sharks-of-qatar/, last accessed 14 March 2023
10. UNESCO: Al Reem, https://en.unesco.org/silkroad/silk-road-themes/biosphere-reserve/al-reem, last accessed 14 March 2023
11. Weber, A.S.: Desert Landscapes and Tourism in the Middle East and North Africa. In: Timothy, D.J. (ed.) Routledge Handbook on Tourism in the Middle East and North Africa, p. 191. Routledge, New York (2019)
12. Weber, A.S.: Desert Landscapes and Tourism in the Middle East and North Africa. In: Timothy, D.J. (ed.) Routledge Handbook on Tourism in the Middle East and North Africa, p. 192. Routledge, New York (2019)

13. Al Mutawa, A.H.: Ecotourism in Qatar, p. 93. Ministry of Environment and Climate Change, Doha (2021)
14. United Nations World Tourism Organization: Sustainable Development of Tourism in Deserts: Guidelines for Decision Makers. World Tourism Organization, Madrid (2006)

Tourism Management

Cultural and Ethnographic Routes as a Tourist Resource for Rural Development: A Theoretical Review

Tania Morales Molina[1](✉), Ximena Morales-Urrutia[2], and Nelson Lascano[2]

[1] Universidad Tecnológica Indoamérica, Ambato, EC 180103, Ecuador
taniamorales@uti.edu.ec
[2] Universidad Técnica de Ambato, Ambato, EC 180104, Ecuador

Abstract. Cultural routes have become an important income-generating resource for the different countries they pass through. At present, this type of route is promoting a series of activities related to the tourism field and that have drawn special attention from its visitors, this has made the governments of the day develop policies aimed at promoting and implementing actions that allow stimulate the development and economic growth of the rural areas in which they are located. The purpose of this research is to carry out a review of the current and evolutionary status of the contributions made for the enhancement of historical and heritage value of these routes. In the present research, a literature review of the main cultural tourism routes and their contributions to the development of the areas through which they run was carried out. In general, the cultural routes have become an important contribution to the growth and economic development of rural areas; and, above all, they have generated jobs and improved the quality of life of the population.

Keywords: Cultural routes · Rural development · Economic growth

1 Introduction

Cultural routes constitute a relatively incipient phenomenon that presents a double dimension. First of all, these routes involve a journey through a geographical territory. On the other hand, they can also be considered as a mental journey, with the significant values, expectations and experiences associated with that journey (Majdoub, 2010). From an economic point of view, cultural routes have become cornerstones of tourism in some regions; being its basis the relationship between tourism and culture, the enrichment of traveler experiences and respect for cultural heritage (Calderón Puerta et al., 2018).

1.1 Cultural and Ethnographic Routes as a Tourist Resource

The concept of cultural routes can manifest itself in different ways, depending on its origins and characteristics. Thus, without any intention of exhaustiveness, we have hiking trails, religious pilgrimage routes, trade routes that were used in the past to transport

C. Montenegro et al. (Eds.): ICMTT 2023, LNNS 774, pp. 313–321, 2024.
https://doi.org/10.1007/978-3-031-43733-5_28

specific goods (salt route, silk route, minerals such as gold, silver or copper), military routes or routes related to certain war episodes, communication routes built in previous historical times and other routes linked to characters or diverse cultural manifestations (Sugio, 2005). The size and density of the tangible and intangible elements that constitute the sustenance of the cultural route vary from one case to another, and the same is likely to happen with the degree to which its authenticity can be verified (Gonzalo, 2011).

Carrera (2006) defines cultural routes as "*spatial routes of heritage content that serve as axes for the dissemination of the culture-territory-identity relationship, the result of a shared and dialectical historical process*". These routes are a good reflection of the evolution of the notion of heritage and its projection in today's society, since heritage ceases to be identified exclusively with "monuments", with objects that have a historical-artistic value, to encompass landscapes, constructions and identities (López Fernández, 2006). In this regard, the Cultural Routes Committee of the International Council of Historic-Artistic Monuments and Sites (ICOMOS) points out that a cultural route is "*Any means of communication, whether land, water, or some other type, that is physically delimited and also characterized by having its own specific dynamic and historical functionality, which must meet the following conditions: reflect the interactive movements of people, as well as the multidimensional, continuous and reciprocal exchanges of goods, ideas, knowledge and values between peoples, countries, regions or continents over a long period of time…*" (ICOMOS, 2015). The richness and attractiveness of these cultural infrastructures is linked to the tangible and intangible heritage that is associated with them: bridges, stations and tunnels, inns and post stops, markets and fairs, defensive and religious constructions, industrial establishments and buildings. Agriculture, landscapes, etc. With the precision that what is relevant is not the separate value of each of these elements, but rather their integration into a route or itinerary with some unifying aspect.

Currently there is a new tourist-recreational use for historic routes and paths. In this sense, Murray and Graham (1997) indicate that cultural routes constitute more than communication routes with signs for visitors, and also try to respond to the needs of tourists by providing links between different experiences that generate a more complete experience. A cultural tourist route can, on the one hand, help in environmental conservation and visitor management (Strauss and Lord, 2001). On the other hand, this type of route is based on a principle of collaboration (Morgan et al., 2009), whereby the grouping of activities, resources and attractions stimulates cooperation between communities of neighboring localities and regions (Mitchell and Hall, 2005); apart from serving to stimulate the economic, social and cultural development of the inhabitants of the localities involved, through tourism (Briedenhann and Wikens, 2004). Likewise, the implementation of cultural tourist routes arouses political and economic interest, since these often imply the development, on the one hand, of cooperative planning agreements and, on the other, of relations within and between the different localities (Rogerson, 2007). Therefore, the planning of tourist routes constitutes a subset of the planning of cooperative tourism (Reed, 1999).

In recent decades, tourist-cultural routes have multiplied throughout the world, particularly in developed countries. This is because they offer significant opportunities for the development of new ancillary products and services (Fagance, 1991), in a context in which tourism trends have led to a change from standardized mass tourism towards more specific and differentiated patterns (Hummelbrunner and Miglbauer, 1994).

Authors such as Guzmán and Gutiérrez (2006) mention the multiple benefits that tourism through cultural routes can bring to rural areas. Specifically, among these benefits we can mention the following: a) it generates a multiplier effect, since it allows activating different economic sectors, creating sources of employment, whether direct (tourism activity) or indirect (production of complementary goods and services); b) stimulates pluriactivity in the population of rural areas, diversifying the income obtained; c) it often provides an incentive (and part of the necessary financing) for infrastructure development, which contributes to the growth of other economic activities in rural areas (Sarasa, 2000); d) it attracts new tourists and contributes to increasing the average length of stay and spending by tourists; e) enables a more integrated product development and commercialization approach; and f) increases the general attractiveness and sustainability of these tourist destinations (Rogerson, 2009). That is, tourism has been identified as a catalyst to stimulate economic growth, increase the viability of underdeveloped regions and improve the standard of living of local communities (Simpson et al., 1998).

1.2 Some Examples of Cultural and Ethnographic Routes in Different Areas of the World; Its Contribution to the Development of Rural Areas

In recent years we have witnessed the multiplication of tourist-cultural routes and itineraries in all corners of the planet (Denstadli and Jacobsen, 2010). Thus, the territory is taking a strategic position in the tourist offer with its multiple characteristics and values. Unlike past stages, in which it was usual for the offer to be constituted by a specific destination, whether it was a place, town, monument or space for pleasure, nowadays the configuration of tourist products referred to is more and more frequent to a complete territory that, under a common denominator, tries to lead visitors following a defined route (Ramírez, 2011). A cultural route is not limited to a series of sites, but rather its richness and attractiveness corresponds to a structured set of assets, both tangible and intangible. A review of the diversity of existing cultural routes throughout the planet allows us to verify that these can correspond to very diverse spatial scales, from the local or regional level to the national level and there are even routes that cross several countries (Majdoub, 2010).

In this regard, a series of cultural tourist routes have been selected with different characteristics in terms of their geographical scale, the unifying element or elements and the activities generated around them, routes that are responding to certain segments of visitors. For this reason, an impressionistic overview of a series of routes that we find in different contexts and areas of the world is offered below.

Wine Route of the denomination of origin Montilla-Moriles

The Montilla-Moriles Wine Route was born at the beginning of 2001, within the framework of the commitments acquired by the City Council of Montilla, as a city of wine, with the Spanish Association of Wine Cities (SAWC). It is made up of three different options. The first runs through Montilla-Moriles and focuses on the two towns that give their name to that denomination, and where the most important wineries are located. The second runs through the cities of Córdoba, Fernán-Núñez, Montemayor, La Rambla and Montilla, and tries to unite the important cultural and heritage legacy of the city of Córdoba with rural tourism resources. The third runs through the cities of Aguilar de la Frontera, Moriles, Lucena and Puente Genil, and the objective of this option is to offer tourists who come mainly from the Costa del Sol a different alternative to the traditional sun and beach destination.

The Valle de Guadalupe wine route

The wine tourism route of Valle de Guadalupe is located 25 km north of the city of Ensenada, in the state of Baja California. This area has important natural resources (landscape and mountains) and heritage resources (winery museums, wine museums, local handicrafts and gastronomy), which make this route a mixture of cultural, environmental and social attractions (Secretariat of Tourism of the State of Baja California, 2011).

Mexico Tequila Route

The Tequila Route emerged as an alternative to improve the tourist competitiveness of the State of Jalisco, based on tequila. This route has a Designation of Origin, and the area in which it is located has been declared a World Heritage Site by the United Nations Educational, Scientific and Cultural Organization (UNESCO). One of the main objectives with the creation of this route was to revitalize the rural areas through which it crosses through the creation of a network of enterprises dedicated to the development and commercialization of rural tourism products and services (Barrera and Bringas, 2009).

The Silk Road

The Silk Road (7,000 km) starts from Xi'an (covers 4,000 km of Chinese territory), through Central Asia, Afghanistan, Iran, Iraq, Syria, to the coast of the Mediterranean Sea. From the 2nd century BC, until the opening of the Sea Silk Road, this route transported large quantities of silk, as well as Chinese inventions (such as paper, newspapers, etc.), while China received the Buddhist and Islamic religions, along with its art and culture, as well as a series of products.

Route of the Quebrada de Humahuaca

The Quebrada de Humahuaca is located in the province of Jujuy in northeastern Argentina. It is an Andean valley formed by the Río Grande along the Cordillera Oriental de los Andes, whose source is located more than 4,000 m above sea level in the Arroyo La Cueva. It stretches for more than 120 km south to Cuesta de Bárcena, where the mountains decrease in altitude and the arid landscape transforms into a lush forest.

According to UNESCO, the valley shows substantial evidence of its use as a major trade route. Regarding the tangible cultural heritage, several interesting physical remains can be found that demonstrate the existence of a road system that runs mainly in the north-south, but also in the east-west direction. In the ravine there are trails, tracks, roads, railways and paved roads, as well as a network of trails that connect cities, inns, places of work and supply, etc., whose courses may have followed those of the ancient Inca routes. (Martorell, 2003).

Camino Real or Estrada Real

The Camino Real or Estrada Real is a 400 km long circuit that connected several places related to mining operations in Brazil. The colonial cities through which the aforementioned road crossed are: Paraty, Rio de Janeiro, Ouro Preto and Diamantina, these last three named as World Heritage Sites by UNESCO. The Estrada Real is in turn divided into three roads: The Diamond Road, which linked Diamantina with Ouro

Preto; the Camino Velho, which linked Ouro Preto with Paraty; and the Caminho do Ouro or Caminho Novo, which improved the links between Ouro Preto and Diamantina with Rio de Janeiro. From this port the gold and diamonds were transferred to Portugal (Tresserras, 2013).

Tempranillo Route

The Ruta del Tempranillo was created as a tourism project in which banditry is considered as a social phenomenon of rural Andalusia in the 19th century, trying to convey the true image of it as a cultural phenomenon. This is how the image of a character appears, half reality half legend, which reflects all the heritage, cultural, social and economic elements of an area and a time, which will now allow, in the 21st century, the economic development of a geographical area. José Pelegio Hinojosa Cobacho "el Tempranillo" is one of the greatest representatives of Andalusian banditry. He received various names, such as King of Sierra Morena, Prince of Andalusia or Andalusian Robin Hood. All this has caused Tempranillo to be the prototype of the romantic Andalusian bandit; that is, the generous bandit who avoided violence in his actions, who stole from the rich and distributed it among the poor (Rodríguez Martín, 1998). But, in addition to the historical documentation about him, there is a rich oral tradition in Andalusia that represents him as the myth of the rebellion of the Andalusian people, transferring the historical knowledge of him to the world of legend and myth (Merinero Rodríguez, 2000).

Tourist-cultural Routes of Teruel

Currently you can find a series of tourist-cultural routes located in Teruel, they have a comprehensive management structure both in terms of heritage and tourism initiatives that can be generated around them. Among the main routes we can mention: Cave Route, Pantoleology Route, Iberian Route of Bajo Aragón, Camino del Cid, Mudéjar Route, Roman Route, Route of the Prisons, Route of the refrigerators and the vaults of the cold, Places of the Memory, Drum and Bass Route (Tresserras, 2013). Three of them are mentioned below.

Path of the Cid

Along this route you can see the path made by Rodrigo Díaz de Vivar, which is described in the Cantar del Mío Cid. What can be highlighted about this route is that it is structured through themed sections, to be able to do it at different times. The main objective of the management of this route is to promote and disseminate the cultural, environmental and tourist resources of the Camino.

Mudejar Route

The Mudéjar Route crosses eight provincial councils (Burgos, Soria, Guadalajara, Zaragoza, Teruel, Castellón, Valencia and Alicante) and was declared a UNESCO World Heritage Site in 1986 due to its great heritage value. Its art and architecture are born from the coexistence between Muslims, Christians and Jews in the Hispanic kingdoms, its main characteristics being the fusion of Arab and Christian elements, using cheap materials such as brick, plaster, wood and ceramics.

Drum and Bass Drum Route

Holy Week is a deeply rooted tradition in Spanish territory and has numerous religious manifestations. In the case of the province of Teruel, celebrations with drums and bass drums take place in many parts of the province. In this way, the Ruta del Tambor y el Bombo del Bajo Aragón, declared of National Tourist Interest, stands out for its traditional touches in the different religious acts and also for the contributions of popular tradition. A heritage has been created around a religious path, where the representative sites are valued thanks to the steps of Holy Week, in which the brotherhoods stop to worship their saint.

The Route of the Dacian Fortresses of Alba and Hunedoara

These are 6 Dacian fortresses, located in the counties of Hunedoara and Alba, called Sarmizegetusa Regia - Grãdiotea de Munte, Costeoti Cetãþuie, Costeoti Blidaru, Luncani Piatra Rooie, Bãniþa and Cãpâlna. For a better assessment, starting in 2006, a series of Dacian Fortresses Festivals were organized (in Alba County, in Cricãu, Piatra Craivei, Cugir), in which both natives and tourists participated. The festival program included equestrian acrobatics, folk performances, and was also aimed at attracting people's attention to the region's tourism potential, by transposing traditional Dacian rituals and customs into real time, such as the restoration of a village Dhaka, with a pottery workshop, a blacksmith shop and a shrine. By promoting the Dacian fortresses, rural tourism was encouraged, focusing on the return to nature and simple traditions.

The Hijaz railway route

The Hijaz Railway was built at the beginning of the 20th century, in order to connect Damascus with the holy cities of Madinah and Makkah, extending to the already established rail links from the Ottoman capital Istanbul to Baghdad and Damascus. It also maintains links with a number of parallel routes that are in close proximity to it, including the pre-Islamic trade route along which frankincense traders moved from Oman to eastern Mediterranean ports and the main Pilgrim routes from northern Egypt. (land route), Turkey and Sham province (present-day Syria and Jordan) (Faroqhi, 1994).

Alpine Way

The cultural route known as "Vía Alpina" is the result of actions undertaken within the Alpine Convention with the aim of contributing to local development, identity and respect for the natural alpine environment (Borsdorf, 2006). The aforementioned route, in addition to being easily accessible, proposes a tourist offer in which culture is mainly promoted, as the main factor along its trails, and the quality of services.

The Alpine route has a total length of more than 5,000 km, its altitudes range between 0 and 3,000 m above sea level. And is divided into 341 stages. The objectives of the project are: the protection of nature (it crosses 9 national parks, 17 natural parks and 22 natural reserves); bring people closer together and preserve the cultures associated with the Alps. In short, this extensive trail is an instrument of local sustainable development, becoming a national and international showcase for the regions it crosses, revaluing natural and heritage resources (Sánchez and Marchant, 2008). This is a concrete example of how it is possible to take advantage of the existence of trails, used for centuries by the inhabitants of the area, from a tourist point of view and become a source of socioeconomic development and integration between nations, through the participation of the actors and the revaluation of the natural and cultural settings associated with it (Pfefferkorn et al., 2005).

The Alpina way project has two major stages. The first was launched between 2001 and 2004, a period in which studies and pilot projects were carried out to improve the quality and sustainability of the tourist offer. With this objective, the signage designs (logo and informative brochures in different languages) and a database of exhaustive technical-tourist information on the description of the natural and cultural heritage, as well as cartography at various scales and an interactive website were built. Which is constantly updated. Studies were also carried out on the quality requirements of the itineraries: maintenance, services, marketing and management; which allowed to build a common quality guide. This was supported by carrying out pilot projects aimed at improving supply, development of complementary services, use of geographic information systems and first actions in the territory for the involvement of local actors. (Borsdorf, 2006).

Between 2005 and 2007, the second phase of the project was launched through financing from the Espacio Alpino. During this period the main priority was to work together with the main local actors in the planning and development of the itineraries and valorization of the natural and cultural heritage. Among the projects is the dissemination of a marketing and communication strategy through the tourism promotion of each country, in addition to carrying out hiking development programs, particularly in cross-border spaces, and environmental education for schoolchildren. (Sánchez and Marchant, 2008).

1.3 Methodology

From the methodological point of view, the qualitative approach is used, which proposes a qualitative and descriptive design, where the researcher accesses a greater coverage of the subject in the bibliographic review process. In this sense, the purpose of the aforementioned approach provides a detailed and documented description of the phenomenon under study.

The information collected for this work is related to the representative heritage and cultural routes, which have been selected for the interest they have aroused by having been studied and named heritage. The purpose of this research was to carry out a review of the current and evolutionary status of the contributions made for the enhancement of historical and heritage value of these routes. Therefore, for this study, an exhaustive review of public documents was carried out, such as: national and international government reports, online magazines, forums, narrations and opinions published on web pages of the agents involved in the process of declaring tourist and cultural routes.

1.4 Conclusions

It is a palpable reality that the development of tourist activities in rural areas is meaning the revitalization of their economies, through a new source of income that complements or replaces the traditional income of small municipalities. The tourist-cultural routes are a source of income not only for the owners of lodging establishments close to the areas, but at the same time they have a pull effect towards other local productive activities, such as restaurants, bars, small businesses, productions agricultural and artisanal, etc. In such a way, it could be said that the practice of this type of tourism not only allows the local population to take advantage of the potential resources of the area, increasing their income and improving their quality of life, but also the tourist who visits them can live a new and differentiating experience to what traditional tourism can offer.

Acknowledgment. We would like to thank the Technical University of Ambato and the Department of Research and Development (DIDE), in Ecuador, for their support.

To Indoamerica University for its support in the research process.

References

Barrera, E.: Las Rutas Alimentarias como instrumento para el Desarrollo Territorial: El caso de la Ruta de la Yerba Mate de Argentina. Revista de la Universidad de Sonora (26) (2009)

Borsdorf, A.: Politishche Rahmenbedingungen alpiner Raumentwicklung un Rauminformation. Das Europäische Raumentwicklungskonzept, die Alpenkonvention und die EFRE-Programme in Relation zu DIAMONT. In: Bordsdorf, A., Lange, S.: Politische, kulturelle und wissenschaftliche Perspektiven der nachhaltigen Raumentwicklung in den Alpen. Erste Ergebnisse des an der LFU Innsburck koordinierten EU-pROJEKTES DIAMONT. Alpine Space: Man & Enviroment **2**, 27–39 (2006)

Briedenhann, J., Wikens, E.: Tourism routes as a tool for the economic development of rural areas. Vibrant hope or impossible dream. Tourism Management **25**(1), 71–79 (2004)

Calderón Puerta, D.M., Arcila Garrido, M., López Sánchez, J.A.: La rutas e itinerarios turístico-culturales en los portales oficiales de turismo de las Comunidades Autónomas españolas. Revista de Estudios Andaluces **35**, 123–145 (2018)

Carrera, G.: Itinerarios y rutas: herramientas para la documentación y puesta en valor del patrimonio cultural. PH Boletín del Instituto Andaluz del Patrimonio Histórico **60**, 52–59 (2006)

Denstadli, M., Jacobsen, J.: The long and winding roads: Perceived quality of scenic tourism routes. Tourism Management (2010)

Fagance, M.: Rural tourism and the small country town. Tour. Recreat. Res. **16**, 34–43 (1991)

Faroqhi, S.: Pilgrims and Sultans: The Haii under the Ottomans. London (1994)

Gonzalo, M.: El patrimonio rural como recurso turístico. La puesta en valor turístico de infraestructuras territoriales (rutas y caminos) en las áreas de montaña del País Vasco y de Navarra. Cuadernos de Turismo (27), 759–784 (2011)

Guzmán, T., Gutiérrez, E.: El turismo rural como agente económico: desarrollo y distribución de la renta en la zona de Priego de Córdoba. CIRIEC-España, Revista de Economía Pública, Social y Cooperativa **55**, 167–192 (2006)

Hummelbrunner, R., Miglbauer, E.: Tourism promotion and potential in peripheral areas: the austrian case. J. Sustain. Tour. **2**, 41–50 (1994)

ICOMOS-CIIC-International Committee on Cultural Routes, International Committe on Cultural Routes. Obtenido de http://www.icomos-ciic.org/ (2015)

López Fernández, M.: Diseño y programación de Itinerarios Culturales. PH Boletín del Instituto Andaluz del Patrimonio Histórico **60**, 20–33 (2006)

Majdoub, W.: Analyzing cultural routes from a multidimensional perspective. Almatourism-Journal of Tourism, Culture and Territorial Development **1**(2), 29–37 (2010)

Martorell, A.: Cultural routes: tangible and intangible dimensions of cultural heritage (2003)

Merinero Rodríguez, R.: El bandolerismo a través de la tradición oral. Anuario Etnológico de Andalucía 1996–1998. Consejería de Cultura de la Junta de Andalucía, Sevilla (2000)

Morgan, J., Lambe, W., Freyer, A.: Homegrown responses to economic uncertainty in rural america. Rural Realities **3**(2), 1–15 (2009)

Murray, M., Graham, B.: Exploring the dialectics of route-based tourism: the Camino de Santiago. Tourism Management **18**(8), 513–524 (1997)

Pfefferkorn, W., Egli, H., Massarutto, A., (eds.): Regional Development and Landscape Change in the Alps. From Analysis and Scenarios to Policy Recommendations. University of Berne, Berne (2005)

Ramírez, J.: Los caminos del patrimonio. Rutas turísticas e itinerarios culturales. Pasos **9**(2), 225 (2011)

Reed, M.: Collaborative tourism planning as adaptive experiments in emergent tourism settings. Journal of Sustainable Tourism, 331–355 (1999)

Rogerson, C.: Tourism routes as vehicles for local economic development in south africa: the example of the magaliesberg meander. Urban Forum **18**(2), 49–68 (2007)

Rogerson, C.: Local economic development and tourism planning in Africa: Evidence from route tourismun South Africa. In: Hottola, P. (ed.) Tourism Strategies and Local Responses un southern Africa. Cambridge (2009)

Sánchez, R., Marchant, C.: Un "Camino Largo" en Sudamérica. Una oportunidad de desarrollo económico y natural sostenible. Biblio 3w: revista bibliográfica de geografía y ciencias (13) (2008)

Sarasa, J.: Aportaciones básicas del turismo al desarrollo rural. Cuadernos de Turismo **6**, 45–59 (2000)

Secretaría de Turismo del Estado de Baja California. Ruta del Vino de Baja California. Gobierno del Estado de Baja California, México (2011)

Simpson, F., Chapman, M., Mahne, L.: Partnership Approaches to Turism and Rural Development in Post Socialist Europe, the Experience of Notranjski Kras, Slovenia. Rural Tourism Management: Sustainable Option International Conference (1998)

Strauss, C.H., Lord, B.E.: Economic impacts of a heritage tourism system. Journal of Retailing and Consumer Services **8**(4), 199–204 (2001)

Sugio, K.: A consideration on the definition of the setting and management protection measures for cultural routes. In: Proceedings of the Scientific Symposium "Monuments and sites in their setting - conserving cultural heritage in changing townscapes and landscapes". 15th General Assembly of ICOMOS, Xi'an, China, octubre de 2005 (https://www.icomos.org/xian2005/papers/4-23.pdf) (2005)

Tresserras, J.: Rutas e itinerarios culturales en Iberoamérica. Cuadernos de Patrimonio Cultural y Turismo **15**, 13–56 (2006)

Ecotourism in Ecuador and the Lack of Cultural Education for Health Benefits

Tania Morales-Molina[1](✉), Ximena Morales-Urrutia[2], Chabely Figueredo-Morales[1,2,3], and Geri Belén Bucheli-Vázquez[1]

[1] Universidad Tecnológica Indoamérica, Ambato, EC 180103, Ecuador
taniamorales@uti.edu.ec
[2] Universidad Técnica de Ambato, Ambato, EC 180104, Ecuador
[3] Instituto Superior Tecnológico España (ISTE), Ambato, EC 180104, Ecuador

Abstract. Ecotourism or ecological tourism is a tourist typology that consists of carrying out recreational activities in natural areas with an environmental sustainability approach, the success of ecotourism in recent times lies in the multiple benefits it brings to our health. The objective of our research is focused on highlighting the lack of cultural education on the subject for health benefits through some strategies. The methodology applied has a mixed approach, documentary-bibliographic type through the review of different scientific articles, documents, books, texts and doctoral theses in order to build the state of the art, at the empirical level, we will work with statistical data from secondary sources in national statistical offices. The results reflect the lack of education that Ecuadorian citizens have on the benefits that ecological tourism brings to health and in the conclusions some strategies are recommended to give solutions to the cultural ignorance of people in the ecological field and to materialize education from early ages to achieve a healthy population, free of stress and improve the quality of life of Ecuadorian inhabitants.

Keywords: Ecotourism · Cultural education · Environmental sustainability · Health

1 Introduction

Ecotourism is tourism that has as its main objective the discovery of the fauna, flora and landscapes of the region we visit. It is usually It is usually carried out in protected natural areas such as Natural Parks or National Parks but not exclusively in them and in the way it is implemented or put into practice it always takes exquisite care that its proposed activities have a low environmental impact and high sustainability.

O.M.T. (World Tourism Organization) in its report "Tourism in the year 2100" that which responds to the needs of current tourists and receptive regions, protecting and magnifying the opportunities of the future. It is represented as a manager of all resources so that economic, social and aesthetic needs can be met while maintaining cultural integrity, essential ecological processes, biological diversity and life support systems. (Ojeda et al., 2019).

C. Montenegro et al. (Eds.): ICMTT 2023, LNNS 774, pp. 322–331, 2024.
https://doi.org/10.1007/978-3-031-43733-5_29

When we talk about this type of tourism we also refer to recreational activities in natural areas with a focus on environmental sustainability, it is characterized by being respectful with the environment and promoting ecological actions, it is a type of responsible travel that not only preserves the green areas of the place, but promotes its ecological development, improving the welfare of the population and the biodiversity of the area of location,despite the fact that nature tourism emerged in the late 1980s, being a relatively early institution, it is the fastest growing tourism class worldwide in recent years. (Suárez, 2020).

Ecotourism is an activity that combines a passion for travel with concern for the environment. Experts who have dealt with the subject suggest that the objectives of ecotourism are sustainability, conservation and the participation of the local community. They also attribute to ecotourism the capacity to achieve the objectives of sustainable development in regions with ecotourism potential. Its growth since the 1980s is due to the economic benefits it provides as nature tourism (Bonfim et al., 2022). This economic incentive, perhaps more than its principles, has given rise to the expansion of this type of "ecological" activity. Some promote it as a solution to the shortage of budget for protected areas, others see it as a basic element of economic growth.

Ecotourism is not adequately understood, being used as a marketing strategy because of its prefix "eco". (Oré, 2020)The term has been used to promote various tourism activities developed in the natural environment and to capture tourists interested in aspects of nature and culture of the regions visited. Ecotourism is part of nature tourism, which presents a great variety of natural offers such as lakes, rivers, mountains, forests, snow-capped mountains, deserts, among others. In these scenarios and according to the motivation and interest of the tourist, different recreational activities are developed such as horseback riding, camping, hiking, diving, climbing, etc. (Rojas et al., 2020).

In Ecuador, the recovery of ecotourism is reflected in the results of the second four-month period of 2022. In August alone, 48,000 visitors chose the destination, representing an increase of 75% over the same period of 2021. being still one of those destinations with hidden places without a large accumulation of tourists and one of those places where you can do absolutely everything: from bathing in the waters of the Pacific to tour the Amazon jungle, diving in the Galapagos Islands or know the "magic" of the Middle of the World. (Morales et al., 2021).

Basic natural resources and ecosystems must be managed in a sustainable manner in order to meet the food demand of the population and other environmental, social and economic needs. Environmental sustainability represents a way to coexist in balance with our environment and prevent a shortage that can put humanity at risk, which makes this topic necessary for every person, organization and government within the planet. (Caviedes and Olaya, 2018).

On the other hand, and taking into account the benefits that ecotourism brings to our health, we can refer that the history of ecotourism in Ecuador goes back to tourists with money in the 90's who were looking for quality medical treatments and found that developing countries offered a great quality of life, quiet and low-cost tourist places. This type of tourism encompasses all forms in which patients travel to other countries in search of health improvement, occurs when people travel from their place of residence in search of improving their state of health in relaxing conditions and nature. (Pérez, 2020).

Ecotourism is an activity that in recent years has gained special importance among those people who are subject to long working days, mainly in big cities where stress, pollution, new pathologies have invaded the life of man. Ecological tourism is a new typology that has evolved into a concept called ecotourism, which is determined by spaces, places and activities dedicated to heal ailments, cure diseases and improve the figure, where treatments and therapies, avoid the maximum use of chemical formulas, using instead natural techniques, being the main element of healing water. Therefore, it is considered important for a better understanding to detail relevant aspects related to mineral-medicinal water such as: the origin, forms of application, classification, benefits, and therapeutic indications, which have for the human body. Finally, a classification of health tourism is made from the point of view of international institutions in charge of regulating these activities, considering the following parameters: concepts, main activities and differentiating characteristics. (Estrada and García, 2020).

To improve ecotourism, guaranteed funds, international promotion, tax relief for entrepreneurs and innovation are some of the keys to achieve "the dream of turning Ecuador into a tourism power related to ecology". Ecuador is one of the most privileged countries in terms of tourism options. (Graciano et al., 2022).

On the other hand, if these activities are not planned or managed according to the characteristics of the ecosystems and the culture of the regions visited, they can cause negative environmental impacts that can destroy the very resources that constitute the tourist attraction and contribute to the sustainable development of the regions, and can offer alternatives to the tourism sector, it must be developed according to the guidelines or principles that support it, i.e., what are its fundamental characteristics. To work on certain precisions of the concept, to elaborate the systematization and diffusion of the principles that support it. In this perspective, the research aims to elaborate a conceptual framework to guide ecotourism according to its principles (Maldonado, 2022).

Sustainable tourism is the development of a responsible recreational activity, which integrates important axes such as conserving natural and cultural resources, sustaining local welfare by ensuring economic viability and raising awareness through environmental education and interpretation among visitors (Mori et al., 2022).

Environmental education is a field of political and pedagogical intervention that promotes comprehensive educational processes aimed at the construction of an environmental rationality. Within this framework, different environmental knowledge, wisdom, values and practices converge and contribute to citizen training and the exercise of the right to a healthy, dignified and diverse environment (Vargas et al., 2022).

Environmental education is a continuous process that promotes sustainability as a social project, understanding that this implies development with social justice, distribution of wealth, preservation and conservation of nature, gender equality, health protection, participatory democracy and respect for cultural diversity. Environmental education, in this sense, seeks a balance between various dimensions, such as social, ecological, political and economic, within the framework of an ethic that promotes a new way of inhabiting our common home (Salas, 2021).

The Ministry of Education, in agreement with the Federal Council of Education, shall provide the necessary measures to provide environmental education at all levels and modalities of the National Education System, with the purpose of promoting values, behaviors and attitudes that are in accordance with a balanced environment and the protection of biological diversity; that tend to the preservation of natural resources and their sustainable use and that improve the quality of life of the population" (Hernandez et al., 2021).

The lack of ecological education is another cause of the destruction of the Amazon. Much has been criticized to the indigenous people for being the major agents causing this destruction. While this may be true on the surface, the real cause of the problem is lack of vision and understanding, i.e. lack of education. Many indigenous people simply cannot understand what is happening in their territories environmentally, and for the few that do, the lack of control of the situation leaves them feeling helpless. Education is the right and responsibility of every Ecuadorian (de Parra et al., 2022).

Ecotourism generates both health and economic benefits for host communities, organizations and authorities that manage natural areas for conservation purposes; it offers local communities alternative employment and income opportunities, minimizes physical, social, behavioral and psychological impacts, creates environmental and cultural awareness and respect, provides positive experiences for both tourists and locals, and provides direct financial benefits for conservation (Cuevas and Iglesias, 2022). (Cuevas and Iglesías, 2022).

It also generates benefits of respect for nature and, therefore, minimal environmental impact. In addition, there is the respect for traditions, local culture and the important generation of income for the inhabitants of the destination visited, it contributes to the conservation of biodiversity. It sustains the well-being of the local population. Includes a learning/interpretation experience. Involves responsible action by tourists and the tourism industry (Barráez, 2022).

Principles of Ecotourism:

Ecotourism is about bringing together conservation, communities and sustainable travel. This means that those who implement, participate in and market ecotourism activities should adopt the following ecotourism principles: Minimize physical, social, behavioral and psychological impacts.

1. Develop environmental and cultural awareness and respect.
2. Provide positive experiences for visitors and hosts.
3. Produce direct financial benefits for conservation.
4. Generate financial benefits for both the local population and private industry.
5. Provide memorable interpretive experiences for visitors that help increase sensitivity to the political, environmental and social climates of host countries.
6. Design, build and operate low-impact facilities.
7. Recognize the spiritual rights and beliefs of Indigenous Peoples in your community and work in partnership with them to create empowerment (Menéndez et al., 2022).

2 Methodology

The methodology will apply a mixed documentary-bibliographic approach through the review of different scientific articles, documents, books, texts and doctoral theses in order to build the state of the art, at the empirical level, we will work with statistical data from secondary sources in national statistical offices.

3 Results

Definition of Ecotourism

Ecotourism or ecological tourism is a type of tourism that consists of recreational activities in natural areas with a focus on environmental sustainability. It is characterized by being environmentally friendly and promoting ecological actions.

It is a responsible type of travel that not only preserves the green areas of the place, but also promotes its ecological development, improving the well-being of the population and the biodiversity of the area. Although nature tourism emerged in the late 1980s, being a relatively early institution, it is the fastest growing type of tourism worldwide in recent years.

Ecotourism Activities

Observing the flora and fauna: Consists of contemplating the variety of living beings that the tourist site possesses, including the species of animals and the diversity of plants that inhabit the area.

Photographing animals and plants: Photographing is an action that does not bother or cause harm to living beings, which encourages users to practice this activity in nature tourism, even if they are not experts in photography.

Participate in flora and fauna rescue programs: These programs aim to avoid, mitigate, or correct the negative effects that can or have had an impact on the tourist area.

Hiking: It consists of walking through forest trails, mountains and any natural environment to discover the virtues that the area possesses. It is the activity that seeks to bring people closer to the natural surroundings of the tourist product.

Explore by bicycle: It is about exploring natural areas using bicycles, which means of transportation is environmentally friendly and is used to travel medium and long distances.

Characteristics of Ecotourism

Ethical tourism: Ecotourism defines itself as a type of ethical tourism.

Respect and conservation of the environment: It is based entirely on respect and conservation of nature.

Ecological actions: Promotes activities that are beneficial to the environment.

Environmental education: Instructs tourists on the most elementary topics to acquire ecological values, knowledge and skills.

Kindness to local populations: Courtesy and respect is not only directed to the environment, but also to the local people living in the area.

Advantages of Ecotourism

Promotes biodiversity conservation.

Requires few non-renewable resources to carry out these activities, and in turn encourages the use of renewable resources.

Increases the environmental education of those involved.

Generates environmental awareness.

Increases ecological culture.

It raises respect for local populations.

Provides positive experiences to tourists and locals.

Generates an economy around the tourist area, providing them with capital and money to carry out environmental conservation operations and tourism growth.

Offers a different approach to minimize environmental impact.

Summaries the benefits, characteristics and textual definition of ecotourism, the main characteristics of which include the conservation and development of the green areas of a specific tourist destination. It is also important to know and raise awareness of this type of tourism as it helps to promote ecological programmed to help improve the biological community.

The ecological balance of many natural areas depends essentially on their economic growth, which is mainly driven by ecotourism, indicating the great relevance of its role in preserving the biodiversity of different tourist destinations (Table 1).

Table 1. The most popular types of passenger cars in Ecuador

Adventure tourism	Nature tourism	Health tourism	Ecotourism
Hang gliding	Mangrove Walk	Medical tourism (surgeries and treatments)	Nature observation
Paragliding	environmental	Wellness Tourism (related to the for welfare, personal relaxation)	Wildlife observation
Skydiving	Mountain Biking	Complementary medicine tourism (alternative medicine)	Ecosystem observation
Autonomous diving		Health maintenance (SPA)	Ecological observation
Free diving	Green Tourism		Interpretive trekking
Descent in rivers	Gentle Nature Tourism		Flora and fauna rescue
kayaquing	Ethnic tourism		Environmental education workshops
Recreational fishing	Sustainable Tourism		Biological education projects
Mountaineering	Socially responsible tourism		Photo safari

(*continued*)

Table 1. (*continued*)

Adventure tourism	Nature tourism	Health tourism	Ecotourism
Canoeing	Tourism Of nature		Sidereal observation
Sport and rock climbing	Cycle touring and mountain biking Sport tourism		Observation of fossils
Mountain biking	Hiking		Observation of natural attractions

The tourism modalities described in the table above refer to activities that are permitted and authorized in 9 provinces of Ecuador, are accepted and considered with greater impact in the same order that we place in the table, in the first place appear the trips that have as purpose to perform recreational and sports activities, associated with challenges imposed by nature, This is what we call adventure tourism. Nature tourism is the tourist activity that is developed without altering the balance of the environment, promoting the conservation of nature and the existing ecosystems by carrying out recreational activities of appreciation and knowledge of nature through interaction with it. Health tourism refers to all types of surgeries and medical treatments to which travelers are subjected, which are performed in clinics and hospitals that provide hotel and lodging services of the highest quality and maximum comfort during hospitalization, and last but not least, Ecotourism, associated with ecological tourism, is the subject of the research.

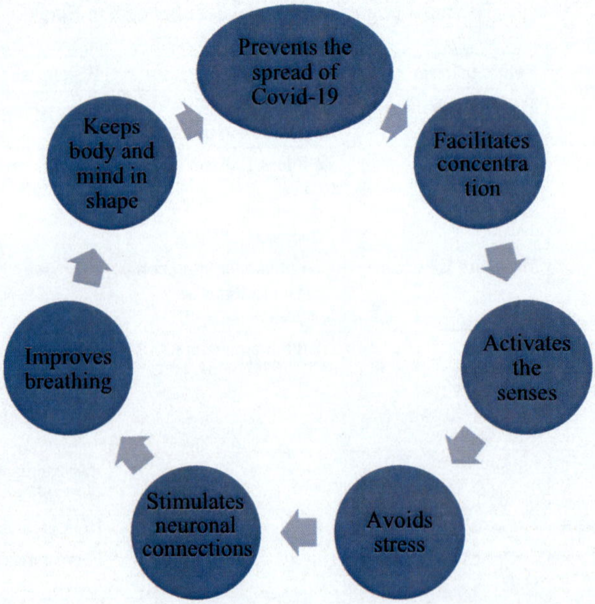

Fig. 1. Health benefits of ecotourism

The purpose of our table is to reflect the order of acceptance of these types of tourism and also to highlight that, due to the lack of culture, environmental and ecological education, hardly the citizens of the country are aware of the benefits that this type of tourism offers to the health and economy of Ecuador (Fig. 1).

Being in contact with nature means that you stabilize your circadian rhythms and improve your mood and health. Ecotourism is booming and its benefits have become a necessity to have a healthy, full and healthy life, synchronize with the earth and its cycles, increase vital energy and positivism. In this modality, the tourist can find and satisfy his search to improve his physical condition, reduce tension, and therefore, benefit his emotional and physical state by overcoming a challenge imposed by nature, where the experience is only between nature and the tourist. Adventure tourism excludes sports competitions or so-called "extreme" activities, where the challenge is against time, or against man himself (Table 2).

Table 2. Strategies to promote ecotourism education in Ecuador

No	Recommendations
1	Encourage the planning of activities associated with educational culture in schools, especially in Amazonian schools
2	Propose the planning or management of the tourism territory as a working tool
3	Local management and consensus of the local population
4	Infrastructures in restaurants, gastronomy and organic farming
5	Ask visitors (market research) to establish goals and objectives
6	Know the natural environment well (Before arriving at the destination, it is essential to get information about the attractions, schedules, guides, etc.)
7	Choose certified companies (Whether for lodging or leisure services, it is important to hire companies that comply with environmental management systems or are certified with the European Charter for Sustainable Tourism)
8	Use adequate material (To observe and learn from nature, it is necessary to have elements that will help us to enjoy the experience (binoculars, GPS, adequate footwear, comfortable clothes and discreet colors...)
9	Moving in a sustainable way
10	Be respectful (Enjoy nature in silence without disturbing or touching animals, plants and flowers)
11	Entering the visitor centers (To observe and learn from nature, it is necessary to have elements that will help us to enjoy the experience
12	Buy local products (Consume products (food, handicrafts, etc.) manufactured in the place of visit and hire services (cabs, guides, etc.) managed by local companies)
13	Help conserve (Learn about non-profit initiatives that help conserve the natural area and contribute to the cause)

Ecotourism or ecological tourism, as already mentioned in the research work is a style of alternative tourism different from traditional tourism, has an approach to tourism activities in which important principles such as sustainability, preservation, appreciation of the environment that welcomes and sensitizes travelers in a pleasant way are privileged, Therefore, in this table we made some recommendations that will be useful to educate us consciously of its main way to achieve to make known and prepare us, to give a better attention to visitors both national and international tourists and give examples in countries with the same destinations with natural attractions.

4 Conclusions

Ecuador's Ministry of Tourism promotes the integration of communities in sustainable tourism activities and complementary alternatives in protected areas. In addition, the country has achieved the commitment of tourism operators in the effective management of environmental processes and best practices, as well as in the regularization of their tourism activities to improve the supply while conserving this natural heritage.

According to the World Tourism Organization, in Ecuador ecotourism is one of the motivations for travel, as both domestic and international visitors are delighted with its benefits both observation and appreciation of Ecuadorian nature, its unique traditional cultures dominant in natural areas. The influence of ecotourism on the environment, includes: Promote as a source of value and enjoyment for tourists, natural ecosystems that help the conservation of the environment, landscapes and biodiversity, thanks to concrete measures that help to that end.

It is necessary to emphasize that it is necessary in Ecuador to have new educational strategies to understand and mitigate from different points of view the knowledge of ecological tourism and its benefits. Environmental education (EE) is considered an elemental tool for all people to become aware of the importance of preserving their environment and be able to make changes in their values, behavior and lifestyles, as well as expand their knowledge to prevent deterioration and promote them actions through prevention and mitigation of existing problems, taking into account the multiple benefits they provide for communities and for the country.

In the study conducted it was found that Ecuadorian citizens, entrepreneurs and tourists do not have a sufficient mastery of the benefits of ecotourism for health, so much so, that it is not considered as the first tourist attractions, however this type of tourism make human life possible, for example, by providing nutritious food and clean water, regulating disease and climate, supporting crop pollination and soil formation, and providing recreational, cultural and spiritual benefits. The natural resources they provide are not only essential for plant and animal life, but also for human survival itself. Carbon sequestration. Ecosystems are the planet's great carbon sinks. In other words, they are essential to the fight against climate change.

Finally, through this research we proposed some strategic recommendations that could serve to increase the knowledge of this activity in a systematic way, which could begin with the students of initial educational units in the Amazonian areas in order to create an environmental educational culture and gradually extend it to other areas of the country.

References

Barráez, D.: Metaverses in the context of virtual education. Revista Tecnológica-Educativa Docentes **2**, 22 (2022)

Bonfim, I., Corbari, S., Mendes, J., Serqueira, B.: Social representations on tourism development in Terena Indigenous Lands. Tourism Research **23** (2022)

Carvache, W., Carvache, M., Carvache, O., Recalde, X.: Preferences for adventure tourism in the development of a tourism package: Santa Elena case, Ecuador. Interamerican J. Environ. tourism **14** (2018)

Caviedes, D., Olaya, A.: Ecotourism in protected areas of colombia: a review of environmental impacts with emphasis on environmental sustainability standards. Luna Azul Journal **46** (2018)

Cuevas, E., Iglesías, S.: The application of the metaverse in the educational ecosystem. Formative Proc. Innov. Edu. Exp. **81** (2022)

de Parra, L., Martínez, V., Losada, H.: Social representations about community ecotourism, environmental problems and relationships with environmental education. Environment & Technology Journal **63** (2022)

Estrada, F., García, M.: Ecotourism in Quintana Roo. Analysis of its Community Projects. El Periplo Sustentable: Journal of Tourism, Development and Competitiveness **113** (2020)

Graciano, R., et al.: Análisis de la factibilidad técnica del centro ecoturístico Puentecillas, Pueblo Nuevo, Pueblo Nuevo, Durango: Analysis of the technical feasibility of the Puentecillas ecotourism center, Pueblo Nuevo, Durango. e-CUCBA, 18 (2022)

Hernandez, C., Prada, R., Aloiso, A.: Strategic technological management: use of the social web 2.0 ecosystem in higher education. Revista Venezolana De Gerencia, 77 (2021)

Maldonado, J.: Construction of a sustainable ecotourism model for Norte de Santander. Aibi J. Res. Manage. Eng. 8 (2022)

Menéndez, O., Ortiz, L., Hormigó, Y.: Postgraduate integrated educational digital ecosystem for education, science, technology and society. Luz ciencia 86 (2022)

Morales, T., Núñez, G., Bucheli, G., Machado, L.: Safety during the fruit and flowers festival and its impact on local entrepreneurs. In: XV Multidisciplinary International Congress on Science and Technology, p. 274 (2021)

Mori, M., Ramírez, O., García, C., Ishuiza, R.: Sustainable tourism based on the evaluation of geosites. Revista Universidad y Sociedad, p. 202 (2022)

Ojeda, A., García, A., Romero, D., Fraga, J.: Ecotourism as a conservation tool in coastal wetlands of Yucatan, Mexico. Cartographic Journal, 155 (2019)

Oré, M.: Nature and culture for the development of ecotourism in the province of Concepción, Junín, Peru. Socialium, p. 143 (2020)

Pérez, B.: Sustainable tourism, ecotourism and the CETS: Sierra Nevada (Spain). HUMAN REVIEW. International Humanities Review/Revista Internacional de Humanidades, p. 18 (2020)

Rojas, E., Castillo, V., Cano, C.: Ecotourism implemented in the globalized world as an alternative for economic and social development. Clío America, 380 (2020)

Salas, H.: Environmental education and its contribution to the care and protection of the ecosystem. Fides et Ratio-Revista de Difusión cultural y científica de la Universidad La Salle en Bolivia, 289 (2021)

Suárez, C.: Valoración económica del ecoturismo el caso del santuario de fauna y flora los flamencos (Colombia). Estudios y perspectivas en turismo, 29 (2020)

Vargas-, H., López, L., Balanta, V.: Sociocultural and environmental components that impact community-based ecotourism. Entramado, 18 (2022)

Villacrés, J.: Ecotourism in Ecuador: Current events and development prospects. Observatory of the Latin American economy, 238 (2018)

Heritage Resources of the Municipality of Tomar in the Context of a Sustainable Cultural Tourism

Eunice Ramos Lopes[1](✉), Célio Gonçalo Marques[1], and Manuela Fernandes[2]

[1] Polytechnic Institute of Tomar, TECHN&ART, 2300 Tomar, Portugal
eunicelopes@ipt.pt
[2] Polytechnic Institute of Tomar, 2300 Tomar, Portugal

Abstract. Faced with the continuous evolution of territories, the management of heritage resources emerges as an instrument that supports the safeguarding and enhancement of place identity. Tourist destinations seek to implement innovative strategies, due to the need to act in the face of competition. The main objective of this paper is to map the existing (material) heritage resources in the municipality of Tomar, to understand their importance and how cultural tourism experiences can be enhanced. The adopted methodology is based on a survey (n = 292). There are some empirical studies that analyze this topic, focusing on specific territories. To that extent, this study constitutes a relevant contribution to the reinforcement of the importance of mapping and managing existing heritage resources in a territory, working within a sustainable tourist and cultural dynamic. The results reveal that the heritage resources existing in the territory and those which can be visited, constitute added value and differentiating tourist attractions, in turn enabling diversification of the supply and the enhancement of heritage.

Keywords: Cultural Tourism · Management of Heritage Resources · Sustainable Tourist Destinations

1 Introduction

In the context of integrated management of heritage resources, cultural tourism is one of the most significant segments of tourism, due to its strong contribution to the evolution of tourism in different regions and countries [1]. Cultural tourism is evidenced to be a lever for the development of tourist destinations, given that all territories implicitly hold culture and heritage resources (material and immaterial). The development of territories involves the relationship that exists between tourism and culture, which, in turn, contributes to the growth of cultural tourism [2].

The practice of cultural tourism can help generate awareness regarding the preservation of heritage, both tangible and intangible, as it brings value to heritage resources that distinguishes and affirms communities individually [3]. Cultural heritage is effectively something that is inherited and therefore affirms the identity of a particular place. Whether of a material or immaterial nature, the legacy that remained as cultural belonging does not lose its symbolic or historical value. Thus, it is up to each person to make

C. Montenegro et al. (Eds.): ICMTT 2023, LNNS 774, pp. 332–341, 2024.
https://doi.org/10.1007/978-3-031-43733-5_30

good use of resources, preserving and restoring them whenever justified, so that historical and cultural memories could persist but also the authenticity of the territory remains. The concept of territory can be used in the political-legal, economic, and cultural context, the latter being used to identify a place. Consequently, there is an association between tourism and territory, as the practice of tourism attributes value to a place, which in turn becomes a tourist destination, but the territory is the main element in the composition of this tourist destination.

Disseminating the tourist resources that are an integral part of a region is important, as it promotes the territories. This increases their attractiveness and, in turn, turns them into tourist destinations. This paper arises from the dynamics of cultural tourism, specifically, the management of heritage resources in a region in central Portugal, implemented as a sustainable form of cultural and tourist attractiveness. The main objective is to map the heritage resources most visited through the constitution of a GIS database.

2 Literature Review

2.1 Sustainable Cultural Tourism

Tourism has a multidimensional, multifaceted, and complex character [4]. This is because tourist activity encompasses several areas and segments, from political, economic, and sociocultural, to environmental in nature [5]. Tourism is in a period of exponential growth, both as a force and as an academic field of study [6]. Even so, there are multiple actors that contribute to the various processes, activities, and results of tourism activity. These encompass interactions between tourists, private organizations, non-governmental organizations, the governments that host them, and the governments of origin (including higher education institutions and other levels of education). These networks deal at different levels with the processes of attracting, transporting, welcoming, and managing tourists as also other visitors [7].

Cultural tourism and the management of heritage resources have been fundamental to leveraging tourist destinations, in a dynamic that involves territorial development. The relationship that exists between tourism and culture is what contributes to the growth of cultural tourism [2]. The practice of cultural tourism, and overall travel with the visiting of cultural heritage in each tourist destination [8], broadened the vision of this segment of tourism. As such, cultural tourism involves visitors experiencing all the 'cultural' aspects of that place, its contemporary lifestyles, food, topography, environment, cities, and villages, as well as many of its historical sites and cultural presentations (Tourism Charter Cultural, ICOMOS, 1976) [9].

Cultural tourism involves the acquisition of knowledge and experiences, assuming the active participation of visitors and tourists and implying the subjectivity of their perceptions and representations [10]. Due to the permanent dialectic between material and immaterial, ancient and modern [11], cultural tourism is subject to transformations, in line with the driving forces of tourist demand.

When visiting specific heritage resources, the experience of tourists/visitors is profoundly unique and subjective, and as such so too is the practice of cultural tourism. This inherence to the diversity of experience derives from the complexity of the very concept of culture [12]. As such, Richards (1996) [13] presents two dynamics associated with cultural tourism: 1) culture as a process, related to a more sociological and anthropological view, in which culture is displayed in codes of conduct, characteristic of a social group; 2) culture as a product, related to a certain critical literary vision, in which culture is understood as the product of an individual or group activity.

Interconnecting with the tourist sector, the tourist components of accessibility, attractions, and accommodation are considered the most important factors for the development of tourism [14]. The existing heritage resources in each tourist destination are a driving force of social and economic development. Allied with heritage resources, cultural tourism encourages travel to differentiate cultural and tourist destinations [15]. However, each tourist destination must establish competitive advantages, inherent to the heritage resources that integrate it, insofar as, in the face of competition, the authenticity of the tourist experiences provided to visitors is a reason for differentiation. This is the only way that cultural tourism, through the integrated management of heritage resources, will be able to add value to the territory where it is located and contribute to the sustainable development of its communities.

2.2 Tourist Destinations and Management of Heritage Resources

It is understood that the management of cultural heritage is a pressing concern for many institutions, professionals, and academics involved in the documentation, restoration, and enhancement of cultural assets. This includes from a tourism perspective of sustainability. The multidimensional nature of tourism makes the context of sustainability complex [16]. In fact, there is growing literature dedicated to the sustainability of tourism, with a focus on the needs of tourists in host tourist destinations.

In this sense, more and more sustainable development strategies have been created, including management and marketing strategies [17]. This dynamic of tourist activity often intersects with cultural and creative tourism where the management of heritage resources requires strategies to balance the cultural and tourist destination and the promotion of this destination regarding the development of events. Recent studies have focused on complementary aspects such as the interaction between events, promotion of the local creative economy, and tourism development [18], to increase the attractiveness of destinations.

Cultural events have played a particularly important role in pursuing the strategic objectives of the attractiveness of tourist destinations. Some authors have worked on this issue of interaction between events and the number of visitors and tourists. Studies ranging from the analysis of small events to the cultural and tourist promotion of large destinations [19], developed for strategic repositioning to manage existing resources.

This process of «monitoring» tourist destinations in relation to the management of cultural resources is increasingly important in the dynamics of tourism and cultural sustainability. Many destinations have faced critical problems related not only to sustainability due to high tourist influx but also related to seasonality and an unbalanced number of tourists [20]. Events play a central role in the strategies that cultural and tourist destinations implement to minimize some critical problems generated by the 'unsustainable' development of tourism. The relevance of holding events for tourism is allied with attracting new tourist segments. This can contribute to mitigating the seasonality factor

[21] which, in turn, can make the spatial distribution of tourists in the territory more balanced. There is, however, a tendency for the sustainability process to lack "monitoring" of the impact of tourism on destinations. The intensity of the impact depends on the number of tourists and the characteristics of the destination, which can be physical, economic, or social [22]. Sustainable planning involves amplifying the positive effects of tourism and minimizing the negative effects [23]. In this process, the scale indicator applied to the study of tourism is relevant when carrying out the impact assessment [24].

The topic of sustainability and sustainable cultural tourism has gained increasing importance, as cultural tourism can contribute significantly to the sustainable development of territories [25]. Tourism and leisure activities combined with the promotion of cultural destinations and the destination's perception of the role of tourist behavior are relevant [26], as it influences the tourist experience [27].

As such, for tourist destinations, the development of actions aimed at protecting the natural environment has been increasing. These actions are aimed at both residents and tourists in the sense of adopting sustainable behaviors applied not only to the destination in general but in relation to the territory's cultural and heritage resources (material and immaterial). Some resources in destinations are in an advanced state of degradation, but with unique characteristics that require good management and conscious behavior on the part of tourists [28]. This has garnered greater attention in recent years due to actions and practices of more conscious and responsible tourism [29]. Development and planning instruments must be a priority in the reaffirmation of destinations/territories, based on the strategic management of heritage resources (material and immaterial).

In this sense, heritage resources combined with cultural tourism have been promoting a strong component of development in the territories, allowing more consistent guidelines to be drawn at economic, social, cultural, and environmental levels. These have been providing experiences for tourists and local populations that interact with the cultural dynamics of the territories [30]. The responsibility of planning in the management of heritage resources is great, requiring tourist and cultural initiatives framed in a type of strategic planning, focused on demand and supply for the development of sustainable tourism [31]. Strategies for the management of heritage resources integrated into a sustainable development concept structure are the factor that will strengthen the development of sustainable cultural tourism. The principles of tourist and community participation in sustainable development are a strategic starting point for integrating sustainability into the management of cultural heritage resources. Moreover, this is also the factor that allows for strategic action regarding challenges such as the environmental crisis and climate change. The bodies that protect heritage resources have been reflecting on these dynamics and drawing up guidelines to minimize these impacts.

This paradigm is interesting as it points to the beginning of a gradual change on these issues involving the attractiveness of tourist destinations and sustainability. Included in heritage discussions is the integration of value and community systems for decision-making related to the management of heritage resources focused on sustainable cultural tourism.

3 Methodology

For the present study, a questionnaire was created with the main objective of understanding the community's perspective in relation to the existing heritage resources in the territory (Tomar municipality, central region of Portugal). The questionnaire was validated by experts in the field.

After previously structuring the questions, the Google Forms tool was used to disseminate the questionnaire.

4 Results

In this study, 292 responses were obtained between November 11, 2021, and February 11, 2022. In accordance with gender, we had 53.8% of women and 46.2% of men. The response to the questionnaire was more frequent in the age group that contains the ages between 26 and 35 years old (26%) and 19 and 25 years old (25%). It is also important to emphasize the ages between 36 and 45 years (21%) since it presents a significant number of respondents. In a smaller number of respondents, there is the age group with people aged over 65 (1%) and 18 or under (2%). Of the total number of respondents, it appears that in terms of academic qualifications, the majority completed secondary education (38.7%), with a significant percentage also having a degree (34.2%). Most respondents are of Portuguese nationality (98%), with only two respondents coming from Angola (1%) and two from Guinea-Bissau (1%). Regarding the question "do you live in the municipality of Tomar", it appears that the majority of respondents indicated positively (61.6%).

There were also respondents from the autonomous regions of Portugal, the Azores (1) and Madeira (2) archipelagos, which indicates positive cooperation on the part of the community regarding the sharing of this questionnaire. It can also count on the inclusion of respondents from other countries, such as the Czech Republic (1); Spain (1), and Switzerland (2). A fundamental question to the purpose of this study is related to the heritage resources already visited by the respondents (Fig. 1).

It is notorious that the most visited heritage resource in the municipality of Tomar (91.4%) is the Castelo de Bode dam, as it is much sought after by visitors and appreciated by locals and neighboring municipalities. This dam is one of the most important Portuguese dams, it is also recognized for its abundant fluvial beaches and for having a unique landscape.

In the second place, the most visited resource (88.4%) is the Convent of Christ, given that it is an ex-libris monumental complex of the municipality, and for being an important landmark in the history of the Knights Templar, it becomes a very sought-after by visitors. In third place (82.9%) are fluvial beaches, also sought after by tourists, due to the quality of their waters and the environment that surrounds them.

However, in the minority and with less demand, is Ponte do Ramilo (6.2%), Cruzeiro and S. Silvestre da Beselga (9.9%), and Capela de S. Bartolomeu (9.9%).

Despite having historical interest and peculiar characteristics, these latter resources experience a lower influx of visitors, due to a cycle of not being known and thus being demoted from cultural promotion, as well as dynamics of visitor distribution of the territory. This can eventually lead to being further disregarded.

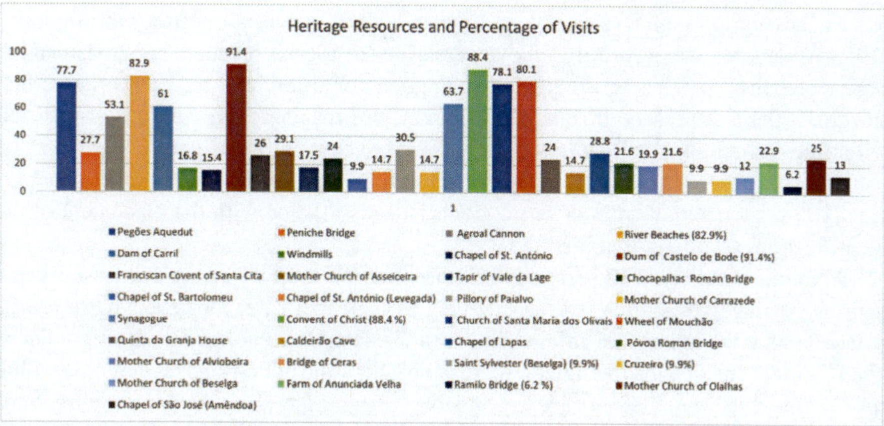

Fig. 1. Heritage resources visited.

On the issue of motivation for visiting these heritage resources (Fig. 2), the answers obtained refer mostly to the option "recommendation from someone" (43.5%).

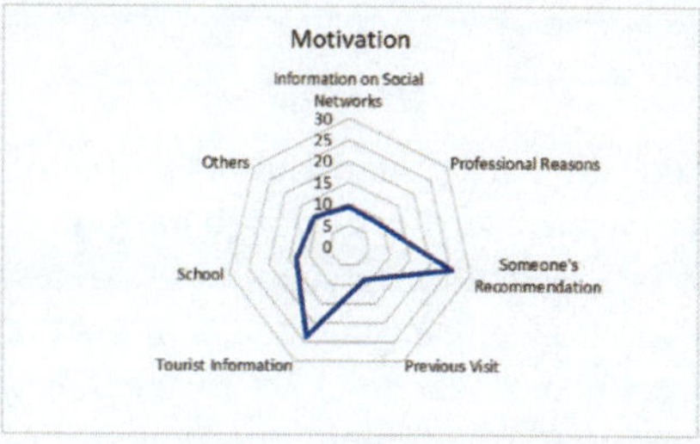

Fig. 2. Motivation for visiting heritage resources.

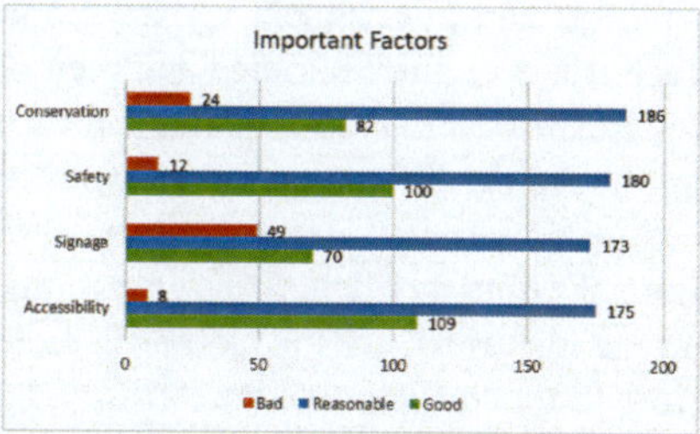

Fig. 3. Important factors to promote heritage

From the responses collected, it is understood that the feedback from those who have already visited a particular heritage resource/tourist attraction, word of mouth, leverages the promotion of these same resources.

The question of tourist experience reflects a strong relevance, as it makes it possible to understand how the target audience interprets some factors intrinsic to the tourist activity (e.g., accessibility, signage, safety, and conservation). On this question, the majority (175 respondents) answered that the accessibility to tourist resources is "reasonably" accessible (Fig. 3). That is, the paving is in good condition, and the environment that surrounds them is preserved. There is a minority (8 respondents) who are dissatisfied with access conditions, which is why they classify accessibility as poor.

Regarding the signs identifying the resources, the signs are placed in a "reasonable" way (173 respondents), with 49 respondents still dissatisfied with the placement of the identification, assuming that it is bad.

Alluding to safety, 180 respondents state that the level of safety witnessed when visiting heritage attractions is "reasonable", as they feel safe. However, there are 12 respondents who do not feel safe when visiting heritage resources, classifying them as "bad". Finally, most consider that the state of conservation of heritage resources is within the "reasonable" parameter (186 respondents), meaning that of the tourist attractions they visited, they consider being preserved. However, there is also an incidence of 24 respondents, classifying the state of conservation of the resources as "bad".

In short, it is possible to attest that in relation to the factor's "accessibilities", "signage", "safety" and "conservation", the municipality of Tomar needs to rethink how to optimize them, to fill in the flaws witnessed by visitors and thus contribute for an enriching tourist experience and to be recommended in the future.

5 Discussion

As can be seen throughout the background and objectives of this study, mapping the existing heritage resources in each territory through integrated and sustainable management of cultural tourism is a meticulous process. It interferes with many factors, such as the preservation, conservation, enhancement, and promotion of heritage resources.

In this sense, safeguarding is essential to ensure the effectiveness of heritage resources, given that there is only attribution of value to a given resource (material or immaterial) when it is recognized as genuine. That is, because there are peculiar characteristics, making it different from other resources. For this, the competent entities, namely the municipalities and government agencies, must ensure the protection of heritage resources, since it is a cultural heritage and memory [32]. Recognizing the potential of heritage resources for tourism in general, and for the cultural tourism segment, constitutes an opportunity for the maintenance, conservation, and promotion of local/regional/national economic activities in the territory where it is located. There are methodologies for analyzing the tourist potential of destinations, based on aspects related to «heritage value», from the historical, artistic, and even architectural perspective and in a dynamic observed from the social actors of the territory. It is necessary, however, to develop operational methodologies that make it possible to streamline the management processes of heritage resources.

The interconnection between cultural heritage management and sustainability is sometimes overlooked. Heritage planning should involve the preservation, conservation, rehabilitation, restoration, and management of heritage resources. It means that one should seek to increase the sustainability of planning interventions by considering the built environment, intangible heritage, cultural diversity, socio-economic and environmental factors, together with the values of the local community. Although the values of cultural and natural resources are often linked to environmental values, currently, there are still few incentives to protect them, much like it was observed in the less indicated heritage resources.

In this context, a planning approach centered on values makes it possible to draw up strategies to deal with the challenges of (cultural and natural) heritage preservation. In addition, heritage planning and management need an integrated link that goes from the identification to the evaluation of preservation policies. These threats make the conservation and management of cultural heritage a complex process, even requiring urgent conservation policies, constant monitoring, protection, and maintenance. In general, the growing importance of cultural heritage management will highlight the importance of sustainability practices in cultural tourism in their relationship with the dynamics of managing heritage resources.

6 Conclusions and Implications

The main conclusions confirm the main objective that had been outlined to achieve in this study. The analysis leads to the conclusion that the mapping of heritage resources contributes to the diversification of the offer and enhancement of the territory. The heritage resource with the most visits was the Castelo de Bode Dam, with the river beaches of this reservoir appearing next as the most visited. The least visited heritage resources were the Ramil Bridge, the chapel of S. Bartolomeu, and the cross of S. Silvestre da Beselga. The main motivation for tourist visits was someone's recommendation and tourist information. In what alludes to the tourist experience, four points were stipulated, accessibility, signage, security, and conservation.

The results obtained made it possible to understand which heritage resources were visited the most, constituting an important step toward the future creation of a GIS database that serves as a promotional lever for the territory. Knowledge presentation, in this sense, cannot be merely informative from a tourist point of view, but also needs to deepen the comprehensive understanding of the territory, allowing visitors to explore other tourist spots that are not as visited as those referenced in higher numbers in this study. They allow for the creation of a useful database that disseminates updated information for visitors who wish to geo-locate the existing heritage resources in the territory, as a basis for the integrated management of resources assets. The limitations of this research relate to the fact that there is no updated inventory of the heritage resources of the municipality of Tomar.

Acknowledgments. Work funded by national funds through the Portuguese National Funding Agency for Science, Research and Technology (FCT) under the project UID/05488/2020 - TECHN&ART. The authors would like to thank Andreia Bernardo for collecting and applying the data.

References

1. Richards, G.: Cultural tourism. In: Blackshaw, T. (ed.) Routledge Handbook of Leisure Studies, pp. 483–492. Routledge, London (2013)

2. Serra, J., Marujo, N., Borges, M.: Turismo Cultural: motivações e experiências dos turistas brasileiros em Évora. In: Cunha-Rocha, S., et al. (eds.) Tópicos Transatlânticos: emergência da lusofonia num mundo plural, pp. 43–53. Universidade de Évora, Évora (2012)
3. Toselli, C.: Algunas reflexiones sobre el turismo cultural. Revista de Turismo y Patrimonio Cultural **4**(2), 175–182 (2006)
4. Mill, R., Morrison, A.: The Tourism System. Kendall Hunt Publishing Company, Dubuque (2002)
5. Torres, N.: Gestão do Património Histórico e Desenvolvimento Urbano Sustentável: Políticas Públicas para o incremento do Turismo em Curitiba. Dissertação de Mestrado, Pontifícia Universidade Católica do Paraná, Brasil (2007)
6. Xin, S., Tribe, J., Chambers, D.: Conceptual research in tourism. Ann. Tour. Res. **41**, 66–88 (2013)
7. Weaver, D., Lawton, L.: Tourism Management, 3rd edn. Wiley, Hoboken (2006)
8. Mckercher, B., Cros, H.: Cultural Tourism: The Partnership Between Tourism and Cultural Heritage Management. THHP, New York, London, Oxford (2002)
9. Tourism Charter Cultural, ICOMOS. The International Seminar on Contemporary Tourism and Humanism, ICOMOS, Brussels, 8–9 November 1976
10. Paiva, O.: Autenticidade e CentrosHistóricos Património Mundial. Contributo para explicar a redundância – uma abordagem multigrupos. Tese doutoramento. Universidade de Coimbra, Coimbra (2016)
11. Mallor, E., et al.: Qué es y cómo se mide el Turismo Cultural? Un estudio longitudinal con series temporales para el caso Español. Turismo y Patrimonio Cultural **11**(2), 269–284 (2013)
12. Vaquero, M.: La ciudad histórica como destino turístico. Ariel, Barcelona (2006)
13. Richards, G.: Cultural Tourism in Europe. CABI, Wallingford, United Kingdom (1996)
14. Camilleri, M.: The tourism industry: an overview. Travel Market. Tour. Econ. Airline Product Springer Nature **1**, 3–27 (2018)
15. Badal, B.P.: Cultural tourism: a review of Karnali. Res. Nepal J. Dev. Stud. **4**(1), 27–43 (2021)
16. Fernández, J., Rivero, M.: Measuring tourism sustainability: proposal for a composite index. Tour. Econ. **15**(2), 277–296 (2009)
17. Pop, I.L., Borza, A.: Factors influencing museum sustainability and indicators for museum sustainability measurement. Sustainability **8**(1), 101 (2016)
18. Ziakas, V.: Planning and leveraging event portfolios: towards a holistic theory. J. Hosp. Market. Manag.Manag. **23**(3), 327–356 (2014)
19. Simeon, M., Buonincontri, P.: Cultural event as a territorial marketing tool: the case of the Ravello Festival on the Italian Amalfi Coast. J. Hosp. Market. Manag.Manag. **20**(3–4), 385–406 (2011)
20. Sainaghi, R., Baggio, R.: Destination events, stability, and turning points of development. J. Travel Res. **72**, 32–46 (2019)
21. Connell, J., Page, S.J., Meyer, D.: Visitor attractions and events: responding to seasonality. Tour. Manag.Manag. **46**, 283–298 (2015)
22. Van der Borg, J., Costa, P., Gotti, G.: Tourism in European heritage cities. Ann. Tour. Res. **23**, 306–321 (1996)
23. Russo, A.P., Van der Borg, J.: Planning considerations for cultural tourism: a case study of four European cities. Tour. Manag.Manag. **23**, 631–637 (2002)
24. Pearce, D.G.: An integrative framework for urban tourism research. Ann. Tour. Res. **28**, 926–946 (2001)
25. Powell, R.B., Ham, S.H.: Can ecotourism interpretation really lead to pro-conservation knowledge, attitudes, and behaviour? Evidence from the Galapagos Islands. J. Sustain. Tour. **16**(4), 467–489 (2008)
26. Ritchie, B., Adair, D.: The growing recognition of sport tourism. Curr. Issue Tour.. Issue Tour. **5**, 1–6 (2016)

27. Ramkissoon, H., Mavondo, F.T.: Pro-environmental behavior: critical link between satisfaction and place attachment in Australia and Canada. J. Tour. Anal. **22**, 59–73 (2017)
28. Ballantyne, R., Packer, J., Falk, J.: Visitors' learning for environmental sustainability: testing short-and long-term impacts of wildlife tourism experiences using structural equation modeling. Tour. Manag.Manag. **32**, 1243–1252 (2011)
29. Lopes, E.R., Simões, J.T.: Economia circular aplicadaao turismo: perceção dos estudantes do ensino superior. In: Estevão, C., Costa. C. (eds.) Turismo – Estratégia, Competitividade e Desenvolvimento Regional: uma visão estratégica no setor do turismo, pp. 207–223. Novas Edições Académica (2019)
30. Lopes, E.R., Simões, J.T., Nunes, M.R.: Turismo Cultural e RecursosPatrimoniais: evolução dos visitantes de um município. In Figueira, M, Cortes, M. (eds.) Turismo Patrimonial: o passado como experiência, vol. 271, pp. 11–42. Pelotas (2020)
31. Lopes, E.R., Simões, J.T.: Turismo cultural e atividade (s) museológica (s). J. Tour. Herit. Res. **3**(4), 107–119 (2020)
32. Martins, G.: Património, Herança e Memória - A cultura como criação. Gradiva, Lisboa (2009)

A Bibliometric Analysis of Guilt in Tourism: The Influence of Gastronomy on Olive Growing

Jessica Ferreira(✉), Elaine Scalabrini, Ana Cristina Silvério, Márcia Vaz, and Paula Odete Fernandes

UNIAG, Instituto Politécnico de Bragança, Campus de Santa Apolónia, 5300-253 Bragança, Portugal
{jessica.ferreira,elaine,ana.silverio,marciafvaz,pof}@ipb.pt

Abstract. Gastronomic development caused many destinations to focus on food as a central element of their tourism products. Studies highlight that tourism development in rural areas will depend on various political, economic, and environmental factors, which is why tourism activities such as oleotourism and gastronomic tourism are essential. To understand the relationship between gastronomy and olive growing from a tourism perspective, this study aims to explore scientific production through a bibliometric analysis. These bibliometric analysis goals are to develop a search equation based on the terms "Gastronomy" and "Olive Oil", to understand the topics' development. Specifically, analysing the evolution of production and citation, it was understood that oleotourism is prominent and relevant in the database trend map. The authors with the highest relevance have worked on topics such as the effect of spice infusion in olive oil, and Italy and Spain are very relevant in this field of research.

Keywords: Gastronomy · Olive Growing · Gastronomic Tourism · Oleotourism · Bibliometric Analysis

1 Introduction

Gastronomic culture cannot be considered an isolated characteristic of a territory [1] but instead as the product of the different cultures and civilisations that populate a territory. Gastronomy is considered a reflection of the cultural identity of a specific geographical area [2]. Gastronomic development, and their growing interest, have led many destinations to focus on food as a central element of their tourism products [3]. This link between the tourism sector and gastronomy stimulates and promotes the consumption of local products, territorial development, and economic sustainability [4, 5].

Hence, it is considered that the gastronomy of destinations has become a strategic tourism product that qualifies the tourism offer, differentiating the experience and adding value and quality to the destination. Further research highlights that tourism development in rural areas depends on several political, economic, and environmental factors [6]. This is the reason for the importance of tourism activities such as oleotourism and gastronomic

C. Montenegro et al. (Eds.): ICMTT 2023, LNNS 774, pp. 342–352, 2024.
https://doi.org/10.1007/978-3-031-43733-5_31

tourism, as the development of these activities affects the rural areas where such activities are inserted [4].

Oleotourism is an expanding typology of tourism. Predominantly found in rural areas where olive trees are the predominant culture (mainly in the countries of the Mediterranean basin: Spain, Italy, Greece, Morocco, Turkey, and Portugal, although cases of oleotourism have also been found in countries such as Argentina, Australia, Chile, Japan, USA, or Lebanon) [7]. Several researchers are dedicated to studying these areas, highlighting that, for various reasons, mills and agro-food companies have decided to diversify their main economic activity and turn to the tourism industry [8]. With this strategy, destinations are opening the way to the growing market demand of potential tourists interested in learning first-hand about olive oil culture: from olive-growing landscapes to its uses in daily life, covering all aspects, from the production process to tasting.

Accordingly, this study explores the existing indexed scientific production through a bibliometric analysis to understand the relationship between gastronomy and olive oil from the tourism perspective. Given the compilation of the SCOPUS and Web of Science (WoS) databases, this bibliometric analysis aims to study a search equation consisting of the terms "Gastronomy" and "Olive Oil" and the various variants of the words. To analyse the most real scientific production possible, the composition of the compiled database did not follow any filtering process, considering all articles open to the research community, i.e., extracting all themes, including tourism.

Specifically, the bibliometric study attempts to understand the themes' development, connection and dynamics considering all the available scientific production. This research presents temporal, geographic and thematic management and development results by analysing scientific production and its impact on the community. For this purpose, tests are performed on the articles' characteristics, reviewing titles, keywords, authors, journals, affiliations and corresponding citations.

2 Framework

Gastronomy is considered a field of knowledge that covers the materials used in food, such as food and beverages, with all the associated cultural aspects. Gastronomy is the primary regional and cultural motivation for gastronomic enthusiasts who visit a place specifically to enjoy its culinary offerings. There are excludable motivations associated with gastronomy or complementary motivations associated with regional culture. Hence, as a secondary motivation (for example, for those for whom, although their primary motivation is not to learn about the gastronomic richness of a destination, they consider gastronomy as a tourism option) to engage in gastronomic tourism [4, 9, 10]. Specifically, gastronomic tourism is defined as a combination of factors such as food, culture, the geography of the destination and the availability of infrastructures to support tourism. This field of research has been extensively explored in science, and several studies have shown that a destination's culture and environmental factors shape its gastronomic identity. Much scientific literature analyses gastronomic tourism from different perspectives: the terrain, the product, motivation and satisfaction [11].

For the territory, gastronomic tourism has been studied in specific regions or countries [12–14]. From another perspective, gastronomic tourism has been investigated based on a specific product, boosting the destinations' crafts and regional culture [15–17]. In 2022, a study on olive oil as a gourmet ingredient in Andalusian cuisine highlighted olive growth as a region's dynamism through the motivation and satisfaction of gastronomic tourism consumption [17]. There has been a considerable growth in gastronomic demand in destinations, combined with the new tourism product proposals that the market has created, among which oleotourism stands out, fundamentally in rural areas as an activity with great potential for the development of olive oil production areas [17, 18].

Oleotourism is developing in offering tourist destinations as a necessary form of special interest tourism within gastronomy, connecting visitors with olive oil and culture in their destinations. This practice of olive oil tourism is usually carried out in areas with a strong tradition in producing this product and which have an ancient olive grove, for the olive tree is among the oldest trees in the world [11]. This type of tourism includes a large variety of motivations that, in addition to the way of consuming it, also conditions the way of evaluating the experience of oleotourism [7]. Specifically, oleotourism refers to tourist and leisure activities dedicated to discovering and cultural enjoyment of the olive tree, olives, olive oil and olive oil-producing areas [11].

3 Methodology

A bibliographic review, or literature review, is a rigorous and broad analysis of current publications in a particular area of knowledge [19, 20]. For a literature review study to be objective, scientific and reproducible, it is imperceptible to define the research objective and specify the inclusion and exclusion criteria of the articles. Thus, the research objective is to identify and analyse the most productive and influential articles, the institutions, countries and affiliations, the authors, and the journals that have contributed to and shaped research on gastronomy and olive growing [19]. Accordingly, it was understood that Bibliometric analysis is rooted in library and information science that use various quantitative techniques to analyse bibliometric data in a field of research [20]. As a focus and observing analysis template, this study follows the guidelines Block and Fisch (2020) stipulated for bibliometric review. Precisely, building a database and collecting data consists of several steps: determining keywords, selecting one or more databases, and searching the selected databases in detail [21, 22].

The framework focused on gastronomy and olive growing using the two scientific reference databases, Scopus and Web of Science (WoS). The choice of these databases arises from the objective of gaining as much depth as possible in the research field. The selection of keywords for the scientific literature search was based on those related to "gastronomy" and "olive growing", focusing on "gastronomic tourism" and "oleotourism". The database search and the subsequent selection of the scientific literature used in this chapter were carried out by establishing an initial time horizon of three years, gradually expanding this time horizon to collect reference research in this field.

As the research topics under study use different words to describe gastronomy and olive growing, the terms that should be included in the research process are discussed in English and Portuguese. To ensure the comprehensive nature of the research, words that refer to gastronomy and the tourism nature of olive growing are considered keywords. However, SCOPUS and Web of Science were chosen as databases because they focus on specific research fields and contain a broader range of journals and databases. Given the diversity of keywords, several publications with irrefutable content are part of the search results. To ensure the validity of the data source, articles published in journals in social sciences were analysed and excluded repeated documents in both databases. Finally, 69 publications were selected, of which 20 publications would be from SCOPUS and 49 publications from WoS. 12 repeated articles were excluded from this compiled database, leaving a final database of 57 articles for analysis.

Recently, bibliometrics has provided the ability to quickly capture the dynamics of publications in specific scientific fields and explore research gaps over time [23]. Specifically, bibliometrics is a method that gives clues about the scientific communication of the relevant field by examining and analysing some characteristics of journals, documents, or other printed publications with mathematical and statistical techniques [22].

This research used bibliometric approaches to conduct various bibliographic analyses of multiple aspects, including authors, citations, journals, affiliations, journals, keywords and countries. R Bibliometrix was used as an analysis tool, the collected articles were analysed, and several clusters were identified. Published articles with similar content [24]. Co-citation refers to the frequency with which two publications are cited together, i.e. their frequency of occurrence in the same reference list. Articles cited together are considered more relevant, and the number of citations indicates their similarity. Based on the reference lists of research findings, the abstracts and keywords of each co-cited article are collected to identify the topic of each cluster. Subsequently, a co-occurrence analysis is provided to extract knowledge from the literature through keywords. Keywords reflect the core of a specific research field and reveal research trends.

4 Results

This section presents the results of the bibliometric analysis of the database under study. The database comprised 57 documents from 1999–2022, with an average age of 3.32. More than 190 authors support this study, 292 authors' keywords and 21.05% international co-authorship. More than 30 sources and 2653 references were also extracted from this database.

A literature coupling analysis was first conducted to reveal the research trend of gastronomy's impact on olive growth under the impact of tourism. Specifically, 14 articles are about tourism; 8 about olive oil; 17 about gastronomy; and 18 about health and nutrition.

The analysis of the annual scientific production showed that there is a constant growth in recent years in research on this theme. With more citations, the article by Giordano, Clodoveo, Gennaro and Corbo [25] on the Factors determining neophobia and neophilia about new technologies applied to the food sector, published in 2018 in the International Journal of Gastronomy and Food Science stands out in the first place in the ranking, 41 citations. Following with 38 citations, the article by the authors Caporaso, Paduano, Nicolettin and Sacchi [26] from 2013, presenting a study on gastronomy, entitled Capsaicinoids, antioxidant activity, and volatile compounds in olive oil flavoured with dried chilli pepper, published in the European Journal of Lipid Science and Technology. In third place in the ranking, with 36 citations, the article on health and nutrition by

Lanza and Ninfali [27] reports a study on antioxidants in extra virgin olive oil and table olives, the connections between agriculture and processing for health choices, published in the journal Antioxidants, in 2020. With a fifth place in the ranking, the article by Jiménez-Carvelo, Lozano, and Olivieri (2019) [28].

Once the articles with the most significant impact in this field of knowledge were analysed, it became necessary to identify the authors with the highest relevance and impact. To study the most relevant authors were analysed the top 5 authors, their h-index (the h-index is a metric to assess the cumulative impact of an author's production and academic performance), the number of articles and the respective citations of the articles. Specifically, Caporaso tops the ranking of most relevant authors with three articles, produced in 2013, in the European Journal of Lipid Science and Technology, on capsaicinoids, antioxidant activity, and volatile compounds in olive oil flavoured with dried chilli pepper; 2016, published in Food Reviews International, on developments, applications, and trends of molecular gastronomy among food scientists and innovative chefs; and 2019, in the International Journal of Gastronomy and Food Science, on the effect of infusion of spices into the oil vs combined malaxation of olive paste and spices on the quality of naturally flavoured virgin olive oils. This author represents an h_index of 3, totalling 65 citations. In the second place, the author Clodoveo presents an h_index of 2, with two articles published in the International Journal of Gastronomy and Food Science, in 2018, with a systematic analysis of determinants of neophobia and neophilia concerning new technologies applied to the food sector.

Moreover, in 2020, an article entitled gastronomic cultural evolution of the Restaurant's virgin olive oil consumption model. In third and fourth place in this ranking are the authors Gambaro and Roascio-Albistur, with an h_index of 2, represented by two co-authored articles from 2018 and 2019. Both articles were published in the International Journal of Gastronomy and Food Science about consumer perception of a non-traditional market on sous-vide dishes, complemented afterwards with the analysis of a focus group and word association. The author Sacchi occupies the fifth place, with an h_index of 2, with two articles. In 2019 the author published an article in the International Journal of Gastronomy and Food Science on the Sensory Profile, phenolic and volatile compounds of an artisanal ice cream functionalised using extra virgin olive oil. Moreover, in 2020, on the contribution of wild edible plants to the Mediterranean diet: an ethnobotanical case study along the Campania coast (southern Italy), published in Economic Botany. It is also interesting to understand that the most relevant authors' affiliations correspond to the University of Naples Federico II from Italy, with 10 articles. Next is the University of Jaén, Spain, with 8 articles. Then, with 6 articles each, the University of Bari, from Italy and the University of Loyola, from Spain. Finally, considering the Top 5 most relevant affiliations, the University of Extremadura from Spain stands out with 5 articles.

In terms of analysing the journals with the highest impact in the database under study. From the analyses performed until now, the International Journal of Gastronomy and Food Science is in the database under study. The International Journal of Gastronomy and Food Science is a peer-reviewed journal that explicitly focuses on the interface of food science and gastronomy, coming from the Netherlands with an h_index of 23. This journal, since 2018, has contributed 18 articles to this database. In 2018 it would correspond to the Q2 quartile, with a SCImago Journal Rank (SJR) of 0.422. Nutrición

Hospitalaria, from Spain, has contributed 6 articles since 2018, with an SJR of 0.34, h_index of 57 and Q3. With two articles since 2020, the journal Antioxidants comes from Switzerland and holds an h_index of 61. In 2020 it also held an SJR of 1.067 and Q2. Finally, there are two Q1 quartile journals with a publication starting in 2013. Specifically, Food Research International from the United Kingdom and the European Journal of Lipid Science and Technology from Germany.

From this analysis, it is concluded that the countries with the most significant dynamics in these topics under study are Italy, Spain and France, both at the level of publication of articles in the field of knowledge of this study but also in terms of the search for authors to reference their studies.

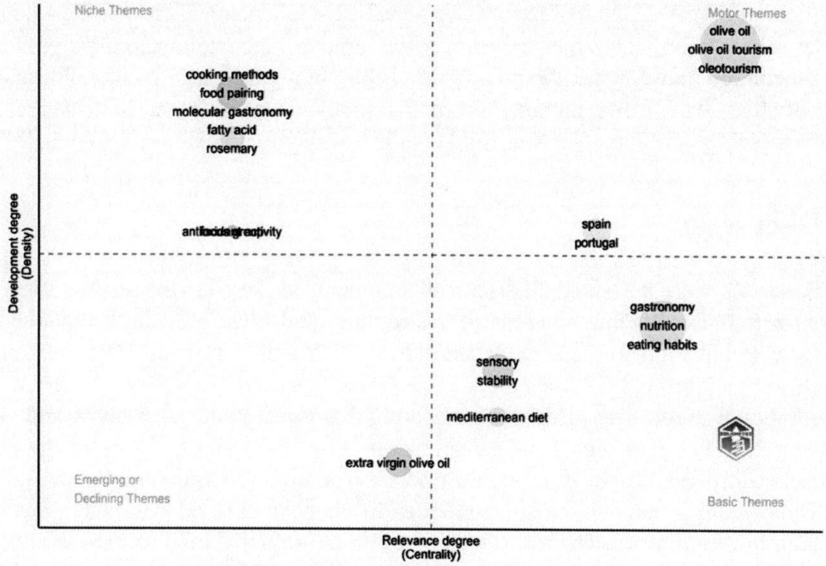

Fig. 1. Thematic map of authors' keywords.

From Fig. 1, it is understood that the thematic map is divided into nine different clusters. With more significant development and relevance, cluster one is highlighted: oleotourism. Cluster one comprises the authors' key. This analysis shows that oleotourism is prominent and relevant in the database trends map. Cluster two: the Iberian Peninsula is analysed with extraordinary relevance and development. This cluster highlights studies with keywords by authors from Spain and Portugal. This analysis transposes the research trend of these countries, not only for their location but also for their characteristics associated with gastronomy and olive growing.

Three distinct clusters are observed with a lower but developing relevance, specifically cluster three: culinary, cluster four: Oil and cluster five: Health. Cluster three highlights gastronomy in cooking through cooking methods, food pairing and molecular gastronomy. Cluster four takes on the quality of fat and the derivatives of fatty acid and rosemary. Cluster five takes on nutrition and healthcare themes through antioxidant activity. From this group of clusters, it is retained that, with less relevance, gastronomy and health maintain research themes in development. With a lower development, there are 3 clusters. With a larger radius, cluster 6: Nutrition explores themes such as gastronomy, nutrition and eating habits. Cluster 7: Sensory presents themes associated with stability and sensory.

Moreover, cluster eight: Mediterranean Diet, takes on characterising the Mediterranean diet. Concretely, it is understood from this analysis that nutrition, sensory and Mediterranean diet are relevant to research topics that need to be developed. Finally, cluster 9: Olive Oil, explores the authors' keywords associated with extra virgin olive oil. I was demonstrating that investigations on this theme need more development and relevance. Studying the co-citation network to understand the relationship between the articles in the database under study was also considered attractive. In particular, the co-citation network relation of the articles reveals five independent clusters. The largest cluster relates two distinct themes, namely the topic of oleotousim and olive growing. The co-citation between this theme is quite visible. In parallel, as a smaller cluster, the topic of olive oil has a citation relationship. In a smaller dimension, three clusters search for the topic's gastronomy, nutrition and cooking, showing no relation with the others.

5 Discussion

In this section were presented all discussed arguments analysed in the results. The evolution of publications and citations in gastronomy and olive growing behaviour has changed. Publication in these fields started in 1999, and only from 2018 presents a significant growth. In 2019, the number of publications increased. Notably, after 2018, the number of publications is higher, indicating that research on gastronomy and olive growing has attracted some scholars' attention. Also, the thematic areas of the 2018 articles mainly focus on health and nutrition, gastronomy, and tourism.

Therefore, the annual scientific production analysis showed constant growth in research in recent years. The top 10 articles with more global citations are articles on health and nutrition, gastronomy, olive oil and tourism. On top was the article by Giordano, Clodoveo, Gennaro and Corbo on the Factors determining neophobia and neophilia about new technologies applied to the food sector, published in 2018 in the International Journal of Gastronomy and Food Science. Following the article by Caporaso, Paduano, Nicolettin and Sacchi, from 2013, presented a gastronomy study published in the European Journal of Lipid Science and Technology. Then, the article on health and nutrition by Lanza and Ninfali reports a study on antioxidants in extra virgin olive oil and table olives, the connections between agriculture and processing for health choices, published in the journal Antioxidants.

Also, it was interesting to note that the article about oleotourism, published in 2017, presents an analysis of what olive oil tourism has to offer in this region by conducting surveys of the companies on the olive oil gastronomic route and their designations of origin to determine their degree of involvement in the development of the "olive oil tourism" tourism product. This article was published by Arjona-Fuentes and Amador-Hidalgo, in the Journal of Tourism Management Perspectives.

Concerning the most relevant authors was highlighted the author Caporaso, who took up the top of the ranking, concerning topics such as capsaicinoids, antioxidant activity, the effect of infusion of spices into the oil and the quality of naturally flavoured virgin olive oils. Then, the author Clodoveo, at the top, presented topics about determinants of neophobia and neophilia concerning new technologies and the gastronomic cultural evolution of the restaurant's virgin olive oil consumption. Following this, the author Gambaro researched consumer perception of a non-traditional market.

Some results were obtained to reveal the research trend of gastronomy and olive growing behaviour under the impact of tourism. Through a co-occurrence analysis, it was analysed five clusters. Concretely, the cluster about olive oil and its relationship with virgin olive is the most featured. With a minor frequency, there is another cluster associated with oleotourism. The authors' keywords related to health and nutrition are divided into three different clusters relating to each other.

Then, based on the research findings, the abstracts and keywords of each co-citation article were collected to identify the topic of each cluster. With more significant development and relevance, cluster oleotourism has been highlighted. This analysis shows that oleotourism is prominent and relevant in the database trends map. The cluster about the Iberian Peninsula is analysed with extraordinary relevance and development. Then, there were clusters about culinary, Oil and Health. Concretely this set of clusters showed that oleotourism, Spain, Portugal, gastronomy, nutrition and health are prevalent keywords in these studies.

6 Conclusions

6.1 Current Issues

Concerning these variables, this study focuses on understanding the relationship between gastronomy and olive oil from the tourism perspective. Looking to explore the existing indexed scientific production through a bibliometric analysis, this study presented some pertinent conclusions. Given the compilation of the SCOPUS and Web of Science (WoS) databases, the bibliometric analysis was made to analyse the most real scientific production possible.

This study concludes that the evolution of publications and citations in gastronomy and olive growing behaviour has changed. Remarkably, after 2018, the number of publications is higher, indicating that research on gastronomy and olive growing has attracted some scholars' attention. The thematic areas mainly focus on health and nutrition, gastronomy, and tourism. The annual scientific production analysis showed constant growth in research in recent years. The articles with more global citations are health and nutrition, gastronomy, olive oil and tourism. Most articles present qualitative methodologies through interviews and case studies.

The authors with more relevance worked on topics such as the effect of the infusion of spices into the oil, the quality of virgin olive oils, the health and culinary tools, and the gastronomic cultural evolution. Also, the production of articles on these themes highlights the University of Naples, Federico II from Italy, and the University of Jaén from Spain. The International Journal of Gastronomy and Food Science Journal and the countries Italy and Spain are very relevant in this field of research.

In the end, to reveal the research trend of gastronomy and olive growing behaviour under the impact of tourism, some keywords' clusters were analysed through a co-citation and co-occurrence analysis. Mainly, there is a relationship between gastronomy, olive oil and virgin olive. Then, there is another relation associated with oleotourism.

Based on the abstracts and keywords of each co-citation article, oleotourism was identified as a cluster with more significant development and relevance. This analysis shows that oleotourism is prominent and relevant in the database trends map. Then, Spain and Portugal were presented with a higher relevance in this topic. Then, the Olive Growing cluster presented that the studies in this field of research are focused on all components that can be related to olive oil.

6.2 Implications and Limitations

As a limitation, this study presents the need to increase the search drivers and complete the database. Future works should exercise in this sense. As future trends for new research, this study may support studies along the lines of investigation.

Acknowledgement. The authors are grateful to the UNIAG, R&D unit funded by the FCT – Portuguese Foundation for the Development of Science and Technology, Ministry of Science, Technology and Higher Education under Project no. UIDB/04752/2020 and UIDP/04752/2020; and to the project "OleaChain: Competências para a sustentabilidade e inovação da cadeia de valor do olival tradicional no Norte Interior de Portugal" (NORTE-06-3559-FSE-000188).

References

1. Ellis, A., Park, E., Kim, S., Yeoman, I.: What is food tourism? (2018). https://doi.org/10.1016/j.tourman.2018.03.025
2. Hernández-Rojas, R.D., Huete-Alcocer, N., Hidalgo-Fernández, A.: Analysis of the impact of traditional gastronomy on loyalty to a world heritage destination. Int. J. Gastron. Food Sci. **30** (2022). https://doi.org/10.1016/j.ijgfs.2022.100585
3. Ab Karim, S., Chi, C.G.Q.: Culinary tourism as a destination attraction: an empirical examination of destinations' food image. J. Hosp. Market. Manag. **19**, 531–555 (2010). https://doi.org/10.1080/19368623.2010.493064
4. Moral-Cuadra, S., de la Cruz, R.A., López, R.R., Cuadrado, E.S.: Relationship between consumermotivation and the gastronomic experience of olive oil tourism in Spain. Sustainability (Switzerland) **12** (2020). https://doi.org/10.3390/su12104178
5. Lee, J.S.H., McCole, D., Holecek, D.: Exploring winery visitors in the emerging wine regions of the North Central United States. Sustainability (Switzerland) **12**, (2020). https://doi.org/10.3390/su12041642
6. López-Guzmán, T., Sánchez Cañizares, S.: Turismo Comunitario Y Generación de Riquezaen Países en vías de desarrollo. Un estudio de caso en El Salvador. REVESCO. Revista de Estudios Cooperativos. 85–103 (2009)
7. Pulido-Fernández, J.I., Casado-Montilla, J., Carrillo-Hidalgo, I.: Understanding the behaviour of olive oil tourists: a cluster analysis in southern Spain. Sustainability (Switzerland) **12** (2020). https://doi.org/10.3390/SU12176863

8. Millán, M.G., Pablo-Romero, M.d.P., Sánchez-Rivas, J.: Oleotourism as a sustainable product: an analysis of its demand in the South of Spain (Andalusia). Sustainability (Switzerland) **10** (2018). https://doi.org/10.3390/su10010101
9. Nicoletti, S., Medina-Viruel, M.J., Di-Clemente, E., Fruet-Cardozo, J.V.: Motivations of the culinary tourist in the city of Trapani, Italy. Sustainability (Switzerland) **11** (2019). https://doi.org/10.3390/su11092686
10. Cruz, F.G.S., Moral-Cuadra, S., Tito, J.C., López-Guzmán, T.: Gastronomic motivations and perceived value of foreign tourists in the city of Oruro (Bolivia): an analysis based on structural equations. Int. J. Environ. Res. Publ. Health **17** (2020). https://doi.org/10.3390/ijerph17103618
11. Oplanić, M., Čop, T., Čehić, A.: Olive tourism: understanding visitors' motivations. Acta Turistica. **32**, 141–167 (2021). https://doi.org/10.22598/AT/2020.32.2.141
12. Batat, W.: The role of luxury gastronomy in culinary tourism: an ethnographic study of Michelin-Starred restaurants in France. Int. J. Tour. Res. **23**, 150–163 (2021). https://doi.org/10.1002/jtr.2372
13. Privitera, D., Nedelcu, A., Nicula, V.: Gastronomic and food tourism as an economic local resource: case studies from Romania and Italy. Geo J. Tour. Geosites. 143–157 (2018)
14. Pavlidis, G., Markantonatou, S.: Gastronomic tourism in Greece and beyond: a thorough review. Int. J. Gastron. Food Sci. **21** (2020). https://doi.org/10.1016/j.ijgfs.2020.100229
15. Cava Jimenez, J.A., Millán Vázquez de la Torre, M.G., Dancausa Millán, M.G.: Enotourism in Southern Spain: the Montilla-Moriles PDO. Int. J. Environ. Res. Publ. Health **19** (2022). https://doi.org/10.3390/ijerph19063393
16. Medeiros, M.D.L., da Cunha, J.A.C., Passador, J.L.: Turismo gastronômico e desenvolvimento regional: um estudo a partir do caso do queijo minas artesanal do Serro. Caderno Virtual de Turismo. **18** (2018). https://doi.org/10.18472/cvt.18n2.2018.1373
17. Millán Vázquez de la Torre, M.G., Sánchez-Ollero, J.L., Dancausa Millán, M.G.: Ham tourism in andalusia: an untapped opportunity in the rural environment. Foods **11** (2022). https://doi.org/10.3390/foods11152277
18. Murgado-Armenteros, E.M., Parrilla-González, J.A., Medina-Viruel, M.J.: What does the olive oil tourist value at the destination? A criterion for olive oil tourism segmentation. Int. J. Gastron. Food Sci. **25** (2021). https://doi.org/10.1016/j.ijgfs.2021.100378
19. Tranfield, D., Denyer, D., Smart, P.: Towards a methodology for developing evidence-informed management knowledge by means of systematic review (2003). https://doi.org/10.1111/1467-8551.00375
20. Martínez-López, F.J., Merigó, J.M., Valenzuela-Fernández, L., Nicolás, C.: Fifty years of the European Journal of Marketing: a bibliometric analysis. Eur. J. Mark. **52**, 439–468 (2018)
21. Block, J.H., Fisch, C.: Eight tips and questions for your bibliographic study in business and management research (2020). https://doi.org/10.1007/s11301-020-00188-4
22. Wu, H., Xu, Z., Skare, M.: How do family businesses adapt to the rapid pace of globalization? A bibliometric analysis. J. Bus. Res. **153**, 59–74 (2022). https://doi.org/10.1016/j.jbusres.2022.08.022
23. Costa, D.F., de Melo Carvalho, F., de Melo Moreira, B.C., do Prado, J.W.: Bibliometric analysis on the association between behavioral finance and decision making with cognitive biases such as overconfidence, anchoring effect and confirmation bias. Scientometrics **111**, 1775–1799 (2017). https://doi.org/10.1007/s11192-017-2371-5
24. Elango, B.: A bibliometric analysis of franchising research (1988–2017). J. Entrep.Entrep. **28**, 223–249 (2019). https://doi.org/10.1177/0971355719851897
25. Giordano, S., Clodoveo, M.L., Gennaro, B. de, Corbo, F.: Factors determining neophobia and neophilia with regard to new technologies applied to the food sector: a systematic review (2018). https://doi.org/10.1016/j.ijgfs.2017.10.001

26. Caporaso, N., Paduano, A., Nicoletti, G., Sacchi, R.: Capsaicinoids, antioxidant activity, and volatile compounds in olive oil flavored with dried chili pepper (capsicum annuum). Eur. J. Lipid Sci. Technol. **115**, 1434–1442 (2013). https://doi.org/10.1002/ejlt.201300158
27. Lanza, B., Ninfali, P.: Antioxidants in extra virgin olive oil and table olives: connections between agriculture and processing for health choices (2020). https://doi.org/10.3390/antiox9010041
28. Jiménez-Carvelo, A.M., Lozano, V.A., Olivieri, A.C.: Comparative chemometric analysis of fluorescence and near infrared spectroscopies for authenticity confirmation and geographical origin of Argentinean extra virgin olive oils. Food Control **96**, 22–28 (2019). https://doi.org/10.1016/j.foodcont.2018.08.024

Enhancing Tourism Through Innovation and Creativity: Exploring the Portuguese Military Tourist Route

João Tomaz Simões[1,2,3](✉), Lígia Mateus[1], and Célio Gonçalo Marques[1]

[1] TECHN&ART – Centre of Technology, Restoration and Art Enhancement, Polytechnic University of Tomar, Estrada da Serra, Quinta do Contador, 2300-313 Tomar, Portugal
{jpsimoes,ligia.mateus,CelioMarques}@ipt.pt
[2] UI&D ISLA Santarem, Santarem, Portugal
[3] Tourism PHD Student, University of Santiago de Compostela, Praza do Obradoiro, S/N, 15782 Santiago de Compostela, Portugal

Abstract. Tourism is an industry that can benefit from innovation and creativity, especially in the field of military tourism. Military tourism offers unique and authentic experiences that combine different tourist segments with the memory, identity, and military history of a place. Portugal has a rich military heritage that can be leveraged for the development of its military tourism industry. To address the need for more structured and developed military tourism products, the Portuguese Military Tourism Association (ATMPT) has developed the Military Tourism Route. This project is a set of thematic routes associated with events and periods of national history that promote and disseminate national tourist and cultural services, as well as heritage assets and equipment that integrate, directly or complementarily, the offer of Military Tourism in Portugal. The Military Tourism Route project includes a digital platform that provides a dynamic program of personalized and educational content, with written, photographic, and audio-visual media. The platform aggregates information about different thematic routes, including heritage assets (material or immaterial), activities and events, travel logistics, and various tourist services. Each asset is directed to a page with more information about it. The platform also provides a set of Experiences, a Highlights section, and a Cultural Agenda with activities and events in course. The project also includes a digital magazine that aims to disseminate the vast and diverse military tourism offer existing in the Portuguese territory, contributing to the consolidation of the Portugal destination brand. The Military Tourism Route is a creative and innovative solution that contributes to the consolidation of a developing tourist segment. By leveraging the human process of creativity, this project offers a space to share information and inspire participative dynamics of tourists in planning their visit adjusted to their interests, needs, and restrictions.

Keywords: tourism · innovation · creativity · Portuguese Military Tourist Route · cultural heritage · sustainable tourism

C. Montenegro et al. (Eds.): ICMTT 2023, LNNS 774, pp. 353–362, 2024.
https://doi.org/10.1007/978-3-031-43733-5_32

1 Introduction

Tourism is a multifaceted economic activity that encompasses several sectors, including hospitality, catering, and transportation [1]. The main objective of these organizations is to satisfy market needs and desires (experiences), while promoting a positive image and operating in a sustainable manner [2]. With the current digital age (knowledge society), it has become increasingly important for these organizations to respond and effectively manage the challenges that arise [3]. Technology, particularly in the form of improved communication, plays a significant role in the tourism industry. It impacts not only the training of human resources, but also the management and sale of services or products consumed by tourists [4]. Creativity and innovation are increasingly valued by these organizations as they understand the potential benefits they bring. With advances in technology, companies are constantly seeking to innovate and stay ahead of the competition [5]. Since creativity and innovation are human processes, they often bring unexpected results, and not only solve problems but also increase the level of competition. Thus, organizations that prioritize creativity and innovation can gain a competitive advantage and ensure their long-term sustainability.

2 Methodology

The Military Tourism Route project was developed by the Portuguese Military Tourism Association (ATMPT) in collaboration with Portuguese municipalities. The project aims to fill an existing gap regarding the structuring of the existing offer in national territory and the development of military tourism products. It is a constantly evolving project that integrates new heritage assets, activities, and events, as well as new projects and products that emerge. The project was implemented using a digital platform that aggregates all the information related to the different thematic routes with a historical background of the theme. The platform provides a set of experiences with suggestions for all tastes according to the thematic routes, a highlights section with recent news, events, and new developments in military tourism, and a cultural agenda with activities and events in course. The digital platform was also accompanied by a promotional video of the territory, a video with the testimony of the President of the Municipality, a teaser, and a video of the Route that includes images of all the territories associated with the theme. The municipalities that integrate the project have access to the BackOffice of the platform where they can directly upload all the information on the assets existing in their territory. The assets uploaded are validated by the ATMPT team before becoming available on the platform, according to their importance and connection to the theme, as direct, indirect, or complementary assets. The project's effectiveness is assessed based on the number of visitors to the different thematic routes and the feedback received from tourists. Additionally, the ATMPT team continuously updates the platform with new information and assets based on the feedback received from tourists and the evolution of the military tourism sector. The Military Tourism Route project was developed as a creative and innovative solution to consolidate a developing tourist segment in Portugal. The project's success can be attributed to its dynamic nature, its use of technology to aggregate and disseminate information, and its collaboration with different stakeholders to promote and develop military tourism in the country.

3 Creativity – Conceptualization and Definition

As an element that "emerges" and is expressed through human actions [6], creativity accompanies the development of society, which is currently in an accelerated context with major technological advances, uncertainty, and instability, known as the Knowledge Society [7]. This society is characterized by its substantial capacity to create more and new knowledge at an accelerated pace [8], enabling more effective and focused responses to innovative processes that, with the help of creativity, promote progress [9]. The importance of studying the phenomenon of Creativity as a capacity or innovative capability of individuals and society began in the 1950s [7]. The interest in the topic resulted in several theses, including: *i*) Csikszentmihalyi's (1988) Systems perspective, which defines creativity as the result of the interaction of three factors: domain, individual, and field: The domain transmits information to the individual, and this produces variations that, through the field, may or may not be incorporated into that domain; *ii)* Amabile's (1996) Componential Model of Creativity states that a product or response will be judged creative to the extent that: a) it is new and appropriate, useful, correct, or of value to the task at hand, and b) the task is heuristic rather than algorithmic; *iii)* Sternberg's (1999) Investment in Creativity theory declares tath creativity exists from the confluence of six resources: *intellectual abilities, knowledge, thinking styles, personality, motivation, and a suitable environment. Iv)*According to Lubart (2007), this is a multiple approach to creativity, a combination of individual and environmental factors is necessary and important, meaning that creativity depends on cognitive, connotative, emotional, and environmental factors, since one component always acts in the presence of others. According to Lubart (1999) *apud* Paletz, Peng [10], eastern views on creativity focus on personal fulfilment processes, compared to Western views that focus primarily on products. This scenario is quite complex, but it is known to be due not only to anthropological aspects such as cultural differences expressed through Art, but also Philosophy, Religion, and Literature. Authors [11] applied the "Four C's Model of Creativity" and determined in the study published in the "International Journal of Experimental Educational Psychology" the perception that Eastern (Chinese) students have of their own creativity and their perception of Western students' creativity. The main conclusion of this work is that Chinese university students clearly consider themselves less creative than Western students, mainly due to the nature of their conservative educational system, although the same study mentions that these perceptions may decrease during the academic journey. In this sense, we can conclude that creativity is a human process (from conception to interpretation) influenced and framed by the surrounding environment over time and therefore manifests and is understood differently, not limited to a single definition as indicated by Robinson (2008 cited in [12]), "the lack of a single widely accepted definition of creativity is undoubtedly due to the wide range of views on the function of creativity". To frame this perspective whenever we mention the term creativity, we delimit the concept to the definition they recognize as the most suitable in the context of this article. In the context of tourism, we distinguish [13], who, in reviewing the literature, grouped the multiplicity of definitions of creativity in the literature into four main areas of scientific approach, which correspond to the "4 P's" of creativity, because, as Richards [12] points out, tourist activity now involves all four approaches: The Creative Person; The Creative Process; The Creative Product; The

Creative Environment. Regarding the characterization of Creativity, after its conceptualization, we can say that it has become a strategy to be followed not only by cities and regions that seek to develop, but also as a strategy in promoting innovation and the development of individual skills (*ibid*), as it requires new, non-traditional ideas and perspectives (without pre-established barriers). By developing Creativity, or being creative, we are exploring new perspectives and possibilities for increasing efficiency, often from vague ideas and guesses. Some studies indicate that the relationship between Creativity applied to organizations seems to offer a fertile potential for knowledge on how to deal with the paradoxical challenges involved in management [14]. For example, Lin (2011, cited in [15] states that the use of employees' creativity can be an organizational innovation in creating competitive advantages. In Table 1 we represent the 4 "P's" of Creativity framed in the tourist activity according to Richards [12]:

Table 1. The "4Ps" of Creativity adapted to tourism.

Item	Description	Reference
The creative Person	The involvement of creative people through the activities of the "creative class	[12]
The creative Process	Travel - use of the creative process in the design of creative activities for tourists (e.g., workshops and masterclasses)	
The creative Product	Tourist attractions (e.g. related to famous authors, painters, etc.)	
The creative Environment	Visits to creative clusters	

Todoroi [9] states that creativity is the result of brain activity that differentiates individuals and can provide a significant competitive advantage for individuals, companies (organizations), society, and especially the Knowledge Society. He also says that highly innovative industries such as the software, computer, and car industries view creativity as a key factor in business success. He believes that natural intelligence and creativity can lead to basic creative activities. Thus, we understand that creativity can contribute to innovation.

4 Innovation – A Dynamic Approach

According to Schumpeter [16], there are five types of innovation: the introduction of a new good, the introduction of a new method of production, the opening of a new market, the acquisition of new raw materials and the implementation of a new form of organization. Schumpeter also highlights that innovation is distinct from invention in that innovation is possible without an invention and an invention does not necessarily lead to innovation. Inventions are not exclusively for industrial use while innovations are amplified by inventions or just brilliant ideas to turn them into useful products.

But in another way, Higgins [17] states that innovation can emerge in two main ways: firstly, as a result of an innovative process; and secondly, as a process that creates new products and services or improves existing ones, or the creation of organizational methods (processes) that have a significant impact on people, groups, organizations, industry or society. As we can see in Fig. 1, Higgins [17] argues that the process of creating innovation is based on creativity, considering it a key element.

Fig. 1. The innovation equation (according to Higgins 1996). **Source.** Higgins [17]

According to Lindič, Baloh [18], companies are actively implementing information technologies in their innovation processes, as technology can play a crucial role in managing knowledge throughout the innovation process, especially when the number of ideas becomes overwhelming. Technology enables the renewal of methods, structural and situational development, and a fast evolution, upgrade, and improvement of production, development, and design of goods. According to Abernathy and Clark (1985) *apud* Hjalager [19], they present a characteristic approach to innovation in the automobile industry that can be framed in tourism. The model presents four types of innovation - Regular, niche, revolutionary, and architectural, consisting basically of two axes. The horizontal axis indicates if a given innovation breaks the process/item, making it obsolete, or, on the other hand, strengthens the existing ones. The vertical axis indicates the knowledge and skills required to create a particular product or service, in some cases, old ideas and qualifications become outdated and need to be replaced while others prove more relevant when adapted and later developed. According to Mota and Oliveira [20], innovation plays an increasingly central role in our society, not only in the pursuit of economic sustainability but also in social development. As Andersen and Kragh [14] state, "in step with the increased importance of innovation as a means of sustaining and developing competitive positions, firms increasingly seek to develop their creative capacity." However, in the tourist industry, as Hjalager [19] notes, "in some important respects, tourism differs from the types of businesses usually targeted for the dissemination of research-based knowledge." As [21] mentions, these words are often used to represent something good or better, leading us to discuss the concept of "better." As he states, "Better" is a slippery, anthropocentric word like "truth" or "beauty." The concept of "better" can have different interpretations and meanings depending on the environment. According to [22], the study of the innovation process has evolved. Early linear and deterministic models, based on the rationality paradigm, have been replaced by contingency models that identify variables that influence innovation in organizations. Today, the complexity paradigm seeks more holistic explanations for this process. Among the various models created to highlight the innovation process, the model created by Rogers dedicated to the diffusion of innovations in agriculture, created in 1962, has become unavoidable (Fig. 2).

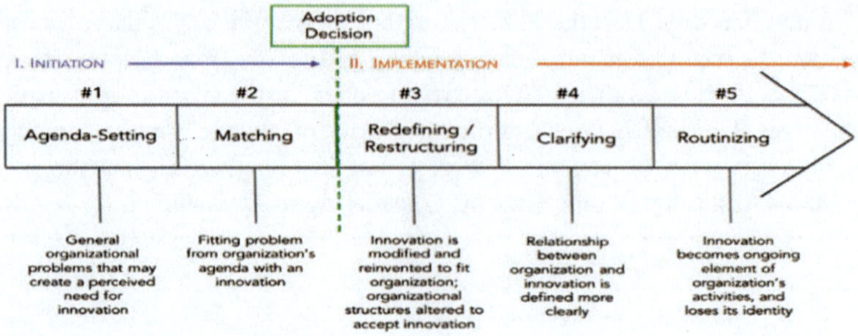

Fig. 2. The Rogers model—The innovative process in organizations. **Source.** Henderson [23]

The model consists of two main phases: the first (initiation) "concerns the definition of the general problems faced by the organization and the study of the environment in order to identify innovation needs (…) decision-makers try to anticipate the effects of innovation in terms of costs and benefits, in order to determine whether it will be introduced or, on the contrary, abandoned" [22]. The second (implementation) relates to the three necessary steps to implement the referred innovation. It is not a static process: "decision-makers try to anticipate the effects of innovation in terms of costs and benefits, in order to determine whether it will be introduced or, on the contrary, abandoned"(*idem, ibidem*). Innovation does not necessarily make something better for everyone, and that there are various forms of innovation. Our position in the development of this article was that the reasons for seeking innovation are to improve or perfect any process or product as a response to the difficulties that arise.

5 Tourism and Innovation

This article chose to treat Tourism as a Human activity because its object form and definition are not definitive, and its status and approach as a "Social phenomenon, expert technique, ethical and aesthetic fact" [24] are still being discussed today. There is also the simultaneous presence of "scientific," "societal," "professional," and "ethical-philosophical" components (at both inter and intra-component levels) (*idem*). We also observe an "Absence of consensus in terms of terminology and concepts and confusion (which also results from intra-academic "political-ideological" struggles but also between academia-industry-politics) among i) Field of study; ii) Scientific field; iii) Scientific discipline; iv) Object of study; v) Form of knowledge and structural dimension of knowledge of reality and its transformation" [24]. For example, Leiper (1979) *apud* Aldebert, Dang [25] states that "(…) the tourism industry is composed of companies that voluntarily commit to coordinating their activities in order to serve tourists. The coordination of activities is the central issue, highlighting the importance of an analysis of actors and their interactions." Another perspective is presented by Ielenicz and Simoni [26], who state that tourism cannot be reduced to just an industry and should be analysed as a set of complex, dynamic areas, and not just reduced to a simple economic sector that generates revenue. This becomes more visible when analysing the components of the tourist system individually, and when we understand the function of each one in the tourist activity chain and the result of their interaction that is balanced when everyone knows their position/category within the same system (*idem*). Out of the various possible ways to analyse and evaluate Tourism, we have chosen to approach it according to the definition of the [27] as an activity that is represented by the movements, activities, and stay of individuals outside their natural environment, for a continuous period not exceeding one year and not less than twenty-four hours, for reasons of leisure, business, and other non-related to the exercise of a paid activity, at the destination. In this perspective, we can indicate that the tourist activity promotes the agents (economic) to currently make available to the traveller a panoply of products that is as vast and varied as the number of existing actors/producers: "The tourist product comprises services from several segments of suppliers: accommodation, transport, catering, entertainment, etc." [19]. Therefore, we automatically associate the tourism industry with restaurants, hospitality, and transportation as fundamental activities for there to be tourism, with the traveller/tourist ultimately consuming other products and services that are not characteristic of the tourist activity. Although, "As Sarmento (2003) refers, the tourist activities that provide the greatest multiplier effects are tourist accommodation and transportation, followed by restaurants and direct support services for the activity" [22]. Silva [28] argues that "tourism is recognized, when properly planned, for its positive effects on the economic, social, heritage, and territorial levels. Many countries have even elected it as a strategic development vector, resulting in an increasing diversification of destinations, both at the level of countries and regional alternatives". He further developed a framework based on Beni (2004) entitled *Tourism and its economic definitions.*

6 Results and Discussion

Nowadays, the innovation and creativity are present in all kinds of Tourism sectors. They're important tools to improve tourist projects and products, in all lines of action. In the military tourism field, their use is no exception. Military tourism has the strength to combine several tourist segments, through the creation of distinctive and authentic experiences, allied to the memory, identity, and military history. According to the recent published 'encyclopedia of Tourism Management and Marketing (2022), *"(…) it is possible to state that military tourism is still very much an open concept, which agglomerates all tourism that is based on the demand for places of historical and cultural interest associated with military history in a given country or region, as well as the associated authentic and distinctive experiences which evoke the history of a place and its community"* [29]. Portugal is a country with a rich and diversified historical and military heritage, spread from the north to the south of the country, both continental and insular. This heritage is an important asset for the development of Portuguese Military Tourism. To fill an existing gap regarding the structuring of the existing offer in national territory and the development of military tourism products, the Portuguese Military Tourism Association (*ATMPT – Associação de Turismo Militar Português)* has created the "Military Tourism Route", a national scope project, developed together with the Portuguese municipalities, composed of a set of thematic routes. The routes associated with events and periods of national history, aim to promote, and disseminate assets, equipment and national tourist and cultural services, which can integrate, directly or complementarily, the offer of Military Tourism in Portugal. It is a project in constant growth, to integrate not only the heritage assets existing in the national territory, but also the activities and events that are being developed, as well as new projects and products that emerge [30]. The aim is to create a space to share all this information, with a dynamic programme of personalised and educational content, with written, photographic, and audio-visual media. Associated to this project is a set of services for the promotion and dissemination of the historical, material and immaterial heritage of the different regions of Portugal, with the objective of divulging all the points of interest and attraction of the territory, whether they are historical, cultural or natural assets, capable of integrating the existing routes. A digital platform has been created aggregating all the information, where it is possible to find the different thematic routes with a historical background of the theme. In each of the thematic routes, the user can search the different types, sub-types and locations: the heritage (whether material or immaterial), activities and events, how to get there, where to stay, where to eat, and the different tourist services available. Each asset is directed to a page with more information about it. The platform also provides a set of Experiences with suggestions for all tastes according to the thematic routes, a Highlights section with recent news, events, and new developments in military tourism, and a Cultural Agenda with activities and events in course. Thus, the platform of the Military Tourism Route opens a virtual window of access to heritage and inspires the participative dynamics of the tourist in planning a visit adjusted to their interests, needs and restrictions. Associated with the digital platform, the municipalities that integrate the project have access to the BackOffice of the platform where they can directly upload all the information on the assets existing in their territory. The assets uploaded are validated by the ATMPT team before becoming available on the platform, according to their importance and connection to the theme, as direct, indirect, or complementary assets – which allows direct assets to be more outstanding on the platform. The services include a promotional video of the territory, a video with the testimony of the President of the Municipality, a teaser and a video of the Route that includes images of all the territories associated with the theme. These audio-visual supports include interviews, reports, guided tours and aerial images taken in the most emblematic places of the different territories with the objective of promoting the visitation. Services also include the digital magazine *VIAGEM NA HISTÓRIA - TURISMO MILITAR* (Journey in History – Military Tourism) with national and international scope, bilingual, with informative and promotional sections in the Military Tourism area, in all its aspects. Open access and published every six months, it is available in digital format. This magazine aims to disseminate the vast and diverse military tourism offer existing in the Portuguese territory, capable of attracting new visitors and adding differentiating experiences, contributing to the consolidation of the Portugal destination brand. The Military Tourism Route is not just an electronic platform, it is the contents, the activities, the dynamics, the operation, and dissemination networks that are associated. This is what makes this project a creative and innovative solution that contributes to the consolidation of a developing tourist segment [31].

7 Concluding Remarks

The creativity is a human process that can be enhanced by technology. It is a distinctive characteristic of each individual and when fostered by the tourism activity, it can generate innovative and efficient products and services in response to market demands, boosting the development of the target territory in a more sustainable perspective. It has various forms of expression and has been the subject of study since the 1950s. On the other hand, innovation is partly composed of creativity and based on new technologies, can reveal itself as a response to social development. It currently plays a central role in our knowledge-based society due to the delicate challenges that have arisen recently, sometimes due to uncontrolled rapid growth. Tourism needs innovative processes, products, and services that are based on creativity to overcome these challenges, as it is performed by millions of people and it is normal that over time, their desires, aspirations, and tastes change, and their satisfaction criteria change. Innovation will have to be practically permanent, in this case, in the Hotel, Restaurant and Transportation industries that make up the tourism activity.

Acknowledgements. Work funded by national funds through the Portuguese National Funding Agency for Science, Research and Technology (FCT) under the project UID/05488/2020 - TECHN&ART.

References

1. Camilleri, M.A., Camilleri, M.A.: The Tourism Industry: An Overview. Springer, Cham (2018). https://doi.org/10.1007/978-3-319-49849-2_1
2. Morrison, A.M.: Marketing and Managing Tourism Destinations. Routledge, London (2013)
3. Saarinen, J.: Sustainable growth in tourism?: Rethinking and resetting sustainable tourism for development. In: Degrowth and Tourism, pp. 135–151. Routledge, London (2020)
4. Giotis, G., Papadionysiou, E.: The role of managerial and technological innovations in the tourism industry: a review of the empirical literature. Sustainability **14**(9), 5182 (2022)
5. Hamel, G.: The why, what, and how of management innovation. In: Harvard Business Review Home – Creativity (2006)
6. Santos, J.F.F.: As cidadescriativas como modelo dinamizador do destino turístico (2012)
7. Oliveira, E.B.P., Alencar, E.M.L.S.d.: Importância da criatividade na escola e no trabalho docente segundo coordenadores pedagógicos. Estudos de Psicologia (Campinas) **29**, 541–552 (2012)
8. Mateo, J.L.: Sociedad del conocimiento. Arbor **182**(718), 145–151 (2006)
9. Todoroi, D.: Information era. Conscience society. Creativity. Informatica Economica **15**, 99–114 (2011)
10. Paletz, S.B., Peng, K., Li, S.: In the world or in the head: external and internal implicit theories of creativity. Creat. Res. J.. Res. J. **23**(2), 83–98 (2011)
11. Wang, B., Greenwood, K.M.: Chinese students' perceptions of their creativity and their perceptions of Western students' creativity. Educ. Psychol. **33**(5), 628–643 (2013)
12. Richards, G.: Creativity and tourism: the state of the art. Ann. Tour. Res. **38**(4), 1225–1253 (2011)
13. Taylor, C.W.: Various approaches to and definitions of creativity. In: Sternberg, R.J. (ed.) The Nature of Creativity: Contemporary Psychological Perspectives, pp. 99–121. Cambridge University, Cambridge (1988)

14. Andersen, P.H., Kragh, H.: Managing creativity in business market relationships. Ind. Mark. Manag.Manag. **42**(1), 82–85 (2013)
15. Rodrigues, A.F.C., Veloso, A.L.d.O.M.: Organizational trust, risk and creativity. Revista Brasileira de Gestão de Negócios **15**, 545–561 (2013)
16. Schumpeter, J., Schumpeter, J.A.: Capitalismo, socialismo e democracia. UNESP ed. (2017)
17. Higgins, J.M.: Innovate or evaporate: creative techniques for strategists. Long Range Plan. **29**(3), 370–380 (1996)
18. Lindič, J., et al.: Deploying information technologies for organizational innovation: lessons from case studies. Int. J. Inf. Manag.Manag. **31**(2), 183–188 (2011)
19. Hjalager, A.-M.: Repairing innovation defectiveness in tourism. Tour. Manag.Manag. **23**(5), 465–474 (2002)
20. Mota, R., Oliveira, J.F.: Combining innovation and sustainability: an educational paradigm for human development on earth. Brazil. J. Sci. Technol. **1**(1), 1–12 (2014)
21. Lagowski, J.: What's So Great about Innovation?, p. 449. ACS Publications, Washington, D.C. (1990)
22. Monteiro, I.P.: Comportamentos do líderinovador no sector do turismo (2008)
23. Henderson, W.D.: Innovation diffusion in the legal industry. Dickinson L. Rev. **122**, 395 (2017)
24. Ferraz, J.: Apontamentos da disciplina de Inovação em Turismo da 5ª edição do curso de Doutoramento em Turismo. IGOT/ESHTE, Lisboa (2014)
25. Aldebert, B., Dang, R.J., Longhi, C.: Innovation in the tourism industry: the case of tourism. Tour. Manag.Manag. **32**, 1204–1213 (2011)
26. Ielenicz, M., Simoni, J.: Reflections on tourism: a systemic view. Caderno Virtual de Turismo **13**(1), 23–37 (2013)
27. OECD, P.: OECD glossary of statistical terms (2008)
28. Silva, J.S.d.S.e.: A visão holística do turismo e a sua modelação (2009)
29. Mateus, L., Marques, C.G.: Military Tourism, in Encyclopedia of Tourism Management and Marketing (Ed. by, D. Buhalis), pp. 213–216. Edward Elgar Publishing Limited, Cheltenham (2022)
30. Mateus, L., et al.: Insignia–a Monitoring and enhancement model for portuguese historical military heritage. In: Conference Chairman (2021)
31. Marques, C.G., et al.: Technological acceptance of a military tourism platform and the challenges of route implementation. In: Carvalho, J.V., Abreu, A., Liberato, P., Peña, A. (eds.) Advances in Tourism, Technology and Systems. Smart Innovation, Systems and Technologies, vol. 345, pp. 351–363. Springer, Singapore (2022). https://doi.org/10.1007/978-981-99-0337-5_30

The Ancestral Practice of Pachamanca for Experiential Tourism: Case Study of the Community of Llangahua, Province of Tungurahua

Mocha-Bonilla Julio Alfonso[1](✉), Chisag Andachi Xıomara Abigail[2], Viteri Toro María Fernanda[2], and Sánchez Guerrero Daniel Oswaldo[2]

[1] Universidad Técnica de Ambato, Facultad de Ciencias Humanas y de La Educación, Unidad Operativa de Investigación y Desarrollo FCHE, Carrera de Pedagogía de La Actividad Física y Deporte, Ambato, Ecuador
ja.mocha@uta.edu.ec

[2] Universidad Técnica de Ambato, Facultad de Ciencias Humanas y de La Educación, Unidad Operativa de Investigación y Desarrollo FCHE, Carrera de Turismo, Ambato, Ecuador
{xchisag3862,mf.viteri,do.sanchez}@uta.edu.ec

Abstract. The community of Llangahua in its retentive preserves Andean knowledge that was transmitted from past times, such as rituals, gastronomy, including ancestral medicine; in a certain way the ancestral knowledge is being devalued due to lack of transmission and interest of the current generations; For this reason, experiential tourism seeks to revitalize that knowledge through a sharing of experiences and activities between the host community and tourists, the study was based on the basic principles of Andean philosophy, which prioritizes the harmony between man and nature, the relationship with the ontological gyro theory. The methodology has a mixed approach, where the qualitative part allowed to describe the ancestral practice of the pachamanca and characterize the experiential tourism, while the quantitative approach was applied with a survey to determine the degree of satisfaction of tourists who accessed this experiential practice; The bibliographic and field research had a non-experimental design - transversal of descriptive scope, a snowball sampling was applied, which allowed the application of the INPC A4 cards, ATLAS of the Food Heritage; the design of the talking map identified the origin of the tourists and primary products for the elaboration of the festive dish. The results highlight that the pachamanca is a traditional dish of the Tomabela culture that is sustained under the doctrines of the ontological gyro theory and evidently experiential tourism would contribute to its revitalization.

Keywords: Pachamanca · Ventillian tourism · Ontological gyro theory · Andean philosophy

C. Montenegro et al. (Eds.): ICMTT 2023, LNNS 774, pp. 363–372, 2024.
https://doi.org/10.1007/978-3-031-43733-5_33

1 Introduction

The valleys and mountains of Ecuador hide among its branches and bushes great knowledge that have been treasured in the memory of its people, such knowledge has surpassed the barrier of time and is preserved by the communities; these places seek to expand their wisdom to their own, national and foreigners in order to preserve their roots; [1] in his research "Analysis of the ancetral culinary technique pachamanca in the Andean zone Llangahua as a gastronomic attraction in Tungurahua" mentions this the case of Llangahua, a community isolated from the city, located between mountains that captivates visitors with its gastronomy and friendly people, this comes from the Tomabela culture and geographically belongs to the parish Pilahuin of the canton Ambato province of Tungurahua.

Llangahua, more than nature, possesses cultural value, tradition and above all gastronomy that characterizes it as unique; however, this last singularity has been affected because the new generations have adapted to the current technologies, strongly avoiding contact with their identity [2].

According to [1] the pachamanca is an ancestral technique proper to the community, but for various difficulties this practice is being lost, however, it is stressed that the identity of a people is non-negotiable, a community that changes its legacy, totally loses its richness, loses the particularity and a possible tourist exploitation.

Tourism in Ecuador represents a potential strategic sector, which has even been very well received in recent years, despite the recent pandemic that hindered the entry of foreign tourists, however this was not an impediment for domestic tourists to start looking for places within their environment [3]. In this sense of searching for places and tourist attractions, Llangahua was discovered [1], due to the pandemic this tourist center obtained greater influx of visitors, where mostly consume activities such as hiking and horseback riding in the community, it is also important to strengthen these activities to diversify the supply of food and beverage service, taking advantage of the culture of the Tomabela people to generate gastronomic experiences [4].

The elaboration of the ancestral gastronomy of a place can be offered as part of an experiential tourism service, where it focuses on making known the activities mentioned by [6], in addition, this activity consists of including the tourist in the experiential experience. The pachamanca is known as a gastronomic tradition of the Tomabela people [5], the article "Pachamanca: the Andean cosmovision through food", mentions that, pacha means earth and manca olla, i.e. earthen pot, since its cooking is under the ground. Therefore, the objective is to disseminate the ancestral practice of pachamanca for experiential tourism in the community of Llangahua, province of Tungurahua.

2 State of the Art

The Andean philosophy is the union and conservation of various beliefs and ideas that were acquired by man, over the years is the experience of the community in their daily life, through their beliefs, dances and rituals that are linked to the Andean world, The community was characterized by having wise men who understood and cured based on natural medicine, with real and visible gods, who have empathy with society, whose basis is reciprocity, correspondence, relationality and complementarity [7], mentions that this theory is based on living well and in harmony, certainly linked to the Andean cosmic coexistence, it implies the relationship between nature and man, where the living being thanks the Pachamama for what he has received. When relating it to the pachamanca, a great analogy can be identified because the community gives thanks to the earth and the different deities for their harvests through the cooking of products underground. Experiential tourism is also related to the Andean philosophy because the tourist at the moment of visiting the community begins to have an approach with the culture and nature, achieving a connection between the two parts, the tourist will learn some knowledge that the community treasures and thanks to this connection the receiving agent begins to revalue its ancestral wisdom that the community possesses.

On the other hand [8], mention that the ontological gyro theory has four edges, among them is intersectionality, this is based on the designation of roles, focused on the pachamanca a person is designated as the administrator, who is in charge of the products for cooking, which are placed by the man in the earthen pit, once the products are sealed, the woman or godmother places a bouquet of flowers in the form of a cross. Under this thinking, the women of Llangahua are identified because they have a certain shy character, since at the moment of interacting they hide among the bushes, they give their ideas to the administrator who is close to them and go unnoticed. Based on this theory, we speak of interculturality, it represents the horizontal decisions that demonstrate their culture, when relating it to the pachamanca it is evident to note that its elaboration is of a festive and collective character, due to the fact that several people cooperate so that this cooking can be elaborated, in addition the cultural character is identified in this preparation because the cooking underground represents the gratitude to the Pachamama for the harvests received and they ask that the next ones are equal or better [9].

Meanwhile, interdisciplinarity aims to respect the knowledge treasured in the memory of the community, either for the elaboration of a product or the realization of an activity [10]. When relating this thought to the pachamanca, it is recognized that its cooking must be under an earthen pit, as an example of the ritual the women cannot put their hands in it and the corn is boiled next to a bonfire near the pit and in clay pots.

Regarding tourism, [10] inferred that tourism is that displacement that is done temporarily and voluntarily, either alone or accompanied, as long as its motive is rest, recreation and food, this moves from their usual area to a different place with the desire to have a new experience without economic gain; the three important aspects of tourism are the voluntary transfer, the unpaid displacement, the gain of going out and disconnecting from everyday life; then tourism has a relationship with the Andean philosophy because it seeks to make known the identity of a community that is treasured in its entrails. In the same way, tourism is conceptualized as the activity that has grown the most in recent years, being a key factor in the creation of income [11], mentions that tourism is transversal because it positively influences the economic sector, in addition to stimulating the creation of infrastructure that indirectly benefits the tourist and directly benefits the receiving community, as is the case of Llangahua, which has ventured into tourism and has generated sources of work, improving the quality of life of the entire community. This activity, seen from the Andean philosophy, is valid because its action maintains a balance in the community by having an economic income and transmitting its knowledge to those who visit it, thus highlighting the sharing between the parties involved.

Experiential tourism is part of the branch of alternative tourism, which emerges as a new choice for tourists who wish to have direct contact with the receiving environment [12]. It focuses on recreational activities, led by people who are part of the receiving area and who are engaged in agricultural, gastronomic and rural activities [13]. For their part [17], mention that this typology goes beyond an experience, because it reflects the stimulation of the heart, mind and senses, this being the fundamental basis of the vivencia as such because feelings play a fundamental role in the dimensions of the tourist and, these opinions will help to improve the management of tourism in that locality; it is related to the Andean theory because it is based on the perspective of optimizing and providing the best alternatives of the locality to tourists. Such population opens the doors of their home in order for the tourist to know and participate in their daily life, aiming at detachment from the routine [14]. Evidently this typology is supported by the Andean philosophy theory because the communities at the moment of opening the doors of their home to tourists are complying with the fundamentals of their vision, which is to share their wisdom with the people who visit their community [15], in turn, emphasizes that experiential tourism is an activity where the host community exchanges its knowledge, resulting in an educational experience between the visitor and the host. In addition, [16] agrees with this definition by saying that experiential tourism is an action of activities that are related based on the customs and traditions of a people. All these thoughts are supported by the theory of Andean philosophy, which highlights the empathy of the Andean man and his interest in preserving and sharing the knowledge treasured in the memory of his community, for this reason the community of Llangahua seeks to incorporate the pachamanca as part of the offer of experiential tourism [18].

3 Methodology

The paradigm was pragmatism, this philosophy is applied in the social and natural sciences because it seeks to relate the theoretical with the practical, through inquiry into the retentiveness of the community, in our case the community of Llangahua [19].

It has a mixed approach; from the qualitative aspect, the study relies on semantics and semiotics to interpret the characteristics of the findings of the qualities of experiential tourism and the festive dish in order to be able to capture it in the talking map. Quantitatively, the consumer's perception of the pachamanca was evaluated in order to determine it as a tourist product [20]. The design was non-experimental-transversal, because the action arises without manipulating the variables, identifying the step by step involved in the elaboration of the pachamanca, without the intention of making any change that affects or distorts its initial elaboration [21].

3.1 Participants

Being a social research based on subjectivity, the representativeness of criteria is sought. The sampling was non-probabilistic intentional by snowball, Baltar, (2020) defines it as a technique that helps to find the main subject; managing to identify the origin of the products for the elaboration of the pachamanca and to specify the information at the time of filling out the instruments and collecting the data. While for the quantitative, there was a sample of 150 respondents, in this case the tourists who visit the place, this instrument is made up of questions divided between knowledge and quality.

The summary Table 1. Shows the study sample.

Table 1. Study sample

Study sample	Frequency
Community Administrator	1
Woman from Llangahua	1
Tourists	150
Total:	152

Table 2. Table crosses between ancestral knowledge and pachamanca flavor.

How do you qualify the ancestral knowledge used in the pachamanca?						
		Moderately bad	Fair	Moderately good	Excellent	Total
How do you qualify the taste of the pachamanca?	Moderately bad	0	2	4	0	6
	Fair	4	2	10	8	24
	Moderately good	0	0	22	26	48
	Excellent	0	0	18	54	72
Total		4	4	54	88	150

3.2 Instrument

The instruments to be used were the cards, survey and talking map, these contributed to obtain information on the study variables. The A4 card of the National Institute of Cultural Heritage seeks to collect knowledge, techniques and practices developed by the place, this card was used in two ways, to inventory the Llangahua community tourist center as an attraction and obtain contacts of the administrative personnel, historical review and geographic location, as well as with this instrument the pachamanca festive dish was described to know the different denominations, historical review, and degree of conservation.

The Food Heritage ATLAS card called "elaborated products" was used to revalue and enhance the culinary richness, which made it possible to inventory the pachamanca dish as such, where it was possible to know its state of conservation, products used and level of ancestry.

The survey was a methodological questionnaire where the information of the tourists' criteria on pachamanca was processed with a quantitative qualification by likert scale that allowed observing: flavor, knowledge, hygiene, service and quality, which were validated by Crombach's Alpha with a value of 0.906, which means that its internal consistency is excellent since it is higher than 0.9 [22]. Finally, the talking map tool facilitates the delimitation and communication of a locality through figures and sketches.

Fig. 1. Talking Map of the Llangahua community

3.3 Collection of information

Field visits were made in order to collect information, the first part was to socialize the community of Llangahua about the objective of the research, after their acceptance and approval we proceeded to expose the survey to be conducted to tourists visiting the area, fill out the A4 sheets of the INPC and ATLAS of food heritage; the following visit is given to know the progress of the document, the amount of filled surveys is verified and the design of the talking map is proposed (Fig. 1) together we worked with all the assistants, where the origin of the products and tourists that arrive to the place after the promotion was detailed; later the information collected is interpreted in the statistical program spss version 24, the results obtained are shown below.

4 Results

Based on the analysis of the A4 card of the National Institute of Cultural Heritage called: "Knowledge and uses related to nature and the universe" the community of Llangahua that belongs to the parish Pilahuin of the canton Ambato province of Tungurahua, located 42 km from the city of Ambato, has an altitude greater than 3800 masl, also has 6312 hectares of protected paramo, which mostly consists of the native forest of polylepis or also called paper tree; peculiarly [23] mentions that this community has roots of the Tomabela culture, same that was characterized by the red poncho in their male clothing, as well as for its typical gastronomy known nationally as the pachamanca; another distinctive characteristic of this community is that being completely away from the city is part of the Fund of Páramos de Tungurahua, same that allowed it to win the contest "Green Award in 2014".

In another sense, the Atlas of the food heritage file called elaborated products, contributed to deepen on the festive dish pachamanca, because thanks to this it was identified that the pachamanca of the community of Llangahua is different from the rest because it has sheep meat, present since the Spanish conquest, but it is in danger because the present generations do not feel interest in investigating and even preserving their ancestral roots[24], in this feeling, [1] exposes a similar feeling and alludes that several knowledge of the community of Llangahua are being lost because there is not the necessary dissemination in the locality.

For his part, [9] agrees with the thought of caring for and revitalizing the knowledge that is treasured in the retentive of the communities, therefore, he emphasizes that the conservation of knowledge (pachamanca) contributes significantly to the development of experiential tourism, because it allows the tourist to be part of a recreational activity, which will not only be based on sharing, but also on learning, respecting and above all valuing the ancestral knowledge of the community.

The pachamanca as such has a great history, but this would not be the only reason for the tourist to visit Llangahua, but also to know and be part of the whole process that involves an experiential tourism, obviously this would imply a connection with the three worlds of Andean thought and by referring to this, we can sustain that indeed experiential tourism and pachamanca could work together and ratify its connection with the community.

Using the spss program, a cross table was made between the ancestral knowledge and the flavor of the pachamanca, in order to identify the criteria of the tourists based on the postures of the Andean philosophy.

Under the analysis of the cross table between ancestral knowledge and flavor of the pachamanca with a population of 150 people surveyed, 88 visitors agree that the flavor is excellent and the knowledge of the cooking process in the mother earth is still preserved in the collective memory of the community. This acceptable result in the flavor and ancestral knowledge through its preparation is a guideline to uncertain the tourist market, as an ancestral living experience. Under the analysis of the spss program, it can be mentioned that the non-parametric Kolmogonv-Smimov test rejects the null hypothesis in the analysis of all the questions, obtaining as a result the parameters of distribution with normality in the closeness of the resulting values to the mean, due to the fact that the resulting statistic is 0.0001 less than 1.

5 Conclusions

From this research work, taking into account the data collection, analysis and procedure, the following conclusions were drawn:

There are several theoretical positions that issue an analytical criterion about the intercultural practice of the pachamanca, which agree that the elaboration of the pachamanca has a festive character cooking, since it is performed in community through sharing; there are several denominations for this festive dish, where it is recognized as an underground preparation, with an approximate cooking time of two hours; Through that process a preparation with a high nutritional value is evidenced, taking into account that this preparation involves products from the Andean moorlands (community of Llangahua), this allows a communion of the human being with mother earth, since a ritual of gratitude for the abundance of the products and the healthy consumption of these, key elements for personal and community wellbeing is achieved.

As for experiential tourism, it belongs to the alternative and educational tourism typology and is characterized by having greater contact with the host community and thus having a direct approach; its activities include rural tourism, nature tourism, and community tourism, which are complemented by recreational activities such as hiking, horseback riding, camping, and tasting local cuisine.

The pachamanca is catalogued as a festive dish from the Andean philosophy and catalogued by tourists under the standards of quality, service, hygiene and knowledge of ancestral knowledge; since the ancestors (Tomabela culture) this practice is maintained as an extremely marked ideology, which is evident when preparing the characteristic ritual, where the man carries out most of the procedure and the woman who generally speaking from a more urban context is "supposed" to be the one who should be in charge of the cooking. is not part of the preparation, being left out of it. Thus, there are different aspects that make the traditional pachamanca more interesting, because it goes beyond a simple tasting, it represents a legacy that undoubtedly would contribute greatly to experiential tourism and in itself to maintain the solid roots of the community of Llangahua.

Since this is a study with preliminary data, it opens the door to new lines of research in the field of study.

Acknowledgement. The authors would like to thank the Technical University of Ambato (UTA) and the Directorate of Research and Development (DIDE) for their support for this work through the PROMOTION OF QUALITY OF LIFE research group RESOLUCIÓN: CONIN-P-169-2020. We would also like to thank the research project: GASTRONOMIC AND INNOVATION TOURISM BASED ON THE USE OF METAVERSES AND VIRTUAL REALITY FOR THE CENTRAL ZONE OF ECUADOR, approved by Resolution No. UTA-CONIN-2023–0050-R.

References

1. Olivo Espinoza, A.X.: Análisis de la técnica culinaria ancestral Pachamanca en la zona andina Llangahua como atractivo gastronómico. Tungurahua [BachelorThesis] (2018). https://dspace.uniandes.edu.ec/handle/123456789/8809

2. Páez-Quinde, C., Torres-Oñate, F., Viteri, M.-F., Porras, M.-E.: M-learning as support tool in the diffusion of the traditional food: Case study ambato. Advances in Intelligent Systems and Computing, 1134 AISC, 327–335. Scopus (2020). https://doi.org/10.1007/978-3-030-40274-7_33
3. Proaño Gómez, Y.A., Viteri Toro, M.F., Torres Oñate, C.F., Chisag Andachi, X.A.: La evaluación y el turismo sostenible, caso de estudio Llangahua. Artículo de Investigación Científica y Tecnológica, 05(1.1), 21. (2022). https://doi.org/10.33262/concienciadigital.v5i1.1.1997
4. Olazabal Arrabal, M. A., Rodríguez Méndez, V., González Fontes, R., Olazabal Arrabal, M. A., Rodríguez Méndez, V., González Fontes. La identidad cultural como recurso local y su integración a la gestión del desarrollo territorial. Retos de la Dirección **15**(1), 27–60 (2021). http://scielo.sld.cu/scielo.php?script=sci_abstract&pid=S2306-91552021000100027&lng=es&nrm=iso&tlng=es
5. Posada Castillo, urelio. Pachamanca: La cosmovisión andina a través de la alimentación. Lienzo, 53–268 (2018). https://scholar.google.es/scholar?hl=es&as_sdt=0%2C5&q=Pachamanca%3A+la+cosmovisi%C3%B3n+andina+a+trav%C3%A9s+de+la+alimentaci%C3%B3n+Aurelio+Castillo+Posadas&btnG=
6. Alencastro, L., Carvache-Franco, M., Carvache-Franco, W.: Preferences of experiential fishing tourism in a marine protected area: a study in the galapagos Islands. Ecuador. Sustainability **15**, 1382 (2023). https://doi.org/10.3390/su15021382
7. Hernández Soto, J. E. La vieja y la nueva filosofía andina. Una crítica a Josef Estermann. Letras (Lima) **93**(138), 137–151 (2022) https://doi.org/10.30920/letras.93.138.10
8. Holbraad, M., Pedersen, M.A.: EL GIRO ONTOLOGICO. UNA INTRODUCCION ANTROPOLOGICA (NOLA EDITORES) (2022). https://www.libreriacanaima.com/libro/el-giro-ontologico_621509
9. Santillán De La Torre, C.N.: VALOR NUTRICIONAL, SENSORIAL Y CULTURAL DE LA PREPARACIÓN DE LA PACHAMANKA, EN EL MUSEO OTAVALANGO, [Investigación, UNIVERSIDAD TÉCNICA DEL NORTE] (2021). http://repositorio.utn.edu.ec/bitstream/123456789/11688/2/06%20NUT%20404%20TRABAJO%20GRADO.pdf
10. Candias, K., Leticia Rojas, M., & London, S. Turismo y Crecimiento en América Latina y Caribe: ¿causa o consecuencia? Economía Coyuntural **5**(3), 99–135. (2020). http://www.scielo.org.bo/scielo.php?script=sci_abstract&pid=S2415-06222020000300005&lng=es&nrm=iso&tlng=es
11. Moreno Freites, Z., Ziritt Trejo, G., Silva, H.: Turismo Sostenible: Percepciones, bienestar ciudadano y desarrollo local. Revista Venezolana de Gerencia **2**, 104–130 (2019) https://www.redalyc.org/journal/290/29063446006/html/
12. Páez Quinde, C., Tamayo, R., Infante Paredes, R., Torres Oñate, C.F., Viteri Toro, M.F.: Tungurahua and the Tourist Brand Within the Perceptive Tourist Development. Advances in Science, Technology and Innovation, pp. 141–146. Scopus (2019) https://doi.org/10.1007/978-3-030-10804-5_14
13. Torres-Oñate, C.F., Viteri, M.F., Infante-Paredes, R., Donato-Moreira, S., Tamayo Soria, R., Núñez-Espinoza, M.: Heritage Cooking as Tourist Motivation: Ambato Case Study. Advances in Science, Technology and Innovation, pp. 109–114. Scopus (2019) https://doi.org/10.1007/978-3-030-10804-5_11
14. Chávez Dagostino, R.M., Andrade Romo, E., Ramos Zúniga, T.V.: Conceptos, enfoques y propuestas sobre el turismo alternativo en bases de datos multidisciplinarias 3(5), 15. 2019/10–24 (2019). https://doi.org/10.47557/FFOP1217
15. Muñoz Ramos, J.J.: El chocolate artesanal en el Turismo vivencial en el cantón Tisaleo [Universidad Técnica de Ambato] (2021). https://repositorio.uta.edu.ec/jspui/handle/123456789/33615

16. Lagua Agualongo, M.L.: TURISMO VIVENCIAL Y EL DESARROLLO SOCIOECONÓMICO [Investigación, Universidad Técnica de Ambato] (2020). https://repositorio.uta.edu.ec/jspui/handle/123456789/31716
17. Meacci Liberatore, L.G.: A senses-based model for experiential tourism. Tourism Manage. Stud. **14**(7), 7–14 (2018) https://doi.org/10.18089/tms.2018.14401
18. Soria Torres, A., Viteri, M., Guerrero, D., Torres Oñate, F.: La melcocha y el turismo vivencial, caso de estudio Baños de Agua Santa. ConcienciaDigital **5**, 1013–1030 (2022). https://doi.org/10.33262/concienciadigital.v5i1.1.2047
19. Martínez Romagosa, P.: La cuestión del método en el pragmatismo contemporáneo. Escritos **29**(62), 123–143 (2021). https://doi.org/10.18566/escr.v29n62.a08
20. Rodríguez, Julio & Reguant Álvarez, Mercedes. Calcular la fiabilidad de un cuestionario o escala mediante el SPSS: el coeficiente alfa de Cronbach. **1**(13), 13. https://doi.org/3 (2020). https://doi.org/10.1344/reire2020.13.230048
21. Hernández Sampieri, R., Fernández Collado, C., Baptista Lucio, M.P.: Metodología de la investigación (Sexta) (2017). elosopanda.com
22. Avila, H.F., González, M.M., Licea, S.M.: LA Entrevısta y la encuesta: métodos o técnıcas de ındagacıón empírıca? Didasc@lia: didáctica y educación ISSN 2224–2643 **11**(3), Art. 3. (2020). https://revistas.ult.edu.cu/index.php/didascalia/article/view/992
23. Fabara Sánchez, F.R., Guamán Rea, J.E.: La ilustración como aporte al fortalecimiento de la cultura indígena de Tomabela (2019). http://repositorio.uta.edu.ec/jspui/handle/123456789/29116
24. Paez Quinde, C., Torres Oñate, C.F., Rivera Flores, D., Chimbo Cáceres, M.: Scopus—Document details—Tourism 3.0 and Indigenous Food Cultures: Case Study Pilahuín | Signed in [Scopus]. Turismo 3.0 y Culturas Alimentarias Indígenas: Caso de Estudio Pilahuín (2020). https://www.scopus.com/record/display.uri?eid=2-s2.0-85096495089&origin=resultslist&sort=plf-f

Natural Parks Image: An Analysis of Tripadvisor Reviews

Elaine Scalabrini[1](✉), Márcia Vaz[1], João Paulo Teixeira[2], Carlos Jesus Rivas Rojo[3], David Alonso[4], Lucía Mestre[5], and Paula Odete Fernandes[1]

[1] UNIAG, Instituto Politécnico de Bragança, Bragança, Portugal
{elaine,marciafvaz,pof}@ipb.pt

[2] CeDRI, UNIAG, Instituto Politécnico de Bragança, Bragança, Portugal
joaopt@ipb.pt

[3] Asociación Ibérica de Municipios Ribereños del Duero, Valladolid, Spain
aimrd@lagunadeduero.org

[4] AEICE-Clúster Habitat Eficiente, Valladolid, Spain

[5] Fundación Santa Maria la Real del Patrimonio Histórico, Valladolid, Spain
lgarrote@santamarialareal.org

Abstract. Natural parks and protected areas have been a significant subject of tourism studies. Understanding the image that visitants have of these areas is important to the planning process, and the analysis of online reviews is a methodology used in different investigations. TripAdvisor is the most used review aggregator in these studies. In this context, this research aimed to analyse the image of three cross-border natural parks based on TripAdvisor reviews. For this, 321 reviews of the *Montesinho* Natural Park (Portugal), *Douro* International Nature Park (Portugal) and Arribes del Duero Natural Park (Spain) were collected and analysed using the machine learning techniques known as Latent Dirichlet Allocation. Two dimensions were determined for the *Montesinho* (scenery and nature), two for *Douro* International (activities and nature) and three for Arribes del Duero (water-based activities, nature, activities). The dimensions were measured on a 1–5 scale It can be concluded that the three parks present a positive image, with averages higher than 4, being 4.47 points (±0.703) for the evaluation of *Montesinho*, 4.48 (±0.671) for *Douro* International and 4.50 (±0.703) for *Arrribes del Duero*. The common dimension in the three parks was nature, which also presented the best average in the three realities. This study presented as a limitation the use only of TripAdvisor for data collection, suggesting the use of other platforms in future studies.

Keywords: Destination image · Nature-based tourism · Marketing digital · TripAdvisor

1 Introduction

National and natural parks are attracting tourism, and the demand has grown in the last few years [1]. This tourism segment offers visitors and residents different opportunities and assists in developing and supporting the region where the parks are located [2, 3].

C. Montenegro et al. (Eds.): ICMTT 2023, LNNS 774, pp. 373–383, 2024.
https://doi.org/10.1007/978-3-031-43733-5_34

In this context, the studies analysing tourism in natural parks have been a constant subject [1, 3–5] in different world places.

In this sense, identifying the opinions of visitors is important in the planning process and in ensuring the sustainability of these spaces. Different methodologies have been used, and with the advancing technologies, the analysis of online reviews is growing, and one of the most used review aggregators is TripAdvisor.

TripAdvisor is the major online tourism reviews (OTR) platform to share tourism experiences in different tourism sectors, such as hotels, restaurants and attractions [3], including natural parks and protected areas. The perceiving cultural ecosystem services (CES) were identified in 48 National Parks in the United States, using TripAdvisor reviews [5]. The same technic was applied to the reality of Tatra National Park, Poland [6].

Recognising the importance of this methodology, this study aimed to analyse the image of three cross-border natural parks based on TripAdvisor reviews. The three parks analysed were *Montesinho* Natural Park and *Douro* International Nature Park, located in Portugal, and Arribes del Duero Natural Park, located in Spain. The natural characteristics and their location were essential for the designation of these areas as the focus of this study. Considering the last ten years of reviews registered on TripAdvisor, 321 reviews were analysed using the software Knime, namely the machine learning techniques known as Latent Dirichlet Allocation (LDA), where the reviews were segmented in dimension based on the words co-occurrence.

Therefore, this paper is structured into five sections, followed by an introduction is presented a theoretical framework in which natural parks concepts and previous TripAdvisor are presented. Next, the methodological process is exposed, focusing on descriptive analysis. The pre-processing of the data and the LDA algorithm modelling are also presented in this section. The findings are presented in section four and finally, the discussion, conclusion, limitations and suggestions for future research are presented in the last section of the paper.

2 Theoretical Framework

This session presents the main concepts related to the subject of this study, namely natural parks, destination image and online tourism reviews, especially TripAdvisor.

The relationship of tourism with natural parks or protected areas is highly evident. The search for activities in natural environments has grown substantially [7] and the visitors´ search for these areas is obvious, especially during the post-pandemic period. One of the concerns regarding these studies is the guarantee of sustainability [4] and discussions of the importance of conserving these environments are also evident [1].

As a tourism destination and, associated with these concerns, natural parks and protected areas have some challenges. In this sense, the emergence of smart destinations and smart tourism can change the way visitors travel and after demonstrates their loyalty [8]. Recommendations are essential in the process of choosing a tourism destination and this can be obtained through word of mouth, in books, and more strongly through the Internet, namely in online tourism reviews (OTR) [9]. Increasingly, Internet reviews influence the decision to select tourist destinations [10]. These reviews are present in different platforms, such as Booking, Google and TripAdvisor.

TripAdvisor, founded in 2000, stands as the world's largest travel website, boasting a massive user base of 390 million unique visitors per month by the end of 2016. With its user-generated content and reviews spanning global destinations and multiple language versions, TripAdvisor offers a valuable resource for information seekers. Its wide-reaching, multilingual platform presents a unique opportunity to address challenges in studying dispersed activities [11, 12]. In this sense, different methods are used to analyse the visitors´ perspectives presented in the reviews of TripAdvisor. An example is the study realised in the Tatra National Park in Poland, the analysis was processed using NVIVO and analysing the most frequent words and word clouds. In this study was analysed 624 reviews and the park presented a positive image.

Other approach, is the use of LDA (Latent Dirichlet Allocation) modeling that has emerged as one of the most commonly used methods. LDA determines dimensions or themes based on the underlying hidden structure within the reviews. By considering the co-occurrence of terms across a large number of reviews, LDA identifies latent constructs that represent the collective sentiment or opinion expressed in the corpus of comments [13]. This method was used to analyse TripAdvisor reviews in touristic attractions [13], hotels [14], and urban trails [15], proving efficiency and innovation in the analysis process, once it is possible to size revisions according to the main elements that arise in each issue. In this context, this was the method chosen to answer the aim of this study, which will be explained in the next section.

3 Methodology

3.1 Fieldwork Under Analysis

In this study three parks located on the Portuguese-Spanish border were selected. The *Montesinho* Natural Park, recognised as a Natural Park in 1979, is in the Northeast of Portugal, covering territories of two municipalities (*Bragança* and *Vinhais*). The park has about 75 thousand hectares and has 9,000 inhabitants in 92 villages, with extensive biodiversity, with species such as the Iberian wolf, the roe deer, and the red deer. Shale is the dominant rock, but granites, ultramafic rocks and small limestone patches can also be found [16].

The *Douro* International Nature Park covers part of the municipalities of *Miranda do Douro, Mogadouro, Freixo de Espada à Cinta* and *Figueira de Castelo Rodrigo* also in the Northeast of Portugal. The border stretch of the *Douro* River is about 122 km long and 86,834.82 hectares, which separates Portugal from Spain. The region's climate has wide temperature ranges, with cold winters and very hot dry summers, and the Park's southern area is integrated into the so-called *Terra Quente* (Hot Earth). The *Douro* international is a fundamental area for the conservation of birdlife of some of the most threatened species, such as the Black Stork (Ciconia Nigra), the Vulture of Egypt (Neophron Percnopterus), the Real Eagle (Aquila Chrysaetos), the Bonelli Eagle (Aquila Fasciata), Falco Peregrinus (Falcon peregrinus), the Red-billed Chough (Pyrrhocorax pyrrhocorax) and the Black Chasco (Oenanthe Leuccura). The valley river´s rugged banks form monumental canyons, where several species of birds, have their nests [17].

The *Arribes del Duero* Natural Park is in Spain where the *Duero* River borders Portugal. In it are formed canyons, with a drop of up to 500 m, considered the deepest and most extensive in the entire Iberian Peninsula. It is a natural area that highlights the rugged beauty of its granite landscape and a rich and varied flora and fauna [18].

3.2 Data Collection and Pre-Processing

TripAdvisor is considered the major platform for review aggregators of tourism products, with over a billion reviews [6, 15], and so different studies use TripAdvisor to understand different aspects of tourism [13, 19–22]. Thus, to achieve the aim of the study, the reviews and ratings platform were used, considering three natural parks located on the Portuguese-Spanish crossborder.

The data, related to quantitative ratings, the titles and the reviews written in all languages were collected between January and February 2023. The period considered for analysis was the last ten years (2012–2022). It was collected 321 reviews, 51 (15.9%) from *Montesinho* Natural Park, 52 (16.2%) from *Douro* International Nature Park and 218 (67.9%) from Arribes del Duero Natural Park (Table 1).

Table 1. Number of reviews

Natural Park	Number of reviews
Montesinho Natural Park	51
Douro International Nature Park	52
Arribes del Duero Natural Park	218
Total	321

Source: Authors' own elaboration.

To prepare and analyse the data Knime Analytics Platform 4.6.4 was used. Different tourism studies used this software [13, 15] that uses modular pipelining for analyses [13]. All the reviews were written in Portuguese or during the collected process were translated into Portuguese to standardise. The pre-processing in the software consists of a punctuation eraser, flowed by the change to lowercase (case converter). The next step was applying the N Chars filters, removing the reviews with fewer than three characters. The following was to apply a stop word filter using common stop words in Portuguese and removing no significant words. The words were stemmed using the Porter algorithm. This process was repeated in the data of each natural park.

The quantitative evaluations were analysed by descriptive statistics (mean, standard deviation and distributions). Based on a previous study [19], score of 4–5 points was classified as a positive image, a score of 3 points as a neutral image and a score of 1–2 points as a negative.

3.3 Data Analysis

Considering different methods used to analyse TripAdvisor review, LDA modelling was selected to analyse the data in this study. The LDA modelling determines dimensions according to the terms co-occurrence and is considered a latent construct of the total of reviews [23]. Therefore, the following steps were performed do the LDA analysis:

i) Determine the number of dimensions.
ii) Extract the dimensions. iii) Appoint the dimensions. iv) Determine the number of reviews in each dimension.

v) Determine the positive and negative image per dimension.

The k-means clustering, namely the Elbow method based on sum square errors was applied to determine the optimal number of dimensions. Two dimensions were confirmed for *Montesinho* Natural Park, two dimensions for *Douro* International Nature Park and three for *Arribes del Duero* Natural Park. According to the LDA algorithm, each dimension contained 10 terms and was named considering previous studies.

4 Findings

As above-mentioned, the dimensions for each natural park were determined using k-means clustering. Calculating the sum squared errors was delimited by the Elbow Method to confirm the number of dimensions. Thus, two dimensions were determined for *Montesinho* Natural Park, two for *Douro* International Nature Park and three dimensions for *Arribes del Duero* Natural Park.

Next, the LDA algorithm was used in Knime to obtain dimensions, their respective terms, and the weight that each term represents in the dimension. The higher weight indicates a more representative term in the dimension. Some terms are repeated along the dimensions. However, their weights are different. The authors determined each dimension's labels based on previous studies and considering the terms with higher weight. Tables 2, 3 and 4 present each Natural Park's dimensions and prevalent terms.

The two dimensions that characterise *Montesinho* Natural Park are scenery and nature (Table 2). The five more representative elements presented in the first dimension are tour, nature, road, stone and village. The second dimension is represented with nature characteristics such as forest reserve, beauty, walk, village, places, tour, vegetation, view, flora and fauna.

There are two dimensions related to *Douro* International Nature Park (Table 3): activities and scenery. The first dimension, activities, includes terms related to the activities that can be realised in the park, such as boat, cruise and walk and the characteristics presented in this area. The second dimension is scnery, which connotes analysis regarding the nature, wine and views of the park.

Table 2. Dimensions of *Montesinho* Natural Park

Scenery	Weight	Nature	Weight
Tour	26	Forest reserve	22
Nature	22	Beauty	20
Road	16	Walk	14
Stone	14	Village	13
Village	13	Places	12
Landscape	13	Tour	12
City	12	Vegetation	10
Calm	12	View	10
Person	12	Flora	9
Visit	10	Fauna	9

Source: Authors' own elaboration.

Table 3. Dimensions of *Douro* International Nature Park

Activities	Weight	Scenery	Weight
Belvedere	32	Nature	22
Boat	26	Wine	18
Cruise	24	Views	14
Beautiful	16	Boat	14
Route	14	Flora	12
Walk	14	Road	12
Explanation	14	Amazing	10
Recommendations	12	Fauna	10
Landscape	12	Village	8
Views	12	Tour	8

Source: Authors' own elaboration.

The Spanish Park was the only one where it was possible to determine three dimensions. The first dimension is named water-based activities, considering the mainly terms related to this variety of activities (boat and cruise). The second dimension is nature, where nature, waterfall, belveder and nature are terms evidenced. The last dimension was related with the different activities' practices in the park such as walk.

After determining the dimensions, the evaluations were divided into positive (4–5), neutral (3), or negative (1–2) based on TripAdvisor quantitative evaluations. Tables 57 present the means, standard deviations, and absolute and relative distribution of the scores for each dimension, determining the image.

Table 4. Dimensions of *Arribes del Duero* National Park

Water-based activities	Weight	Nature	Weight	Activities	Weight
Boat	128	Views	79	Walk	37
Recommendations	51	Boat	76	Belvedere	30
Fauna	51	Belvedere	74	Landscape	28
Cruise	41	Show	48	Views	25
Flora	37	Amazing	41	Boat	24
Walk	34	Nature	35	Cruise	23
Show	31	Waterfall	31	Like	22
Explanation (guide)	29	Walk	30	Nature	20
Wine	28	Wonder	28	Amazing	19
Vulture	28	Dam	27	Charm	18

Source: Authors' own elaboration

The global mean and standard deviation of *Montesinho* Natural Park (Table 5) was (4.47 ± 0.703) points, where 92.2% was positive (scores 4 and 5 points). The evaluation of each dimension follows the same trend, and the dimension with the highest mean was Nature (4.70 ± 0.470) points. This dimension presented a 100% of the analysis in the positive scores (4–5). A positive evaluation written in August 2019 by a visitor from Lisbon (Portugal) evidences the nature potential of the park:

> "The vast *Montesinho* Natural Park is of enormous beauty. There are dozens of places to visit, countless corners to discover, fauna and flora of unparalleled richness".

Although with a low percentual in negative image (1–2 score), the dimension related to the scenery was the one that presented a review with a 2 score. This review was written by a Portuguese visitor in June 2016 and was related to difficulties he had while walking through the park.

Table 5. Evaluations of *Montesinho* Natural Park

Score	Global $\overline{X} \pm \sigma$		Dimension 1: Scenery $\overline{X} \pm \sigma$		Dimension 2: Nature $\overline{X} \pm \sigma$	
	4.47 ± 0.703		4.32 ± 0.791		4.70 ± 0.470	
	n = 51	%	n = 31	%	n = 20	%
1	0	0	0	0	0	0
2	1	2.0	1	3.2	0	0
3	3	5.9	3	9.7	0	0
4	18	35.3	12	38.7	6	30
5	29	56.9	15	48.4	14	70

Source: Authors' own elaboration

Related to the evaluation of *Douro* International Nature Park (Table 6) the global mean is 4.48 points (±0.671), representing a positive image, also corroborated with 90.4% with scores 4 and 5 points. Dimension Activities presented a slightly smaller (4.44 points ± 0.698), representing 88.9% of positive evaluations (4 and 5 points). Dimension Scenery presented a mean of 4.52 points (±0.653), with 92% of positive image evaluations. It is significant to highlight that this park has not presented any negative evaluations in the analysed period.

Arribes del Duero National Park presented the highest global mean (4.50 ± 0.917). In this reality is presented the dimension with the parks' highest mean (nature, 4.80 ± 0.431) and the minimum mean (activities, 4.29 ± 0.1.170). The park presented a score of 1 point (3.7%), possibly related to the higher number of reviews. One of the negative evaluations was written by a Singapore visitor in January 2020. It is worth noting that evaluations may be influenced by previous expectations of the park and also by the time spent on the visit.

> "My friend and I made a diversion from Salamanca to the Mirante, as I had read somewhere that this was the best place to get a great view in a short time. What a mistake!!! First of all, the scenery leading to the viewpoint was horrible - it looked like a post-apocalyptic landscape with gravel and rocks and not much beauty. When we arrived at the viewpoint still with some hope (after passing through miles of electricity pylons and towers), the dam and river were totally brown and stagnant! Total waste of time".

On the other hand, positive comments value the natural aspects of the park, such as the one made by a Spanish visitor in August 2020.

> "It is one of Spain's great unknown factors. It is 140 kilometres long, mostly along the Douro and the River. I spent a week getting to know it and didn't see everything. You have to enjoy its landscapes, visit its viewpoints, get to know its villages and above all enjoy the Douro, where it is essential to take a cruise (somewhere it

Table 6. Evaluations of *Douro* International Nature Park

Score	Global $\overline{X} \pm \sigma$		Dimension 1: Activities $\overline{X} \pm \sigma$		Dimension 2: Scenery $\overline{X} \pm \sigma$	
	4.48 ± 0.671		4.44 ± 0.698		4.52 ± 0.653	
	n = 52	%	n = 27	%	n = 25	%
1	0	0	0	0	0	0
2	0	0	0	0	0	0
3	5	9.6	3	11.1	2	8
4	17	32.7	9	33.3	8	32
5	30	57.7	15	55.6	15	60

Source: Authors' own elaboration

Table 7. Evaluation of *Arribes del Duero* Natural Park

Score	Global $\overline{X} \pm \sigma$	Dimension 1: Water-based activities $\overline{X} \pm \sigma$		Dimension 2: Nature $\overline{X} \pm \sigma$		Dimension 3: Activities $\overline{X} \pm \sigma$	
	4.50 ± 0.917	4.34 ± 0.999		4.80 ± 0.431		4.29 ± 1.170	
n = 218	%	n = 79	%	n = 81	%	n = 58	%
1 8	3.7	4	5.1	0	0	4	6.9
2 1	0.5	0	0	0	0	1	1.7
3 13	6.0	6	7.6	1	1.2	6	10.3
4 48	22.0	24	30.4	14	17.3	10	17.2
5 148	67.9	45	57	66	81.5	37	63.8

Source: Authors' own elaboration

reminded me of the Norwegian fjords that I visited not long ago). Another thing to see are the local industries: oil mill, olive press, cheese factory, bakeries with wood oven, etc. It is worth buying their products. The dams are very interesting, highlighting the immense engineering work of the Almendra dam".

5 Discussion and Conclusions

Technology is an essential factor for marketing and for determining the image of tourism destinations, TripAdvisor is an important tool that helps understanding the perception of visitors in relation to tourism destinations. Being one of the evaluated items the natural parks in OTR, this study aimed to analyse the image of three border parks (Portugal and Spain) through TripAdvisor reviews.

Different studies demonstrate the positive destinations evaluation [8, 13, 24] from the analysis of TripAdvisor, using text mining. Using the same methodology of these previous studies, the present study can positively assess the cross border national parks. The three parks presented a similar positive evaluation. Comparable studies have also revealed this positive image of natural parks such as USA [5] and on Poland [6]. In these two cases, the evidence of a positive evaluation was through word clouds that highlighted the positive feelings towards the experience in these places.

Based on the methods applied, namely the LDA with the use of KNIME, the some conclusions remarka demonstrate that the analysed parks present a very positive evaluation in TripAdvisor when considering the mean ratings obtained (4.47 points $\pm$ 0.703 for *Montesinho* Natural Park; 4.48 points $\pm$ 0.671 for Douro International and, 4.50 points $\pm$ 0.917 for *Arribes del Duero*) It was also evidenced that visitors value the natural aspects since the dimension of nature was presented in two cases and the activities practised in these parks are also an aspect to be considered.

Other conclusion is that the characteristics of each park influenced the dimensions of the analysis. For example, in the Spanish case, there is an emphasis on aquatic activities, which is due to the park's characteristics. Similar results were found in other studies, such as the case of US National Parks in which one of the dimensions identified was water-based activities [5]. A practical implication of these findings is that these results can help managers and planners identify and minimise the weaknesses as well as maximise strengths in their marketing strategies to attract demand to the parks.

The just use of TripAdvisor can be presented as a limitation of this study. In this sense, in future studies is suggested to extend the analysis to reviews of other OTR´s. Nature-based tourism, specially in natural and national parks became focal points in the pos pandemic period. In this sense, other suggestion for further studies is to apply the method of this study in other realities.

Acknowledgments. This project is being carried out within the scope of the DuraDOURO - El Duero como recurso perdurable del paisaje y el territorio de la frontera, 0785_DURADOURO_2_E, financed by FEDER, through the POCTEP. The authors are also grateful to the UNIAG, R&D unit funded by the FCT – Portuguese Foundation for the Development of Science and Technology, Ministry of Science, Technology and Higher Education. "Project Code Reference: UIDB/04752/2020 e UIDP/04752/2020".

References

1. Sæþórsdóttir, A.D., Wendt, M., Ólafsdóttir, R.: Tourism industry attitudes towards national parks and wilderness: a case study from the icelandic central highlands. Land **11**(11), 2066 (2022)
2. Haq, S.M.A.: Multi-benefits of national parks and protected areas: an integrative approach for developing countries. Environ. Socio-Economic Stud. **4**(1), 1–11 (2016)
3. Dumitras, D.E., Mihai, V.C., Jitea, I.M., Donici, D., Muresan, I.C.: Adventure tourism: Insight from experienced visitors of Romanian national and natural parks. Societies **11**(2), 1–11 (2021)
4. Trišic, I., Privitera, D., Ristic, V., Štetic, S., Milojkovic, D., Maksin, M.: Protected areas in the function of sustainable tourism development - a case of delibrato sants special natural reserve. Vojvodina Province. Land **12**, 1–20 (2023)

5. Kong I., Sarmiento F.O., Mu L.: Crowdsourced text analysis to characterize the U.S. National Parks based on cultural ecosystem services. Landscape and Urban Planning 233(June 2022), 104692 (2023)
6. Niezgoda, A., Nowacki, M.: Experiencing nature: physical activity, beauty and tension in tatra national park - analysis of TriAdvisor reviews. Sustainability **12**, 1–15 (2020)
7. Scalabrini, E., Correia, A., Nunes, A., Veloso, C., Vieira, E., Ferreira, F., et al.: Outdoor Tourism Demand Segmentation: a Case Study From North of Portugal. GeoJournal Tourism Geosites **44**(4), 1342–1349 (2022)
8. Kladou, S., Mavragani, E.: Assessing destination image: An online marketing approach and the case of TripAdvisor. J. Destin. Mark. Manag.Manag. **4**(3), 187–193 (2015)
9. Sparks, B.A., Bradley, G.L.: A "Triple A" Typology of Responding to Negative Consumer-Generated Online Reviews. J. Hospitality Tourism Res. **41**(6), 719–745 (2017)
10. Pantano, E., Priporas, C.V., Stylos, N.: 'You will like it!' using open data to predict tourists' response to a tourist attraction. Tour. Manage. **60**, 430–438 (2017)
11. Spalding, M., Parrett, C.L.: Global patterns in mangrove recreation and tourism. Mar. Policy **110**(April), 103540 (2019)
12. Mkono, M., Tribe, J.: Beyond reviewing: uncovering the multiple roles of tourism social media users. J. Travel Res. **56**(3), 287–298 (2017)
13. Taecharungroj, V., Mathayomchan, B.: Analysing TripAdvisor reviews of tourist attractions in Phuket. Thailand. Tourism Manage. **75**, 550–568 (2019)
14. Kalnaovakul, K., Promsivapallop, P.: Hotel service quality dimensions and attributes: an analysis of online hotel customer reviews. Tourism and Hospitality Research, 1–21 (2022)
15. Taecharungroj, V.: An analysis of tripadvisor reviews of 127 urban rail transit networks worldwide. Travel Behav. Soc. **26**, 193–205 (2022)
16. Turismo Câmara Municipal de Bragança: Parque Natural de Montesinho [Internet]. https://turismo.cm-braganca.pt/paisagens-e-biodiversidade/parque-natural-de-montesinho. Accessed 28 Feb 2023
17. Visit Portugal: Parque Natural do Douro Internacional [Internet]. https://www.visitportugal.com/pt-pt/content/parque-natural-do-douro-internacional. Accessed 28 Feb 2023
18. Turismo Castilla y León: Parque Natural Arribes del Duero [Internet]. https://www.turismocastillayleon.com/es/espaciosnaturales/parque-natural-arribes-duero. Accessed 28 Feb 2023
19. Ali, T., Marc, B., Omar, B., Soulaimane, K., Larbi, S.: Exploring destination's negative ereputation using aspect based sentiment analysis approach: case of Marrakech destination on TripAdvisor. Tourism Manage. Perspectives **40**(January), 100892 (2021)
20. Banerjee, S., Chua, A.Y.K.: In search of patterns among travellers' hotel ratings in TripAdvisor. Tour. Manage. **53**, 125–131 (2016)
21. Chu, M., Chen, Y., Yang, L., Wang, J.: Language interpretation in travel guidance platform: text mining and sentiment analysis of TripAdvisor reviews. Front. Psychol. **13**, 18 (2022)
22. Colantonio, C.: TripAdvisor reviews: a new genre in the tourism discourse? Textos en Proceso **8**(1), 42–59 (2022)
23. Guo, Y., Barnes, S.J., Jia, Q.: Mining meaning from online ratings and reviews: Tourist satisfaction analysis using latent Dirichlet allocation. Tour. Manage. **59**, 467–483 (2017)
24. Gitto, S., Mancuso, P.: Improving airport services using sentiment analysis of the websites. Tourism Manage. Perspectives **22**, 132–136 (2017)

The Gamification of Children's Games to Revitalize the Cultural Heritage: Case Study Fiestas de las Flores y Frutas Ambato

Viteri-Toro María Fernanda[1,1](✉), Sánchez-Guerrero Daniel Oswaldo[1], Falcón-Salazar Diana Alexandra[2], and Castro-Acosta Washington Ernesto[3]

[1] Universidad Técnica de Ambato, Facultad de Ciencias Humanas y de la Educación, Unidad Operativa de Investigación y Desarrollo FCHE, Carrera de Turismo, Ambato, Ecuador
{mf.viteri,do.sanchez}@uta.edu.ec

[2] Facultad de Dirección de Empresas, Universidad Regional Autónoma de los Andes, Carrera de Turismo, Ambato, Ecuador
ua.dianafalcon@uniandes.edu.ec

[3] Universidad Técnica de Ambato, Facultad de Ciencias Humanas y de la Educación, Unidad Operativa de Investigación y Desarrollo FCHE, Carrera de Pedagogía de la Actividad Física y Deporte, Ambato, Ecuador
we.castro@uta.edu.ec

Abstract. Ambato, cantonal head of Tungurahua, is a city that stands out for the celebration of the Festival of Flowers and Fruits (F.F.F.), a cultural legacy of national renown. The origin of this festivity comes from the historical bases that took place on August 5, 1949. Day in which the inhabitants of the city suffered a psychological, physical and natural situation in which an earthquake, with epicenter in the Pelileo canton, devastated the city and consequently the people of Ambateo [1]. This date is essential because since then the locals from Ambato, are recognized as people of strong character who do not let themselves be defeated in the face of adversity and above all very cheerful, mentioning the latter brings to mind several festivities held by the city, but the one that has more prominence is the FFF, which is being lost due to the new inclinations of society. For this reason the researchers seek to revitalize this festival in children through gamification in children's games. Therefore, the methodology has a qualitative approach, which allowed characterizing the variables, under the survey instruments and cards, the information was obtained through a bibliographic and field research, based on a non-experimental design - transversal, with a non-probabilistic sampling by convenience, data that allowed the design of a didactic document. The results show that the didactic documents contribute significantly to the conservation of the identity of the children.

Keywords: Gamification · FFF of Ambato · Didactic document · Children's games

C. Montenegro et al. (Eds.): ICMTT 2023, LNNS 774, pp. 384–393, 2024.
https://doi.org/10.1007/978-3-031-43733-5_35

1 Introduction

The development of society over time has allowed individuals of an entity to participate in social transformations, in addition to these inherit a cultural legacy that forges the identity of a group from generation to generation, preserving the cultural expressions of the past to revive it in the present and transmit them in the future.

[2] mentions Tungurahua, a province located in the center of the Ecuadorian country, where more than 50 years ago it suffered a terrible earthquake before and after it.

This event not only affected the city of Ambato, but also several cantons of the province were destroyed, however, this city was the hardest hit by the natural disaster. Faced with this, the inhabitants of Ambato sought a way to express their will of acceptance before the exteriorization of Pacha Mama, in addition to showing gratitude because the same granted fertility to the lands to develop agricultural activities, so that such manifestation was demonstrated in the development of the first Flower and Fruit Festival [3].

In that first edition, this popular festival emerged as a consolidation of different ideas that citizens contributed. All the different ideas were consolidated in the Municipality, and these ideas were supported by the Cantonal Agricultural Center. This last entity, was the precursor of the organization of the first festival, since the members of the center with their dynamism turned their entity as the transversal axis for the development of the celebration [4].

Being a festivity developed by an agricultural center, this commemoration was circumscribed on February 17, 1951, with an agricultural, livestock and industrial motif, so from there part of that one of the main elements of the solemnity, residing radicates the allegorical floats decorated with flowers and fruits regarding the designated theme [5].

Additionally, another of the identifying elements of the festival are the dance troupes, where the main actors are students from various educational institutions, as well as international dance groups [6]. These dance groups usually accompany the floats with a choreography that carry the queen, who is chosen weeks in advance. The queen represents an important symbol of the celebration, whose function is to be the representative of the festivity and later she extends her role and adopts a philanthropic role for a determined period of time. These elements are present in the parade of the confraternity.

Likewise, one of the traditions celebrated during this period of festivity is the "Blessing of the flowers, fruits and bread" [7]. This tradition takes place one day before the parade, where these elements are taken as representative symbols of the city. The blessing takes place in the main atrium of the Cathedral church, where the entire population is invited to attend this Catholic event, in which it can also be observed that, in the atrium, each year decorations are made that exceed 5 m high, which has as its theme religious allegories and Christian problems that occur in today's society. The main characteristic of these ornaments is that their realization is based on fruits and flowers of the city, where the goodness that the lands of the countryside receive is exposed, so these ornaments are exhibited from the beginning to the end of the festival [8].

In this sense, the ambateños are the main actors of the festivity, who annually attend the celebration of this tradition and participate in the parades and festivals that are held, which generally coincide with the carnival festivities in the country. Thus, it is stated that the local people have a high level of knowledge about the commemoration, including that the scope of knowledge of this event reaches a national scope and on several occasions its renown has become international [9].

However, this festivity from its beginnings to the present has been transformed, starting from the fact that today, the axis of organization falls on the Permanent Committee, the same entity that is responsible for selecting the themes of the festivities of each edition and consequently that the floats are in accordance with the theme [10]. Since there are no protocols that contemplate the parameters so that the designers of the floats can expose their proposals, at present there is evidence of a novelty in the concepts related to the main theme of the solemnity, an act that has generated a waste of the naturalness of the celebration, corroborating the social and cultural transformations of the Festival of Flowers and Fruits [11].

Despite this, the aforementioned event has not been an impediment for the people of Ambato to participate in the festivities, although with the social changes and devaluation of the culture, mainly by the younger population, the essence of the festival has been lost and there is even a lack of knowledge of the origins of this festivity. Therefore, the objective is to motivate the gamification of children's games to revitalize the Festival of Flowers and Fruits of Ambato.

2 State of the Art

Faced with this cultural devaluation, gamification has been applied to recover cultural heritage. Understanding this as a learning method that is related to the playful games, strengthening the educational aspect that by improving a skill and potentiating knowledge will allow the learner to have better learning results, since gamification has as one of its objectives to generate an influence on the actions of individuals, who by their sense of curiosity constantly receive learning subconsciously [12].

By applying ludic resources to generate knowledge in students, a tool can be generated that promotes the preservation of popular festivities through dynamic learning, mainly in the younger population who will be responsible for disseminating cultural expressions [13].

This is how it is reaffirmed that; the ludic components are a strategic media base that seeks to disseminate the benefits of the teaching-learning process, which additionally serves as a strategy that fosters communication among learners and this in turn strengthens cognitive development, same that will develop skills to infer an assumption [14].

Therefore, gamification focuses on the use of ludic resources, i.e., games, and other tools that allow for forging experiences that create meaningful and attractive learning and generate a feeling of motivation to continue with the study process [15]. By developing learning based on dynamics and games, creativity is developed, the same aspect that when strengthened with playful processes for learning, where there is a liberal dynamic, potentiates intelligence through a process of social relations [14].

Thus, by applying new methodologies to consolidate a meaningful learning process, the brain begins a process in which it starts to release dopamine, which activates positivism in the attitude of individuals, which is evidenced through emotions, feelings and other expressions that facilitate the training process, especially when these methodologies employ tools such as games, which take a person out of everyday life and activates the aforementioned process [16].

Consequently, gamification as a tool to recover cultural heritage seeks to encourage the revaluation of this through playful games that create awareness in learners so that they have a complete knowledge of the historical context, in this case, of the indicated festivity itself that will come to position itself in the memory of the deserving in an eloquent way [17].

Thus, to apply gamification in the recovery of cultural heritage, relating it to the Festival of Flowers and Fruits, the activity of an interactive game is contemplated [18]. An interactive game that can be enjoyed by people of all ages with a cultural context is monopoly. This game will be adapted to the context of the festivity, where starting from the main square, the players will have the objective of acquiring the greatest amount of goods. Thus, in this cultural edition, the goods that can be purchased are the entities organizing the celebration, floats, relevant buildings that are protagonists of the earthquake and emblematic of the city where the celebrations take place, flowers and fruits.

When implementing this game as a gamification strategy, it is important that the actors who will carry out the activity consider that, for the development of the game, there must be a dynamic environment, i.e., a correct playful space. In addition to the fact that it is beneficial to reinforce the activity with prizes or badges to have better results [19].

Finally, it is important to understand that, in a society with constant transformations, gamification has taken great boom and is of great contribution to the development of learning activities, where there is a relationship between the learner and the educator, and a process that strengthens not only the educational welfare of the learner, but also contributes to their personal, emotional and in this aspect, cultural development.

3 Methodology

The research methodological proposal applied in this study is based on pragmatism because it seeks to relate the FFF of Ambato (theory) with games for children (practice). The design was descriptive, retrospective and experimental, focused on the dissemination of knowledge of the Fruit and Flower Festival through specific gamification resources for children. For the development of this, 3 stages were proposed, starting from the collection of information, design of the playful material and ending with the evaluation of the experimentation with a pretest and post-test [20].

3.1 Participants

The social research is based on subjectivity, where the characteristics of the FFF of Ambato were interpreted. The sampling was non-probabilistic by convenience, choosing arbitrarily the number of individuals for the study, i.e., 20 children were identified in a range of 9 to 12 years of age, who performed a pre-test and a post-test [21].

The summary table 1. Shows the study sample.

Table 1. Study sample

Study sample	Frequency
Children from 9 to 12 years old	20
Total:	20

3.2 Instrument

The instruments were interviews and cards, tools that allowed to discern and deepen the variables, the interviews is the technique used to collect data [22]. Was designed a pretest and post-test were designed where the children's retentive was investigated to identify how much they know about the patrimonial festivity.

On the data sheet allows you to collect real information through a document [23], therefore, to characterize the party the data sheet called of Intangible Goods of the National Institute of Cultural Heritage, it was created by the authors to acquire reliable and structured information that reflects the sensitivity to change over time.

The instruments were evaluated with a rubric designed based on the learning hierarchy of Bloom's taxonomy.

3.3 Collection of Information

The research group met with 20 children between 9 and 12 years old with their legal representatives to explain about the experimentation, ask them to issue an authorization and commit to the use of the didactic document for at least 30 min a day. A pretest was applied to the investigated subjects and the gamification material was given to them, after two weeks the group met again and the posttest was applied.

4 Results

It started from the premise of sensitivity to the high change of the heritage events and the collective memory around the heritage festival, as a result of the information from the inventory cards, which led us to propose a playful teaching model that reaffirms the sense of identity and belonging to the festival in the children of the canton.

It began by assigning a name to the children's didactic document, followed by the elaboration of the contents, which started with the writing of information in a summarized and easy manner to assimilate: Parades of festivities, Election of the queen, Parade of the confraternity, Night round, Blessing of fruits, flowers and bread, Tour of the flowers [24]; a game was made in the form of a snake which contains questions about the events of the festivity, in this phase is given the layout of the material where the photographs are transformed into illustrations.

For the development of the structuring, the didactic document [25], was divided into three parts, the first is the informative part, which explains details of the festival, its history and the main events; the second was intended for creativity through gamification of painting, blank illustrations are presented for the child to paint and recognize the details, emphasizing the use of fruits, flowers and bread for all the decorations of the festival and ended with an evaluation tool, where a game that showed the children's learning is performed. A non-probabilistic sample was taken at the discretion of the research group of 20 children, residents of the Ambato canton, who contributed with the taking of two related samples, a pretest and a post-test, in which it is evident that the gamification resource has an impact on the dissemination of the identity of the heritage festival de la Fruit y de las Flores through the learning of each of its components and history.

Next, the variables before and after the use of the didactic document are detailed (Figs. 1 and 2), the pretest and post-test variables will be described, which are analyzed in a frequency and distribution parameters will be taken.

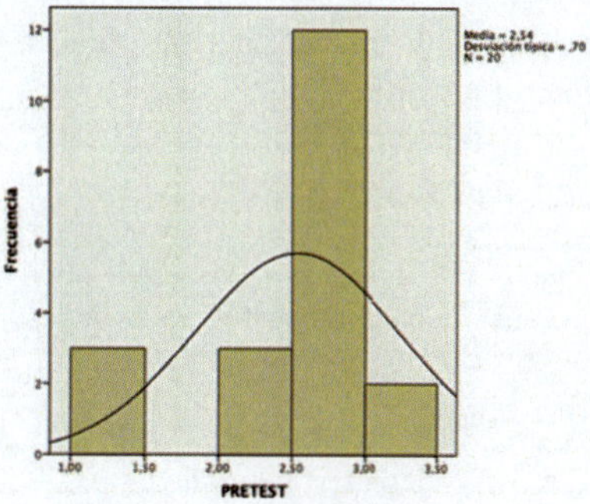

Fig. 1. Pretest variable

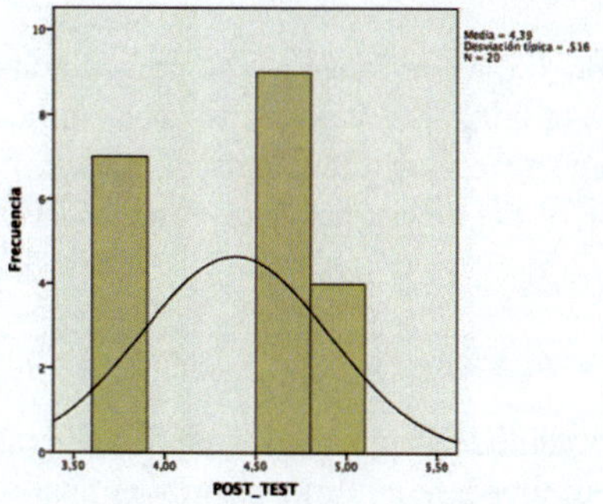

Fig. 2. Post-test variable

As shown in the normality curve (Fig. 1), it can be observed that in the pretest there is asymmetry to the right, thus having non-normally distributed data; while in the post-test there is evidence of a standard deviation of 0.516, resulting in symmetry in the data obtained.

Next, in Table 1, the normality test based on the Shapiro-Wilk statistic, which contrasts the normality of data from a pretest and post-test, used in a number of samples is less than 50, will be specified.

Table 2. Test for Normality

	Shapiro-Wilk		
	Statistical	Gl	Sig
PRETEST	0,734	20	0,000
POST-TEST	0,831	20	0,003

According to the normality contrast under the Shapiro Wilk statistic (Table 1), the result is a significance of 0.000 in the pretest and 0.003 in the post-test; thus, rejecting the null hypothesis of normality because the values obtained are less than the statistic (0.005), i.e., the contrast values are normal or parametric and the alternative hypothesis is accepted (Gamification presents a normal distribution as a tool for the dissemination of tourism heritage: Fruit and Flower Festival).

For greater statistical consistency, the nonparametric Wilcoxon test for signed ranks of related samples was applied as shown in Table 2.

Table 3. Summary of hypothesis test

Null hypothesis	Test	Sig.	Decision
The median of the differences between PRETEST and POST_TEST is equal to 0	Wilcoxon test for signed ranks of related samples	0,000	Reject the null hypothesis

In the previous section, the mean range of the pretest and post-test samples is compared and when the value of 0.000 is obtained, it is determined that there is a difference between them, which leads us to affirm the alternative hypothesis that gamification is a tool for the dissemination of the tourist heritage: Fruit and Flower Festival (Table 3).

5 Conclusions

According to the results obtained by applying the pretest and posttest it was found that the use of playful resources to preserve cultural identity is positive, since the development and use of playful material in children helps to understand the functionality of any event. And heritage history, revalue and preserve the Fruit and Flower Festival, demonstrating the hypothesis that gamification is a tool for the revaluation of tourism heritage.

Gamification seeks to influence the behavior of individuals, in this case, those who, through creativity and curiosity, receive information in the subconscious and in the same way produce new skills and knowledge, so that the use of game material is an effective tool to participate.

Acknowledgement. The authors would like to thank the Technical University of Ambato (UTA) and the Directorate of Research and Development (DIDE) for their support for this work through the PROMOTION OF QUALITY OF LIFE research group RESOLUCIÓN: CONIN-P-169-2020. We would also like to thank the research project: GASTRONOMIC AND INNOVATION TOURISM BASED ON THE USE OF METAVERSES AND VIRTUAL REALITY FOR THE CENTRAL ZONE OF ECUADOR, approved by Resolution No. UTA-CONIN-2023-0050-R.

References

1. Freire Vaca, S.J.: Estudio socio histórico de la prensa de Ambato respecto al terremoto de agosto 1949», bachelorThesis, Riobamba, 2022. Accedido: 1 de marzo de 2023. [En línea]. Disponible en. http://dspace.unach.edu.ec/handle/51000/8633
2. Páez Quinde, C., Tamayo, R., Infante Paredes, R., Torres Oñate, C.F., Viteri Toro, M.F.: Tungurahua and the Tourist Brand Within the Perceptive Tourist Development. Adv. Sci. Technol. Innov., 141–146 (2019). https://doi.org/10.1007/978-3-030-10804-5_14
3. Rueda Ramírez, M.A.: Ilustraciones para microrrelatos basados en tradiciones, costumbres y origen de La Fiesta de las Flores y las Frutas de Ambato, para la Casa de la Cultura Ecuatoriana Benjamín Carrión Núcleo de Tungurahua, March 2018, Accedido: 1 de marzo de 2023. [En línea]. Disponible en: http://repositorio.ucsg.edu.ec/handle/3317/10615
4. Lascano Cevallos, A.D., Castillo Pérez, E.I., Mena Méndez, D., Vayas Ruiz, E.C.: La Fiesta de la fruta y las flores en el escenario local de Ambato: análisis de la comunicación publicitaria (2012–2017). URBS Rev. Estud. Urbanos Cienc. Soc. **8**(1), 61–77 (2018). Accedido: 1 de marzo de 2023. [En línea]. Disponible en: https://dialnet.unirioja.es/servlet/articulo?codigo=6416873
5. Gaune, R.: El jesuita como traductor. Organización, circulación y dinámicas de la Compañía de Jesús en Santiago de Chile, 1593–1598, p. 195, 10 de febrero de 2023. [En línea]. Disponible en: Edición electrónica https://journals.openedition.org/histcrit/10741
6. Guerrero Muepaz, M.N.: Trajes de comparsas de la Fiesta de la Fruta y de las Flores: Un recorrido histórico del año 2009 al 2018, bachelorThesis, Universidad Técnica de Ambato. Facultad de Diseño y Arquitectura. Carrera de Diseño de Modas., 2019. Accedido: 1 de marzo de 2023. [En línea]. Disponible en: https://repositorio.uta.edu.ec:8443/jspui/handle/123456789/30123
7. Rosero Guamán, Y.S.: La fiesta de las flores y las frutas como estrategia de diseño interior de espacios efímeros para la difusión cultural en la ciudad de Ambato, bachelorThesis, Universidad Técnica de Ambato. Facultad de Diseño y Arquitectura. Carrera de Diseño de Espacios Arquitectónicos., 2022. Accedido: 1 de marzo de 2023. [En línea]. Disponible en: https://repositorio.uta.edu.ec:8443/jspui/handle/123456789/36017
8. Supe Llanganate, J.R.: «Las fiestas tradicionales y la gastronomía en la provincia de Tungurahua», bachelorThesis, Universidad Técnica de Ambato. Facultad de Ciencias Humanas y de la Educación.Carrera de Turismo y Hotelería, 2019. Accedido: 1 de marzo de 2023. [En línea]. Disponible en: https://repositorio.uta.edu.ec:8443/jspui/handle/123456789/30212
9. Martínez Mesías, J.P., Vásconez Gavilanes, R.: Impacto económico de la fiesta de la fruta y de las flores de Ambato. Analítika Rev. Análisis Estad., n.º 8, pp. 41–50, 2014, Accedido: 1 de marzo de 2023. [En línea]. Disponible en: https://dialnet.unirioja.es/servlet/articulo?codigo=5004626
10. Flores-Carrillo, D.G.: Criterios morfológico-estéticos: diseño de los carros alegóricos en la Fiesta de la Frutas y de las Flores en Ecuador. *Ágora Heterodoxias*, vol. 5, n.º 1, Art. n.º 1, jul. 2019, Accedido: 1 de marzo de 2023. [En línea]. Disponible en: https://revistas.uclave.org/index.php/agora/article/view/2457

11. de los Á. Palomo Landa, J.: «Análisis histórico de la Fiesta de la Fruta y las Flores en el desarrollo sociocultural de la sociedad ambateña.», bachelorThesis, Quito: UCE, 2021. Accedido: 1 de marzo de 2023. [En línea]. Disponible en: http://www.dspace.uce.edu.ec/handle/25000/24125
12. Pascuas Rengifo, Y.S.: Experiencias motivacionales gamificadas: una revisión sistemática de literatura (2018). https://www.scielo.org.mx/scielo.php?script=sci_arttext&pid=S1665-26732017000300063&lng=es&tlng=es (accedido 1 de marzo de 2023)
13. Rodríguez-Miranda, R., et al.: Aprendizaje a través de estrategias lúdicas: una herramienta para la Educación Ambiental. Rev. Cienc. Ambient. **56**(1), 209–228 (2022). https://doi.org/10.15359/rca.56/1.10
14. Stefani, G., Andrés, L., Oanes, E.: Transformaciones lúdicas: Un estudio preliminar sobre tipos de juego y espacios lúdicos. Interdisciplinaria **31**(1), pp. 39–55, jun. 2014, Accedido: 1 de marzo de 2023. [En línea]. Disponible en: http://www.scielo.org.ar/scielo.php?script=sci_abstract&pid=S1668-70272014000100003&lng=es&nrm=iso&tlng=es
15. Barbecho-Benalcázar, M.M., Uyaguari-Guamán, J.L., Torres-Toukoumidis, Á., Barbecho-Benalcázar, M.M., Uyaguari-Guamán, J.L., Torres-Toukoumidis, Á.: Descriptive analysis of the game as a tool to learn about cultural heritage: case study. Estud. Pedagógicos Valdivia **46**(3), 33–44 (2020). https://doi.org/10.4067/S0718-07052020000300033
16. Barceló Hernando, A.: «La gamificación como herramienta de difusión del patrimonio cultural en el Grado en Turismo: el caso de Fuentidueña (Segovia)», en El uso de las TIC en la Innovación Docente, 2020, ISBN 978–84–1324–571–3, págs. 109–129, 2020, pp. 109–129. Accedido: 1 de marzo de 2023. [En línea]. Disponible en: https://dialnet.unirioja.es/servlet/articulo?codigo=7592040
17. Corrales Serrano, M., Garrido Valverde, J.: Uso del patrimonio y aprendizaje de la historia de la ciudad: Una experiencia de gamificación en la alcazaba de Badajoz. UNES Univ. Esc. Soc. **9**, 30–40 (2020). Accedido: 1 de marzo de 2023. [En línea]. Disponible en: https://dialnet.unirioja.es/servlet/articulo?codigo=7741439
18. Castillo Mora, M.J., Escobar-Murillo, M., Barragán-Murillo, R., Cárdenas-Moyano, M.: La Gamificación como herramienta metodológica en la enseñanza, vol. 7, n.º 1, p. 16, enero de 2022. [En línea]. Disponible en: https://doi.org/10.23857/pc.v7i1.3503
19. Ivanjko, T.: «La gamificación en el ámbito del patrimonio. Crowdsourcing con un diseño lúdico», Tabula Rev. Arch. Castilla Ón, n.º 21, pp. 177–196, (2018). Accedido: 1 de marzo de 2023. [En línea]. Disponible en: https://dialnet.unirioja.es/servlet/articulo?codigo=6843742
20. Hernandez Sampieri, R.: Metodología de la Investigación, 6.ª ed. Mexico: osopanda.com (2017). [En línea]. Disponible en: https://www.uca.ac.cr/wp-content/uploads/2017/10/Investigacion.pdf
21. Hernández González, O.: Aproximación a los distintos tipos de muestreo no probabilístico que existen. Rev. Cuba. Med. Gen. Integral, **37**(3) (2021). Accedido: 1 de marzo de 2023. [En línea]. Disponible en: http://scielo.sld.cu/scielo.php?script=sci_abstract&pid=S0864-21252021000300002&lng=es&nrm=iso&tlng=es
22. Piza Burgos, N.D., Amaiquema Márquez, F.A., Beltrán Baquerizo, G.E., Piza Burgos, N.D., Amaiquema Márquez, F.A., Beltrán Baquerizo, G.E.: Métodos y técnicas en la investigación cualitativa. Algunas precisiones necesarias. Conrado **15**(70), 455–459 (2019), Accedido: 24 de febrero de 2023. [En línea]. Disponible en: http://scielo.sld.cu/scielo.php?script=sci_abstract&pid=S1990-86442019000500455&lng=es&nrm=iso&tlng=es
23. Diáz, M.: «El videoanálisis, evolución a las fichas de observación de clase», 30 de junio de 2021. https://www.codimg.com/education/blog/es/fichas-observacion-clase (accedido 1 de marzo de 2023)

24. Guarderas, A.B.: Ambato: La fiesta de las flores y las frutas - Clave! Turismo Ecuador. *Clave!*, 17 de enero de 2018. https://www.clave.com.ec/ambato-en-carnaval/ (accedido 1 de marzo de 2023)
25. Quinga Nacimba, Y.C., Maurera Caballero, S., Guijarro Vinces, J.: Qué tipo de material didáctico es empleado para el desarrollo del lenguaje en Educación Inicial?|Revista Cognosis. ISSN 2588-0578, vol. 7, n.º 1, 2022. Accedido: 1 de marzo de 2023. [En línea]. Disponible en: https://revistas.utm.edu.ec/index.php/Cognosis/article/view/4529

Author Index

C. Montenegro et al. (Eds.): ICMTT 2023, LNNS 774, pp. 395–396, 2024.
https://doi.org/10.1007/978-3-031-43733-5